Smart Grids and Buildings for Energy and Societal Transition

Smart Grids and Buildings for Energy and Societal Transition

Benoît Robyns
Laure Dobigny
Dhaker Abbes
Benoit Durillon
Hervé Barry
Christophe Saudemont

iSTE

WILEY

First published 2024 in Great Britain and the United States by ISTE Ltd and John Wiley & Sons, Inc.

ISTE Ltd
27-37 St George's Road
London SW19 4EU
UK

www.iste.co.uk

John Wiley & Sons, Inc.
111 River Street
Hoboken, NJ 07030
USA

www.wiley.com

Library of Congress Control Number: 2024939666

British Library Cataloguing-in-Publication Data
A CIP record for this book is available from the British Library
ISBN 978-1-78630-736-1

Contents

Foreword by Thierry Magnin

Technoscience at a Time of Ecological Urgency

The use of fossil fuels such as coal, oil and gas has enabled the development of industry, with all its benefits, but it has also contributed significantly to global warming via the increase of greenhouse gases, which is currently a source of profound anxiety. The climate, once thought to be relatively stable on a generational scale, despite daily and annual fluctuations, has become a scientific and even a social object. Since the mid-19th century, there has been a concerning acceleration in the rising global average temperature of the Earth's surface. In this book, one author notes that, compared to 0.6°C between 1860 and 2010, the global average temperature increase was already 1.2°C in 2023 (average increase over the period 2014 to 2023). Various scenarios predict that, by 2100, this temperature will have risen by between 2 and 7°C unless we effectively implement changes in production and consumption patterns.

Heatwaves are not the only manifestations of climate excesses, with disruption also affecting rainfall surpluses (floods) and large deficits (droughts). Climate change is not only reflected in extreme events but also in profound changes to the major biochemical cycles that sustain life on Earth.

The climate crisis is thus an indicator of a much deeper crisis. The relationship between humans and nature (a topic which has profound existential resonances) is being called into question. The scale of human development is such that some equate anthropogenic actions as equivalent to a geological force, even though the resulting notion of the Anthropocene is itself controversial. What is more, it is the countries that emit the most CO_2 that enjoy the highest life expectancy. This indicator is often indicative of the quality of life of which the poorest countries are deprived.

However, if humanity is responsible for climate degradation, this means that it also has the capacity to modify this harmful trajectory, to initiate a "great turnaround" that is being played out at the level of individuals, local communities, countries and the planet as a whole.

The ecological emergency calls for far-reaching transformations in energy consumption and production patterns, which in turn implies profound changes in mindsets and lifestyles. The survival of our planet is under threat, and our responsibility to future generations is at stake. The UN's Sustainable Development Goals, which consider not only climate and environmental degradation but also poverty and social inequalities, are valuable benchmarks in the search for solutions that will enable us to make a success of the energy and societal transition. It is also clear that substituting renewable energies for fossil fuels is not enough to achieve climate objectives when factoring in prevailing consumer trends.

This is the complex context in which the present book is set. It follows on from *Smart Users for Energy and Societal Transition*, published in 2023 and coordinated by Benoît Robyns. It continues the interdisciplinary approach of the first book, traversing topical technoscience and societal issues, in terms of smart grids and smart buildings to accelerate the necessary ecological transition. In this sense, the book presents a didactic and well-supported overview of the issues, including global warming, energy consumption, the balance of living ecosystems and societal challenges, against the backdrop of the sharing economy.

In the introduction to this book, we read that while cities are home to 50% of the world's population, they consume 75% of the energy produced and emit 80% of CO_2 emissions! This highlights the importance of designing smart buildings in conjunction with smart energy networks.

Smart grids, smart buildings, smart users, self-generation and self-consumption are the major elements that weave their way through the book, not forgetting the qualitative analysis of the word "smart" when we speak of buildings, digitized technologies or users concerned with the common good. Chapter 2 examines the meaning of the term, its effectiveness in meeting environmental challenges, the imagination it conjures and the impasses involved. This chapter guides the "architecture" of the following chapters, which attempt to take better account of users in their interaction with technologies and the use of digital technology, in the context of energy networks and increasingly smart buildings.

In the same vein, it is particularly significant that the book is based on the four typical scenarios devised by ADEME[1] to meet the climate challenges and achieve carbon neutrality by 2050. Each of these four scenarios combines, in very different ways, the use of technology, economic vision and lifestyle choices in the fight against global warming. Different forms of intelligent living are thus proposed to the ethical choices of societies and their leaders. In this sense, this book is particularly enlightening when it comes to situating technological innovations in terms of their contribution to the fight against global warming, wherein users and their lifestyles are considered "in the same research movement", not forgetting the underlying economic model.

It is worth noting the seriousness of the approach proposed in this book, which brings together technoscience and societal issues: the necessary technical precision, which some might find very dry at first glance, is always at the service of the effectiveness of solutions combining technologies, economics and lifestyles. It is not simply a question of offsetting the weight of technology with few ethical regulations to achieve a socially acceptable result. It is more a question of transforming the mindset, so that technical development is truly at the service of a society's lifestyle choices, and the appropriation of techniques by users becomes central. This notion of appropriation is particularly well discussed within the book. We are still used to thinking that it is human beings who must adapt to technological evolution. The authors of the various chapters in this book show that, when technology can be appropriated by the user, without the latter having to be an experienced technician, humans are no longer subject to technological constraints that are beyond their control. The user becomes increasingly central to technological development, in a process whose intelligence is no longer solely technical or technoeconomic.

As biology and neuroscience are showing in ever finer detail, living organisms are the seat of a dynamic balance between robustness and vulnerability. This balance determines its capacity for adaptation and resilience, which can be severely penalized by the acceleration and scale of environmental change. Therefore, we fear a collapse, and this is also why we need to imagine energy and societal transitions that will enable humankind to rediscover the dynamics of life in the service of our common home, the inhabited Earth of today and tomorrow.

In this sense, the city of tomorrow – a smart city that combines quality of life and balanced ecosystems – is the focus of new outlooks, of which this book provides pertinent examples. Techniques, uses and imagination are all examined in the implementation of smart grids and smart buildings. Studies go as far as proposing tangible solutions for today and tomorrow, including full-scale demonstrations.

1 Agence de l'Environnement et de la Maîtrise de l'Énergie: the French Environment and Energy Management Agency.

Most of the work presented has been carried out as part of an industrial research chair held by one of the engineering schools of the Université Catholique de Lille (Junia), in conjunction with the private sector, and as part of the Université Catholique de Lille's Live TREE energy and social transition program. This practical, demonstration-type approach is also what makes this book so original, with new, concrete solutions that help to better control the consumption and production of electrical energy, in an approach that is both technological and sociological.

For example, this book provides models for forecasting production (photovoltaic) and consumption of electrical energy based on digital artificial neural networks, with application to a block of buildings at the Université Catholique de Lille. We take into account the various actors (producers, consumers, storage operators, network managers) involved in energy management strategies, not forgetting those in fuel poverty. Various actor profiles are introduced into the models, in line with the notion of flexibility among residential consumers. Finally, these approaches are situated in relation to the four ADEME scenarios mentioned above.

A strategy for supervising a local smart grid, integrating as far as possible the implication of stakeholders, is developed. The conditions for self-consumption within a human community are specified. These are referred to as "renewable energy communities", in line with ADEME's Scenario 2, "territorial cooperation". The management of energy exchanges explores blockchain technology.

Each chapter presents the development of methods and models, simulations and feedback. Chapter 7 focuses on the concept of sustainable and desirable housing, thanks to digitally controlled smart buildings, with particular attention on indoor air quality. This gives rise to the idea of "connected, supportive and humane buildings", applied in the new Rizomm building and the renovated ESPOL building at the Université Catholique de Lille. Chapter 8 is devoted to the tangible development of demonstrations, helping designers to confront the reality of the field, which is highly interdisciplinary, particularly with regard to techniques and human behavior. The transition is both energetic and social!

The contributions in this book help to redefine what can be called genuine technoscientific progress today, in the face of the ecological crisis and the rise of artificial needs fanned by our consumer society. As Pope Francis says in his encyclical *Laudato si*[2], linking environmental and human ecology: "It is not enough to balance, in the medium term, the protection of nature with financial gain, or the preservation of the environment with progress. Halfway measures simply delay the inevitable disaster. Put simply, it is a matter of redefining our notion of progress. A

2 Pope Francis, *Loué sois-tu*, Bayard, Paris, 2015.

technological and economic development which does not leave in its wake a better world and an integrally higher quality of life cannot be considered progress". The priority given to quality of life in the choice of energy production and use is an example that goes in this direction, along with the criteria of the world's habitability and the balance of ecosystems that respect nature.

Finally, it is particularly significant to note, through the pages written by the engineers and researchers involved in the writing of this book, how fruitful their dialogue with researchers in the Humanities and Social Sciences can be. It fosters collective ingenuity and broadens the sense of responsibility and civic commitment of the engineer in the 21st century: so many new ways to embody this ancient and noble profession!

Thierry MAGNIN
President of the Humanities Department, UC Lille

Foreword by Audrey Linkenheld

Smart Grids and Smart Buildings Tested by a City

Today, more than ever, we need to move away from fossil fuels if we are to combat climate change and move towards carbon neutrality. This implies an in-depth transition of our society and all economic sectors, firstly by drastically reducing our energy consumption, and also by massively developing renewable energies and electrifying uses that currently rely on fossil fuels, such as the heating of buildings or for mobility.

On its own scale, the Métropole européenne de Lille (MEL)[1] adopted an ambitious Climate Plan in February 2021, which aims to make the territory carbon neutral by 2050 by setting targets to reduce energy consumption by 16% between 2016 and 2030 and to multiply by 2.3 local renewable energy production over the same period.

As a result, the national electricity system will have to integrate decentralized production resources more and more, while at the same time satisfying new uses. This is possible, as shown by the 2021 RTE[2]/AIE[3] study on the technical conditions for integrating a high proportion of renewable energies into the electricity mix and the Futurs énergétiques 2050[4] study by RTE. It will nevertheless be necessary to significantly develop the flexibilities of the national power system: storage facilities, load shedding, etc.

1 The European Metropolis of Lille.

2 Réseau de transport d'électricité: France's Transmission System Operator.

3 Agence internationale de l'énergie: International Energy Agency.

4 Energy Pathways to 2050, available at: https://assets.rte-france.com/prod/public/2022-01/Energy%20pathways%202050_Key%20results.pdf.

Energy-saving and energy-efficient solutions, particularly in buildings, are also mechanisms for combating fuel poverty, as well as for reducing users' exposure to the fluctuating costs of fossil fuels, constitute major challenges for ensuring the acceptability of the energy transition.

In addition to technological solutions, energy-saving actions by citizens, as well as simple gestures to shift the use of appliances away from peak consumption, are essential to ease the pressure on the power system, as demonstrated by feedback from the winter of 2022–2023.

The book you are holding in your hands highlights the role that smart grids and smart buildings can play in improving the flexibility of the electric system, making buildings more efficient and integrating a higher proportion of renewable energy, particularly photovoltaic. The authors show that these tools reach their full potential when combined with individual or collective self-consumption of renewable electricity. Self-consumption approaches also have the virtue of enabling citizens, communities or local authorities to become directly involved in the production of local renewable energies, which makes a major contribution to their societal acceptability.

This book provides a practical overview of how to effectively operate a smart grid in conjunction with a smart building and photovoltaic production. However, beyond these technical aspects, its originality and interest lie in the multidisciplinary approach adopted, integrating the human sciences to focus on user behavior. Their support is fundamental if we are to move towards a more sufficient and resilient model of society. The second major advantage of this book is that it systematically draws on experience gained from full-scale demonstrations.

The use of such demonstrations is crucial in preparing for the widespread deployment of smart grids, as they enable theories to be tested against real-life situations involving those users who do not necessarily spontaneously behave optimally. For example, the MEL has piloted the "So MEL So Connected" program, implemented between 2017 and 2021 and funded by ADEME, which has enabled several demonstrations to be set up in order to test smart grid solutions.

The program was coordinated by the MEL, with ENEDIS[5] (MEL's public distribution system operator) acting as technical director. The project was also supported by several partners involved in one or more experiments: Dalkia[6],

5 ENEDIS, a fusion of "énergie" and "distribution", the new name for ERDF (Électricité Réseau Distribution France).

6 An energy service company, subsidiary of the EDF Group.

EDF[7], Intent Technologies, the Laboratoire Lille Economie Management (LEM)[8] of the Université de Lille[9], the Junia Grande école d'ingénieurs[10] and the FGES[11] faculty of the Université Catholique de Lille[12].

The program has enabled the development and testing of innovative services aimed at a wide range of actors, from manufacturers to private individuals. These services have been tested in four use cases, to demonstrate not only their technical feasibility, but also the relevance of the associated business models.

The first case study was applied to the Institut Catholique de Lille's historical block (FGES Rizomm building), in order to test an individual self-consumption solution on a smart tertiary building, with fine control of surplus electricity fed into the public distribution grid. The lessons learned from this demonstration are described in detail in this book.

The second use case study was made up of various devices enabling the consumption of public facilities to be controlled. Shading devices were tested on the Salomé wastewater treatment plant and on the Ansereuilles drinking water plant (two-hour delay of certain high-consumption processes). The experiment demonstrated the effectiveness of these systems, which contribute to the flexibility of the power grid by shaving off part of the demand during periods of high consumption. Another component involved using the functionalities of Linky smart meters to better control the city's 1,150 street lighting points. Thanks to the data sent back by the meters, it was possible to carry out a full-scale test of an IT platform developed by ENEDIS to automatically identify faults and plan interventions. This platform is still in operation and is being rolled out by ENEDIS to other local authorities.

So MEL So Connected has also enabled the installation of a demonstration site combining electric vehicle charging stations and on site photovoltaic shading, combined with batteries. An optimization mechanism has been implemented to control these three components. It makes the most of photovoltaic production, by arbitrating between self-consumption, storage and injection into the public grid, as well as modulating charging power according to the number of vehicles connected. In this way, the system reduces the impact of recharging on the electrical grid, by

7 Electricité de France S.A.: A French multinational electric utility company owned by the French state.

8 Lille Economics and Management.

9 The University of Lille.

10 Junia Graduate School of Science and Engineering.

11 The Faculty of Management, Economics and Sciences.

12 The Catholic University of Lille.

lowering the peak power demand. This demonstration has been installed in the relay parking lot at the Saint-Philibert metro station in Lomme and has been operational since 2018.

Finally, the program includes support for people in fuel poverty, through the introduction of innovative services based on the functionalities of Linky smart meters. For example, workshops have been organized to help people better understand their energy consumption, including using the data provided by Linky, and to learn how to control it. A remote display of household electricity consumption was also proposed.

In the end, the So Mel So Connected program yielded an abundance of lessons. Firstly, the technical relevance and viability of the various demonstrations were established. In particular, modules enabling equipment to communicate with each other and with the public electricity grid were tested. The setting up of a platform centralizing all the energy consumption data of the metropolitan equipment and buildings taking part in the experiment, cross-referenced with meteorological data for the area, demonstrated the potential of this grouping to optimize equipment control and identify new uses (in particular, the possibility of erasing consumption for the benefit of the electricity grid). The experimentation also enabled us to acculturate the agents and professionals managing the equipment or buildings, as well as the users, which is a sine qua non condition for success.

Feedback from this program will feed into local projects. Firstly, functionalities such as power reduction and fine control of public lighting will be increasingly integrated into public facilities. The communes of Lille and Lambersart have signed up to the Eff'Acte program run by the FNCCR[13], which aims to support the reduction of electricity consumption in tertiary buildings. Eight municipalities in the Métropole will also benefit from the FNCCR's Lum'Acte program, which addresses the issue of efficient public lighting. The lessons learned from So Mel So Connected will also be invaluable for the continued deployment of photovoltaic solar panels and electric vehicle charging stations on public property. Finally, the various So Mel So Connected partners are continuing to work on smart grids, moving on to larger-scale projects involving other actors. The authors of this book give several examples.

The So Mel So Connected program has therefore achieved its main objective, namely to encourage the development of smart grids at a larger scale.

---

13 Fédération nationale des collectivités concédantes et régies françaises: National Federation of French Local Authorities.

However, we know, and this book confirms this, that if we are to succeed in the ambitious energy transition that lies ahead, technological solutions – of which smart grids and smart buildings are a part– must go hand in hand with a profound change in our behavior and the support of the actors involved, so as to ensure that no one is left by the wayside, and be allocated the scale of investments and transformations they require.

Local authorities have a major role to play in both these areas. They are at the heart of the Plan climat de la Métropole européenne de Lille[14] and the Plan lillois pour le climat[15] of the City of Lille. One of the flagship actions has been the creation in 2021 of the HCmC[16], which brings together representatives of municipalities, citizens, the socio-economic world, as well as the scientific world. The HCmC is a forum for exchange with civil society, whose purpose is to encourage the implementation of actions that contribute to achieving the objectives of the Climate Plan. The issue of behavioral change is therefore central to the HCmC, which has worked on all its components: energy sufficiency, mobility, consumption of goods, food, societal acceptability of renewable energies, etc.

For the winter of 2022–2023, the Métropole has acted to encourage energy sufficiency, not only for its own assets, but also for local businesses, municipalities and private individuals. This has resulted in the publication of a take-home booklet on eco-actions. The Métropole also supports the DECLICS[17] local citizen challenges, since 2015, on water and energy, and since 2023, on food, which give volunteer participants concrete keys to reducing their water and energy consumption and moving towards healthier, more sustainable food. The 2022–2023 edition of the DECLICS water and energy program, which has just been completed, saved nearly 200 MWh and 850 m^3 of water. In the field of transport, the Mobility Plan currently being revised will include a section on behavioral changes, with a view to encouraging soft mobility and public transport over the use of private cars. The ecobonus scheme currently being tested by the MEL, which provides financial rewards to individuals who refrain from using their private vehicles on certain major roads during rush hour, is part of the same dynamic.

Supporting private individuals in the energy renovation of their homes is also crucial in our region, where 30% of tenants in the private sector are in fuel poverty, and 40,000 homes are considered potentially unfit for habitation. Over 10 years ago,

14 The Climate Plan for the European Metropolis of Lille.

15 The Lille Climate Plan.

16 Haut conseil métropolitain pour le climat: High Metropolitan Climate Council.

17 Défis citoyens locaux d'implication pour le climat et la sobriété: Local Citizen Challenges for Climate and Usage Sufficiency.

the MEL set up its public AMELIO[18] service to advise and support metropolitan residents – whether homeowners, tenants or co-owners – in improving their housing. This service is supported by Maison de l'Habitat Durable (MHD)[19], a facility managed by the City of Lille and the Métropole, where everyone can find all the information they need to build, renovate or live better in their own home. By 2022, AMELIO had helped more than 12,000 low-income or very-low-income households with over 8,600 housing projects. This represents almost EUR 33 million in assistance from the Métropole, the Région, the Département and the Agence nationale de l'habitat[20]. Thanks to this work, homes consume less energy while being healthier and more comfortable, which reduces energy bills and helps combat climate change.

At the Maison de l'Habitat Durable, the Solaire en Nord[21] association, with the support of the Métropole, offers advice and training to individuals wishing to install solar panels on their roofs. This initiative complements the launch of a solar cadastre in May 2022, a free online tool enabling all metropolitan residents to simulate a roof's capacity to produce solar energy.

The city's six district heating networks, 66% of which are supplied by renewable and reclaimed energy sources, are another key element in our climate and energy policy, whose development also helps to alleviate pressure on the electricity system.

The Métropole européenne de Lille also supports local communities with a range of financial and technical tools. For example, a contribution fund dedicated to energy and low-carbon transition was set up in 2021, with a total envelope of EUR 25 million over five years to provide financial support for energy renovation and renewable energy production projects on communal assets. Several years ago, the Métropole also set up a shared energy consulting service for municipalities with fewer than 15,000 inhabitants, enabling them to pool qualified engineering expertise on the technical aspects of renovating their assets. A new, complementary dynamic has also been underway since the beginning of 2023 with municipalities of more than 15,000 inhabitants, to encourage feedback and the spread of efficient solutions.

Businesses also need support. In 2023, the Métropole launched transformation contracts, offering awareness-raising, personalized advice and help with sustainable building, which can finance the energy renovation of premises. The Métropole also facilitates contacts between companies with solarizable surfaces and photovoltaic developers. A call for private initiative, launched in the autumn of 2022, has

18 "Energy-efficient renovation of your home by the Métropole Européenne de Lille".

19 Home of Sustainable Housing.

20 France's National Housing Agency.

21 Solar in the North.

identified almost 106 hectares of land that could potentially be fitted with solar panels, which should lead to the installation of 35 MWp of photovoltaic power. Finally, in March 2023, the Métropole signed a renewable heat contract with ADEME, under which it is responsible for coordinating and managing a subsidy scheme for renewable heat production projects. The contract sets a production target of 25 GWh over three years.

The example of actions implemented in the metropolitan area shows how innovative technical solutions need to be deployed in tandem with accelerated behavioral change and support for local stakeholders. By presenting in a clear and instructive way the way in which smart grids and smart buildings can contribute to the energy transition, while taking into account the actual behavior of users, this book is perfectly in line with this approach.

Audrey LINKENHELD
Vice-President for Climate, Ecological Transition and Energy, MEL
First Deputy Mayor of Lille

Introduction

In the summer of 2022, a series of extreme climatic phenomena, uncharacteristic for that season, occurred all over the planet. This surprised meteorologists, as no weather model was previously capable of imagining or anticipating these extreme weather events. The culprit is climate change, caused by unbridled human activity that consumes energy and raw materials, leading us towards ecocide and the sixth mass extinction, dramatically reducing biodiversity.

In early 2020, a vortex of wet smoke 1,000 kilometers in diameter persisted for 13 weeks, traveling 66,000 km across the Southern Hemisphere. This new kind of specimen was spawned by another monster: a megafire that ravaged 5.8 million hectares of bush in Australia after 35 months of drought, exacerbated by global warming. When these ultra-convective plumes encountered the cool, moist air masses, they formed approximately 30 massive cloud formations known as "pyrocumulonimbus". These intruders disrupted the dynamics and chemistry found at very high altitudes (35 km high), strengthening the winds around Antarctica by around 40 km/h and destroying 20% of the Antarctic ozone layer[1], which was particularly large in 2020 and 2021 [NOU 22].

A deadly cloud of scorching humidity gripped regions located between a desert and an overheated sea, or an overheated irrigation zone, leading to intense evaporation. Humidity-saturated heat at 35°C prevents the human body from evacuating heat and leads to certain death in less than six hours, although this can be

1 The ozone layer, located between 20 and 30 km above sea level, filters out the sun's most dangerous UV-B rays. They have a harmful effect on all living organisms, both terrestrial and aquatic, as they can alter the DNA of cells. This layer therefore protects living organisms, but ozone is also a greenhouse gas; its increase near the ground will increase the greenhouse effect, contributing to increased climate change. The average concentration of ozone increases with the consumption of fossil fuels, and with the number of forest fires.

fatal as early as 31°C. This phenomenon, accentuated by global warming, leads to an increase in latent heat[2] throughout the atmosphere, particularly affecting sub-tropical coastal areas, such as India and Pakistan, but which in future could also affect the American Midwest, Southeast Asia and the coastlines of the Mediterranean. This humid heat can affect the survival of warm-blooded animals and humans during extreme heatwaves [NOU 22].

The frequency and intensity of mega-hurricanes are on the increase, taking advantage of an ocean that is getting hotter as a result of global warming, an ocean that fuels these hurricanes. The frequency of Category 3 and above cyclones (wind speeds in excess of 180 km/h) is set to double by 2050 in most regions of the world, with maximum straight line wind speeds likely to increase by more than 20% [NOU 22].

At an altitude of 3,000 meters and a thickness of around 2 km, atmospheric rivers develop. When they meet a mountain or rising air front, all this water vapor rises, cools, condenses and triggers a deluge. These phenomena, which originate over tropical oceans, threaten many of the world's coastal regions and play a role in the floods that hit Western Europe, particularly Great Britain and the Iberian Peninsula. Global warming is exacerbating this phenomenon, since the laws of thermodynamics imply that a rise of 1°C allows the air to contain 7% more water vapor [NOU 22].

Droughts are on the increase, but there is a new phenomenon on the rise: flash droughts, in which water disappears from an area so quickly (in just five days) that no-one has time to react. The intensification of this phenomenon, which has affected the American Great Plains but could also affect the Amazon Basin and Southeast Asia, seems to be caused above all by a formidable coupling between the aridity of the soil and the atmosphere: low humidity in the soil and vegetation reduces the flow of evaporation into the air, while this drier air reduces the chances of precipitation – a feedback loop that is being reinforced by global warming [NOU 22].

All these phenomena occur on the back of the accelerated melting of glaciers and sea ice, raising sea levels by up to 3 meters by 2100 (depending on the scenario), endangering populations living near the coasts. In addition, global warming is

2 Latent heat causes a material to change state without altering its temperature (e.g. liquid water turns into steam at 100°C). When latent heat is absorbed or released, it causes instability in the atmosphere, which can lead to severe weather. The change in latent heat alters the temperature of objects when they come into contact with warmer or colder air. Available at: www.greelane.com/fr/science-technologie-math%C3%A9matiques/science-4132464.

altering ocean chemistry, including acidification, affecting marine life, as is already visible in corals.

"Without adaptation measures, and under a scenario of 3 degrees Celsius (5.4 degrees Fahrenheit) global warming by 2100 [compared with the pre-industrial era before 1750–1800], 90,000 Europeans could die from extreme heat annually", according to the European Environment Agency (EEA). However, if this warming is limited to the 1.5°C targeted by the Paris Agreement (COP 21), the number of deaths will be divided by 3. The EEA also points out that climate change favors the emergence and transmission of infectious diseases (including mosquitoes that carry malaria and dengue fever).

But even if the Paris Agreement targets are met, tropical regions could still experience dangerous heatwaves for several months of the year, with temperatures that are unbearable for the human body. A temperature of 39.4°C is considered dangerous for the human body, and 51°C extremely dangerous. Figure I.1 shows the number of days when temperatures will exceed the mortality threshold in different regions of the planet in 2100 in the RCP 8.5 scenario, which simulates a business-as-usual situation, that is, with no change in our current lifestyles. By 2050, the global average temperature will have risen by around 2°C, and by 2100 by around 4°C. In this scenario, greenhouse gas emissions would not stabilize until 2250! This would encourage the displacement of populations, with all the geopolitical imbalances and precariousness induced by uncontrolled migration.

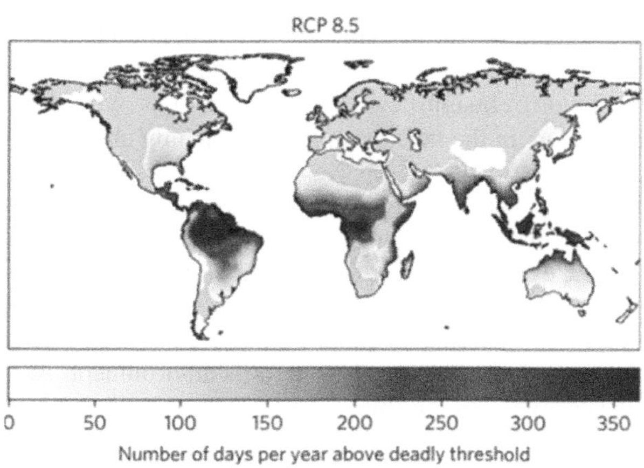

Figure I.1. *Number of days when the temperature would exceed the mortality threshold in different regions of the planet in 2100 in the RCP 8.5 scenario, which simulates a "business as usual" situation (Source: 5th report of the Intergovernmental Panel on Climate Change (IPCC) published in 2014). For a color version of this figure, see www.iste.co.uk/robyns/smartgrids.zip*

According to INSERM[3], every degree rise in temperature increases the risk of suicide, a risk that will be further increased by global warming.

Therefore, as the planet heats up, it is our bodies, our intellects and our institutions that will be placed under pressure. Moreover, the list of harmful effects induced by global warming could go on and on, as not all of them are necessarily known yet. What is more, an action or event that is a priori positive in terms of limiting greenhouse gas emissions may in fact have a negative effect! Here is an example: in 2020, while global CO_2 emissions fell as a result of Covid-19, the concentration of methane, another powerful greenhouse gas, soared in the atmosphere. Although methane has a short lifespan – only about 10 years in the atmosphere – it is the second most abundant greenhouse gas in the atmosphere, behind CO_2, and has a much higher warming potential than the latter; it is responsible for 20% of global warming. Although the origin of methane emissions is linked to fossil fuels, agriculture (mainly ruminant farts and burps), waste treatment, forest fires, etc., once emitted into the atmosphere, methane is largely oxidized by photochemical reactions. In particular, hydroxyl radicals (OH), present in infinitesimal quantities in the atmosphere, nevertheless account for 85% of methane destruction. This chemical species is produced by pollutant emissions, in particular nitrogen oxides (Nox). These Nox, which are largely responsible for air pollution in major cities, are mainly emitted by combustion engine vehicles. In 2020, it is important to note that these engines ran much less as a result of the pandemic [POU 22].

"The Earth is far more powerful than we imagined, and the human species is far more insignificant than we expected", says American economist Jeremy Rifkin [RIF 22, authors' translation]. Insignificant, vulnerable, but with what capacity for adaptation and resilience[4] in the face of rapid, far-reaching change?

"Every living thing has a structure that gives it coherence and a kind of functional unity, guaranteeing a "robustness" that enables it to retain a certain invariance over time", explain Vincent Grégoire-Delory and Thierry Magnin:

> The robustness of a living being thus defines its ability to maintain itself in the face of disturbances linked to its environment. At the same time, however, every living thing can be influenced by the effects of

3 Institut français de la santé et de la recherche médicale: the French Institute for Health and Medical Research.

4 *Le Robert* dictionary offers several definitions of resilience. In psychology, resilience is the ability to overcome traumatic shocks. In ecology, it is the ability of an ecosystem, species or individual to return to normal functioning or development after suffering an ecological disturbance.

> the external environment, thanks to the "structures" that accommodate these external influences (vulnerability). Plasticity [of the brain and of living beings as a whole] and adaptability are thus two essential characteristics of living beings. In this sense, we can apply the adjective "vulnerable" to malleable living organisms [vulnerability being an essential space for their growth], independently of any fragility linked to disease or deficiency. However, this tension between robustness and vulnerability can encourage a certain "fragility" in living things [GRE 22, authors' translation].

Living things therefore have a capacity for adaptation and resilience, which is nevertheless put to the test, and can be weakened by the speed and scale of change. The more pessimistic predict a collapse (which can be used to describe endangered species); a more positive reaction speaks to adaptation and resilience encouraging us to imagine and initiate large-scale transitions (energy, societal, agricultural, digital, etc.), and also revolutions (pacifist in our lifestyles, for example), and the reconstruction of ecosystems destroyed by humanity and harmed by global warming.

The planet is experiencing a serious ecological crisis, caused by humankind's unbridled consumption of energy and materials, which is having an irreversible impact on climate and biodiversity. The consumption of fossil fuels since the 18th century has undeniably enabled the development of industry, transport and living standards, particularly in the most industrialized countries, but it has also contributed to the increase in greenhouse gases (GHGs) in the atmosphere, which are causing global warming. This phenomenon is continuing, and if emissions of these gases are not rapidly and drastically reduced, global warming will have impacts on the planet and our lifestyles that will become increasingly significant and difficult to live with. Fossil fuels are the main source of greenhouse gas emissions (CO_2, methane), but not the only one.

Energy and societal issues are linked, as energy plays a very important role in our lifestyles, whether it is the energy we consume directly (for heating, lighting, food, transport, running all our electrical and digital appliances, etc.), the energy needed to produce the products we consume (to extract and transform materials, for cultivation, livestock farming, etc.) or through the impact of our lifestyles, which, for example, lead to deforestation that reduces natural sinks, favoring its storage in the atmosphere, and also through the impact of our lifestyles which, for example, lead to deforestation that reduces natural CO_2 sinks, encouraging its storage in the atmosphere.

The buildings in which we live, work and shop, as well as those used for leisure and sport, are the main CO_2 emitters.

The technologies used to build and operate them play a major role in the scale of these carbon emissions, but the way in which they are lived in and used also contribute to these emissions. These buildings are equipped with devices that enable them to better understand and control energy. They can become energy producers and store energy, thanks to the new materials used to construct these buildings, but also increasingly thanks to new energy and digital technologies, while ensuring the comfort of their occupants. They are becoming smart buildings.

Buildings are becoming intelligent, as are energy networks (smart grids), with increased user involvement. These buildings interact with energy networks, integrating new practices of energy self-production and self-consumption. All these systems are interconnected by information systems, leading to the convergence of smart buildings, smart grids, the Internet of Things and the Internet of People.

Buildings and housing in general, combined with energy networks, are becoming more environmentally friendly and low-carbon, making them the building blocks of tomorrow's cities. If we combine this with the quality of the living environment, with the return of nature to the city (or even urban agriculture), low-carbon transport such as public transport (transport is another major emitter of CO_2), spaces and organizations that encourage community life, and the democratization of information and decision-making (participatory democracies), enabling residents to be partners in the city and not just consumers, we have the ingredients for a smart city. This evolution of cities is important, because while cities today occupy 2% of the world's surface, they are home to 50% of the world's population, consume 75% of the energy produced and emit 80% of CO_2 emissions [ROB 19, ROB 23]. The challenge of making cities more sustainable, carbon-neutral and resilient is thus considerable.

For French economist Gaël Giraud:

> Tomorrow's France will be meshed with small urban centers whose buildings will themselves produce energy to be shared on smart grids, surrounded by organic polycultures, woven with short circuits, and linked by rail. This will be the end of large-scale distribution, zoning, stacked warehouses and supermarkets [VIO 20, authors' translation].

Smart buildings, smart grids, smart cities – these are concepts with many promising technologies for combating climate change, especially if they are inhabited and used by smart users[5] [ROB 23]. We could optimize everything with

5 Smart users, the subject of [ROB 23]. According to *Le Robert* dictionary, an intelligent person is one who has the faculty of knowing and understanding, who is, to a variable degree, endowed with intelligence. According to Wikipedia, intelligence is the set of processes found

suitable algorithms, and artificial intelligence, with a huge amount of data (big data) [GIO 21], could give instructions to all the smart components of the system and thus achieve the optimal behavior (of materials and humans) to save the planet! It is a Big Brother dream that the technosciences could soon make a reality! While technological solutions will be part of the solutions of the future, it is becoming increasingly clear that these solutions will not be optimal and may even have negative effects for the planet, that is, if society does not take ownership of it; the ownership necessary for it to evolve, to enrich it with its own creativity and make it its own. Only in this way can these solutions be desirable and economically viable, since human rationality is not the same as machine rationality, and individual and social acceptability is not at all automatic [ROB 19, ROB 23]. We have arrived at the crossroads of technosciences and societal issues.

To make a success of the energy and societal transition, we need to get the various scientific disciplines to work together, to develop the interdisciplinary approaches between the humanities, social sciences and engineering sciences that are essential for a successful transition that belongs to everyone.

After taking stock of the necessary transition, this book proposes interdisciplinary approaches to help buildings, energy networks and their users and stakeholders contribute to the energy and societal transition. It covers:

– the importance of understanding energy production (technical systems) and consumption (users) together, in order to achieve energy sobriety objectives, that is, to question and design correlative techniques, uses and imaginaries in the implementation of smart buildings and smart grids;

– the importance of forecasting energy production and consumption;

– modeling the involvement of actors in a local electrical smart grid with a view to energy supervision;

– self-consumption within a local renewable energy community;

– the potential of smart buildings, particularly in terms of user comfort and energy efficiency;

– the contribution of full-scale demonstrators, and the acceptability and appropriation by users of the potential of electrical load shedding.

The interdisciplinarity of this book is ensured by the diverse expertise of its authors: engineers and researchers in the field of electrical grids and their cross-fertilization with buildings (Benoît Robyns, Dhaker Abbes, Benoit Durillon,

in systems, more or less complex, living or non-living, that enable us to understand, learn or adapt to new situations. In English, smart means intelligent, astute (Cambridge dictionary).

Christophe Saudemont), a sociologist (Hervé Barry) and a socio-anthropologist (Laure Dobigny). It follows on from the book *Smart Users for Energy and Societal Transition* [ROB 23], but can also be read independently.

After summarizing the climate challenges, Chapter 1 presents four typical scenarios devised by ADEME to achieve carbon neutrality by 2050. These scenarios suggest a number of possible solutions, with varying degrees of technological maturity, social acceptability and desirability, which are discussed: sobriety and prosperity (the best energy is the energy we do not consume!), the link between ecological transition and spirituality, ethical questions and political issues, forking researchers to introduce smart energy networks, smarter buildings in a desirable habitat, and the crossover between energy networks and buildings. The book's methodological contributions are summarized at the end of Chapter 1. The developments presented are positioned in relation to ADEME's 2050 scenarios, as well as in terms of their contribution to the UN's sustainable development goals.

In Chapter 2, socio-anthropologist Laure Dobigny discusses the limits of both an approach focused on technical change (energy efficiency) and one focused solely on changing practices. Demonstrating the intrinsic link between technical systems and uses implies thinking about techniques and their uses together, that is, that the appropriation of techniques by the user plays a central role in determining uses and thus the achievement of sobriety practices. It questions the "effectiveness" of the *smart* concept (smart city, smart grid, smart building) in meeting environmental challenges, as all these technical developments are based on the same imaginary of an ever-increasing distance between technical systems and their users. With this in mind, in Chapters 4 through 8, attempts are made to take better account of the user, via his or her actor profile and/or by enabling him or her to act on his or her environment by interacting with the technology, in the context of energy networks and buildings becoming smarter.

Chapter 3 shows the importance of predicting energy production and consumption, particularly intermittent renewable energies such as solar and wind power. Forecasting methods for predicting photovoltaic production using satellite images, cameras and artificial neural networks are developed. Artificial neural networks are also used to predict the consumption of tertiary buildings. These methods are being tested on the Université Catholique de Lille smart grid demonstrator [ROB 23]. These production and consumption forecasts make it possible to make better use of self-generated renewable energies by maximizing their self-consumption, for example, by modulating consumption (making it flexible) to synchronize it with the means of production where possible. These forecasts also enable greater flexibility with regard to the public electricity distribution grid, with a view to limiting any constraints on this grid and promoting the consumption of low-carbon, less expensive energy.

Smart grids will really deploy their potential for innovation by promoting interactions between the various actors in the power system (producers, consumers, storage operators, network managers), who themselves become "electrically" more intelligent actors. These actors have very different consumption and/or production profiles, and very different economic and societal objectives and/or constraints. As a result, new sets of actors may emerge, making it possible to target new economic models, but also to meet energy and climate challenges by promoting the development of renewable energies (REs). One of the challenges of this evolution is to ensure that all actors win, not forgetting those in fuel poverty [ROB 19].

Since understanding and integrating energy behavior into a network is one of the main challenges of the energy issue, this question will be tackled in Chapter 4 from an interdisciplinary angle. It is at the crossroads of electrical engineering, microeconomics and sociology that this exploratory work is situated, with the aim of ensuring efficient network management while respecting the sensitivities of each stakeholder in these new smart grids. This chapter attempts to answer the following questions, with the aim of bringing energy flexibility to the power system in a residential context: what are the involvement profiles of network actors? How can these profiles be modeled? How can these profiles be used for energy management?

Based on the residential stakeholder profiles identified and modeled in Chapter 4 (thrifty, eco-sensitive, technofiles, moderate indifferent opportunists, disengaged), Chapter 5 proposes an answer to the question: how can these profiles be used for energy management? An energy supervision strategy for a local residential smart grid, with the overall aim of bringing energy flexibility to the power system while integrating stakeholder involvement, is developed in two stages: a predictive supervision stage the day before and a real-time supervision stage. The game of actors is modeled using Game Theory, widely used in economics. Optimization of the actors' objectives (economic gain, renewable energy, comfort) uses dynamic programming. Real-time adjustment is achieved through a fuzzy-logic-based supervision stage, based on a methodology developed by the Laboratoire d'électrotechnique et d'électronique de puissance de Lille (L2EP)[6] [ROB 15, ROB 19]. This supervision is tested and analyzed using numerical simulations.

In order to maximize self-consumption within a local renewable energy community, Chapter 6 proposes a techno-economic energy exchange model using an approach combining Game Theory, the Alternating Direction Method of Multipliers (ADMM) optimization algorithm and a (distributed, secure and decentralized) blockchain. Actors exchange energy in a decentralized way, without intermediation. Exchanges are automated according to photovoltaic (PV) production and consumption via smart contracts, taking into account cost, PV self-consumption and

6 Lille Laboratory of Electrotechnics and Power Electronics.

the comfort of the actors. Highly differentiated actors are taken into account: permanent consumers, transient consumers such as electric vehicles (EVs), consumers/producer, producers, storage operators, public distribution system operators, etc. Through utility functions inspired by the developments presented in Chapter 4, actor profiles are taken into account. The electrical grid of the Université Catholique de Lille's historical island demonstrator, developed through its Live TREE Energy and Societal Transition Program [ROB 23], serves as a case study for the application of the energy management methods developed in this chapter. The network links four tertiary buildings, two PV power plants, a battery storage system and six EV charging points. An example of the implementation of energy exchange management via blockchain technology is developed. The energy consumption of different blockchain technologies is evaluated and compared with more conventional centralized management.

Chapter 7 defines the concept of the smart building, illustrating the services it can offer thanks to the vast amount of data it can collect and manage, particularly in the case of tertiary buildings. These digital data, managed, for example, using a building information modeling (BIM) digital twin, can offer new services to users and to the energy networks to which the building is connected (buildings become nodes in a smart energy network) and help to integrate the building more effectively into its environment. Several examples concerning the use of data in building management are presented. These involve the dynamic energy optimization of a building using weather data and potentially data provided by a BIM model, taking into account room occupancy and user comfort. The interest of managing indoor air quality (IAQ) to ensure user comfort is then developed. Finally, the contribution of blockchain technology to smart buildings is illustrated by two examples. However, while the deployment of digital technology offers many new opportunities for the development, management and enhancement of buildings, it must not overshadow the environmental impact of buildings, which must be drastically reduced in the future. To explain this evolution towards more environmentally virtuous buildings, two examples of tertiary buildings, one renovated, the other built, at the Université Catholique de Lille, are presented and compared. The user has a role to play in a building, contributing to its environmental performance while ensuring its comfort. This dimension is illustrated by the results obtained from a socio-technical approach to managing a building that was not initially designed as "intelligent". Sufficient building and low-tech approaches are also important ways of reducing the climatic impact of buildings. The case of residential buildings is used to illustrate the value of these approaches.

Implementing the energy and societal transition will require full-scale experimentation with a wide range of technological, economic and local societal organization solutions, while accepting trials, errors and adjustments. We need to validate the viability of technologies, their efficiency and their impact on the environment, as well as their social acceptance (which is by no means automatic), and find new economic models and design new legal frameworks.

Chapter 8 presents an approach to the development of demonstrations of new solutions to better control the consumption and renewable production of electrical energy, via self-generation and self-consumption of energy, electric mobility and the flexibility of electrical loads in buildings. The approach is both technological and sociological. Therefore, as part of the assessment of the potential for electrical load shedding in the demonstrator site, the possibilities for load shedding being provided by a socio-technical context, we first look at developments in the philosophy prevailing in building design; this philosophy sets out the respective responsibilities of technical and human factors in the proper energy operation of a building. Next, a socio-technical approach is taken to the potential for energy savings in the context of commercial buildings, with a focus on hot water tank savings and electric vehicle charging.

The majority of the work presented in this book (Chapters 3–8) was carried out as part of the Smart Buildings as Nodes of Smart Grids (SBnodesSG) industrial research chair held by the Junia Engineering Graduate School with nine companies and the financial support of the MEL, as part of the Université Catholique de Lille's Live TREE Energy and Societal Transition Program [ROB 23].

References

[GIO 21] GIORGINI P., MAGNIN T., *Vers une civilisation de l'algorithme ? Un regard chrétien sur un défi éthique*, Bayard, Paris, 2021.

[GRE 22] GREGOIRE-DELORY V., MAGNIN T. "L'humain troublé par sa vulnérabilité : exemple en sciences et technologies du vivant", *MSR*, pp. 41–54, 2022.

[NOU 22] NOUYRIGAT V., "Méga phénomènes climatiques. Ce qui nous attend", *Epsilon*, no. 17, pp. 43–57, 2022.

[POU 22] POULIQUEN F., "Climat : Pourquoi donc la concentration de méthane dans l'atmosphère a bondi en 2020 ?", *20 minutes*, 2022.

[RIF 22] RIFKIN J., BOUCHER Y., "La terre est beaucoup plus puissante qu'on l'imaginait, l'espèce humaine est beaucoup plus insignifiante que prévu", *L'atlas éco de la Voix du Nord*, pp. 1–11, 2022.

[ROB 15] ROBYNS B., FRANCOIS B., DELILLE G. et al., *Energy Storage in Electric Power Grids*, ISTE Ltd, London and John Wiley & Sons, New York, 2015.

[ROB 19] ROBYNS B., DAVIGNY A., BARRY H. et al., *Electrical Energy Storage for Buildings in Smart Grids*, ISTE Ltd, London and John Wiley & Sons, New York, 2019.

[ROB 23] ROBYNS B., LENGLET C., BARRY H. et al., *Smart Users for Energy and Societal Transition*, ISTE Ltd, London and John Wiley & Sons, New York, 2023.

[VIO 20] VIO-DURY P., NOYON R., *Face aux chocs écologiques*, Marabout, Paris, 2020.

1

From Transition Challenges to Smart Grids and Smart Buildings

1.1. Introduction

The consumption of fossil fuels (coal, oil, gas) since the 18th century has undeniably enabled the development of industry, transport and improved living standards, particularly for those living in the more industrialized countries; however, it has also contributed to the increase of greenhouse gases in the atmosphere, responsible for the warming of our planet's climate. This phenomenon is continuing, and if these gas emissions are not rapidly and drastically reduced, global warming will have an enormous impact on the planet and our lifestyles, which will become increasingly consistent and difficult to live with. Fossil fuels are the main source of CO_2 emissions but are not the only culprits.

In [ROB 23], connections between energy and societal issues were highlighted. Section 1.2 of this chapter summarizes the climate challenges. Section 1.3 presents four typical scenarios devised by ADEME to achieve carbon neutrality by 2050. These narratives suggest several possible solutions that are technologically more or less mature, and socially more or less acceptable, sought after or desirable, which are discussed in the remainder of this chapter.

Sufficiency appears to be a major area for development, but it is not the stuff of dreams. Section 1.4 identifies different types of sobrieties, introduces the concepts of smart sharing and wise sharing and discusses the link between human sufficiency and nature's prosperity and the potential drivers and motivators of sufficiency such as meaning, values, spirituality, etc.

The ecological transition raises numerous ethical questions of both a private (individual) and public (institutional) nature, which present a major challenge for our democracies today: to combine ethics and politics in a way that respects individuals and protects their interests, as well as those of future generations. After summarizing some of these ethical issues, the question of governance to ensure the energy and societal transition in view of its urgent nature is discussed in section 1.5, highlighting the need to reinvent our democracies. The difficulty for politicians to take account of the urgency of climate change and to act to reduce our societies' impact on the climate is a source of eco-anxiety, particularly among the younger generations.

Research must make an essential contribution to the success of the energy and societal transition. To do this, it needs to move out of the laboratory and into the real world, which is interdisciplinary, by testing innovations on a real scale and in real operating conditions, via demonstrations. Section 1.6 develops this necessary shift in the world of research towards trans-disciplinary research in the engineering sciences, the humanities and the social sciences and underlines the importance of systematically assessing the societal and environmental impact of research results. The UN's Sustainable Development Goals (SDGs) are benchmarks for assessing this societal and environmental impact, as are ADEME's four 2050 scenarios. The results presented in this book are based on research work that is generally interdisciplinary, often carried out in partnership with companies or local authorities and in many cases tested by means of demonstrations. They will be positioned in relation to the UN's SDGs and ADEME's scenarios.

Section 1.7 introduces the importance of the energy issue in terms of the transition, with energy consumption needing to be reduced and renewable energies to be increased, and introducing the notion of the smart grid. New modes of local self-production and self-consumption of energy represent a revolution adapted to the development of intermittent renewable energies (such as photovoltaics and wind power). Sharing this energy in collective self-consumption or within local energy communities raises new technological, economic, sociological and ethical questions: can this sharing be smart sharing or wise sharing? The nature of the digital dimension associated with the smart dimension, as well as a few other ethical and sociological concerns, are raised.

Section 1.8 introduces the concept of the smart building, after outlining some 2050 prospects for the building sector and positioning these prospects in relation to ADEME's 2050 scenario. The low-tech approach is also introduced, as it is of particular interest in the digital buildings of the future.

Section 1.9 intersects the concepts of the smart building and smart grid, showing that both will be actively interrelated in the future, with buildings becoming active nodes (producer, consumer, storage, flexible, integrating electric mobility) of future smart grids.

Section 1.10 summarizes the case studies and methodological contributions of each chapter in the book.

Section 1.11 draws attention to some of the issues raised by artificial intelligence (AI).

1.2. Climatic challenges

For hundreds of thousands of years, the concentration of carbon dioxide (CO_2) in the Earth's atmosphere has remained stable due to a balanced carbon cycle: CO_2 emitted was practically equivalent to CO_2 absorbed, which corresponds to carbon neutrality. However, since human activity has increasingly involved deforestation and fossil fuel combustion, what is emitted exceeds what is absorbed, and CO_2 concentration in the Earth's atmosphere is increasing, leading to a rise in the global temperature [GRA 16].

Due to human activity, the concentration of greenhouse gases has exploded since the pre-industrial period (1750–1800). The concentration of CO_2, the main greenhouse gas, has increased by more than 30% since the pre-industrial era. The combined effects of all greenhouse gases (CO_2, methane, nitrous oxide, carbon monoxide, ozone, etc.) are now equivalent to an increase of over 50% in CO_2 since that period [ROB 21, ROB 23].

Between 1860 and 2010, the average temperature of the Earth's surface rose by 0.6°C, and by 1°C if we extend the period to 2018 [INT 18] and 1.2°C in 2023[1]. Various prospective scenarios predict that by 2100, this temperature will have risen by between 1.5 and 7°C if current energy sources and consumption habits are not modified (an estimated increase of 4°C in France and 7°C in the Arctic). This considerable increase would be accompanied by a rise in sea levels of between 20 cm and 1 m, depending on the area of the planet concerned. While climate change appears irreversible, it is possible to slow it down by significantly reducing greenhouse gas emissions, which requires strong and urgent action.

Natural CO_2 sinks such as soils, trees and oceans are only capable of absorbing just under half of the CO_2 produced by human activity (in 2000). In order to stabilize CO_2 concentration at its current level, we would therefore need to immediately

1 See: https://wmo.int/fr/news/media-centre/lomm-confirme-que-2023-bat-le-record-mondial-de-temperatures.

reduce emissions of this gas by 50–70%. While such a drastic reduction is impossible, action is urgently needed, as we are faced with a cumulative problem. Indeed, given that the lifetime of carbon dioxide in the atmosphere is of the order of a century, it will in any case take several generations for CO_2 concentrations to stabilize at an acceptable level. In 2018, the Intergovernmental Panel on Climate Change (IPCC)[2] estimated that in order to limit global warming to between 1.5 °C and 2 °C by 2100, CO_2 emissions must fall by 45% by 2030 compared with 2010 and by 91% by 2050 [INT 18].

CO_2 is produced by the combustion of all fossil fuels: oil, gas and coal. CO_2 emissions from coal are around twice as high as those from natural gas, with those from oil in between [ROB 21, ROB 23].

Global warming and, more broadly, human activities are having an impact on biodiversity. The rate of species extinction is estimated to be 50 to 560 times higher than the extinction rate for stable biodiversity; this is known as the sixth extinction. Moreover, 11% of our greenhouse gas emissions are due to deforestation and land-use change, with uses that store less carbon, such as palm oil [GRA 16].

Livestock farming is also a major emitter of greenhouse gases, accounting for around 9% of global CO_2 emissions. The primary source of these emissions is the production and transport of food, which requires agricultural land and contributes to deforestation. The second source of emissions is ruminant gastric fermentation. According to the IPCC, beef farming emits five times more CO_2 worldwide than pig or chicken farming.

Figure 1.1 shows the evolution of annual mean global surface temperature, observed and simulated by the IPCC over the period 1850–2020. The black line represents the observed temperature. The brown line represents the temperature evolution simulated by the IPCC model, taking into account human and natural factors. The green line represents the evolution of this temperature considering only natural factors such as solar and volcano activity. The bold-colored lines show the model averages, and the colored shading shows the probable limits of the simulation

2 The IPCC was set up in 1988 and published its first report in 1990. Two years later, the United Nations Framework Convention on Climate Change was adopted, with the aim of stabilizing greenhouse gas emissions. The COPs, or Conference of the Parties, are the annual meetings of the members of this Convention. In 1990, under the Kyoto Protocol, industrialized countries agreed to reduce their emissions by 5% between 2008 and 2012 (compared with 1990 levels). However, these agreements were scarcely respected. It was not until 2015, with the Paris Agreements, that the parties made a strong commitment. They call for the rise in average global temperature to be kept below 2°C compared with pre-industrial levels or ideally 1.5°C. Only a tiny proportion of signatories are meeting these commitments.

range. This figure clearly shows the decisive influence of human activities. Over the same period, the increase in atmospheric CO_2 has risen from 280 ppm (or parts per million) in 1850 to 400 ppm in 2020, following a curve similar to that of observed mean temperature ([ROB 23] and Figure 1.1).

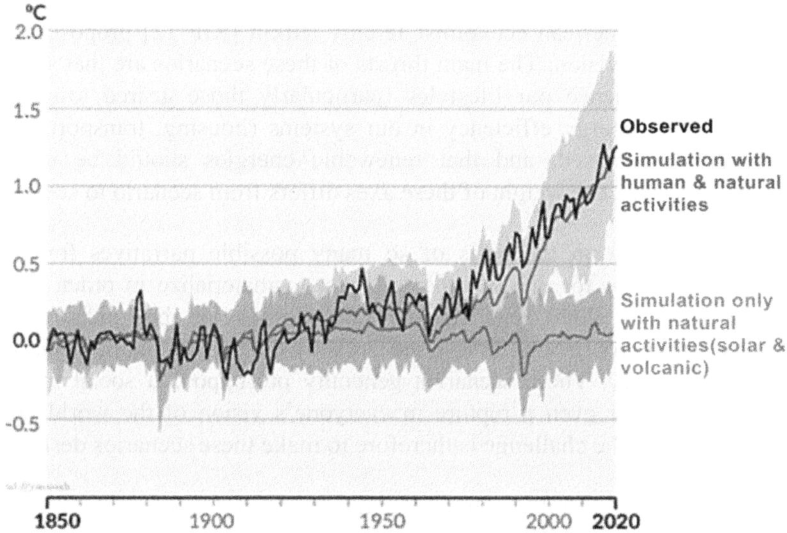

Figure 1.1. *Change in average annual global surface temperature observed (in black) and simulated by the IPCC[3] over the period 1850 to 2020, with simulation taking human activities into account (in brown) or not (in green). For a color version of this figure, see www.iste.co.uk/robyns/smartgrids.zip*

1.3. Four inspiring social and climate scenarios

In order to limit the temperature rise by 2100 to 2°C, we need to aim for carbon neutrality by 2050. Carbon neutrality does not simply mean that we no longer emit carbon at all, but that the carbon still emitted is either offset by the development of carbon sinks (forests) or that this carbon is captured and stored so that it is not released into the atmosphere.

With the aim of achieving carbon neutrality by 2050, various organizations are proposing scenarios based on the evolution of technologies, the economy,

3 IPCC Report 2021: Summary for Decision-makers. See: https://resumegiec.wordpress.com/2021/08/11/ipcc-report-summary-for-decision-makers/.

governance and territories, as well as our lifestyles. In France, three organizations have proposed scenarios for 2021: négaWatt, ADEME and RTE. The negaWatt [NEG 21, ROB 23] and ADEME [ADE 21] scenarios present proposals that can be widely shared worldwide.

Greenpeace [GRE 13], the World Commission on the Economy and Climate [THE 15] and the American economist Jeremy Rifkin [RIF 12] propose scenarios with an international vision. The main thrusts of these scenarios are that sufficiency should strongly influence our lifestyles (particularly those geared towards mass consumption), that energy efficiency in our systems (housing, transport, industry, etc.) should be increased, and that renewable energies should be used more extensively. However, the weight of these axes differs from scenario to scenario.

These scenarios form the basis of so many possible narratives for our next 30 years; it is realistic for one or other of them to materialize in order to achieve carbon neutrality by 2050, but are they desirable? One big unknown is the acceptability and appropriation of these scenarios by the population, particularly those calling for sufficiency. These scenarios generally presuppose a social mutation, a change in behavior or even a rupture in everyone's vision of the world and their relationship with it. The challenge is therefore to make these scenarios desirable.

In the remainder of this section, we introduce and discuss ADEME's four transition scenarios for 2050, to serve as a reference point for the methodological and experimental proposals and the various developments presented in the rest of this book.

Figure 1.2 summarizes ADEME's S1 and S2 transition 2050 scenarios, while Figure 1.3 summarizes ADEME's S3 and S4 scenarios [ADE 21].

Let us start with the fourth S4 scenario, entitled "Restoration Betting"[4], as it is the closest to current lifestyle trends. The lifestyles of the early 21st century are preserved, but the proliferation of goods consumes a lot of energy and materials, with potentially strong impacts on the environment. Society places its trust in the ability to manage, and even repair, social and ecological systems with more material and financial resources to maintain a habitable world. This exclusive reliance on technologies is a gamble, given that some of them are not yet mature. This gamble is based on, among other things, the capture and storage of CO_2 already present in the atmosphere, thanks to technologies that are not yet technically and economically mature on a large scale. This scenario also predicts an increase in the importance of

4 Proposed translation for: *Pari réparateur*.

digital technology, the ubiquity of the Internet of Things (IoT)[5] and AI [ROB 23], and data centers consuming 15 times more energy in 2050 than in 2020 [ADE 21].

This S4 scenario is a business-as-usual scenario, close to the current trend in our society, with a high risk of not achieving the necessary carbon neutrality by 2050, while leading to depletion of materials and other resources before 2050. Although this scenario relies heavily on innovation, research and development, it does not appear realistic from the point of view of the 2050 target and even less desirable. Indeed, we might well ask whether focusing on adaptation and repair might not entail ratifying the continued economic development of activities that generate greenhouse gases. François Gemenne then speaks:

> of "maladaptation", that is to say a "process of adaptation which results directly in an increase in vulnerability to climate variability and change and/or in an alteration of current capacities and opportunities and 'futures of adaptation'". This is the case when these projects increase greenhouse gas emissions, increase pressure on the most vulnerable, cost too much (compared with more effective policies), or commit countries to technological paths on which they will become dependent. Today, we have a problem with the concept of growth: we have always associated it with the idea of having more, doing more, and so growth is necessarily inseparable, for the moment, from greenhouse gas emissions. The big challenge will be the idea of associating growth with the idea of doing better, not more [VIO 20].

The third scenario, S3, entitled "Green Technologies", is based on technological development to meet environmental challenges, rather than on behavioral changes towards greater sufficiency. Metropolises are growing. Technologies and digital technology, which enable energy efficiency and the optimization of material resources, concern all sectors. The best technologies are widely deployed and widely accessible to solvent populations.

This scenario predicts that data centers will consume ten times more energy in 2050 than in 2020. However, the energy would be decarbonized, and the CO_2 still emitted would be captured and stored [ADE 21]. Numerous innovations, R&D projects and demonstrations [ROB 23] are being developed in line with this scenario. While this scenario seems realistic for reaching the 2050 target, we will need to assess the carbon impact of the technologies we develop with ever greater precision, in order to minimize it, while integrating the socio-technical, economic and ethical dimensions.

---

5 The Internet of Things concerns objects (appliances, buildings, etc.) connected to the Internet, enabling communication between them and digital algorithms.

In this scenario, the transformation of jobs towards more specialized ones is very likely, with the risk of accentuating inequalities if "green technologies" are difficult to appropriate and not necessarily affordable. This scenario, which is essentially aimed at solvent individuals, therefore runs the risk of increasing social inequalities, and as such is undesirable.

ADEME's second scenario is entitled "Territorial Cooperation"[6]. Society is being transformed by shared governance and territorial cooperation. Non-governmental organizations, public institutions, the private sector and civil society are finding pragmatic ways of cooperating to maintain social cohesion. To achieve carbon neutrality, society is betting on a gradual but steady evolution of the economic system towards a sustainable path combining sufficiency, efficiency and renewable energy. Consumption of goods is becoming measured and responsible, and sharing is becoming widespread [ADE 21]. Meat consumption is halved, and housing renovation is favored over new construction and adapted to the number of occupants. Land is no longer artificially developed, and nature is preserved and even protected. Personal mobility is reduced by 17%, while encouraging the development of medium-sized towns.

This scenario relies on the sharing economy (or economy of functionality), the circular economy, second-hand approaches and therefore second lives for appliances, clothing, etc. To ensure that everyone is aware of their environmental impact, the carbon footprint of each and every one of us should be measured on a regular basis. This scenario seems both realistic and desirable for reaching the 2050 target. One major obstacle will be the highly developed individualism of Western societies.

The first scenario, entitled "Frugal Generation"[7], relies on major alterations to the ways in which we travel, heat, eat, buy and use equipment. It makes it possible to achieve carbon neutrality without involving carbon capture and storage technologies, which are unproven and uncertain on a large scale. New consumer expectations and, above all, new consumption practices are rapidly taking shape. The growth in energy demand, which is depleting resources and damaging the environment, is being interrupted by behavioral, organizational and technological innovations. The transition is mainly driven by frugality through constraint (increasing user costs, restrictive regulations, etc.) and sufficiency [ADE 21].

---

6 Proposed translation for: *Coopérations territoriales*.

7 Proposed translation for: *Génération frugale*.

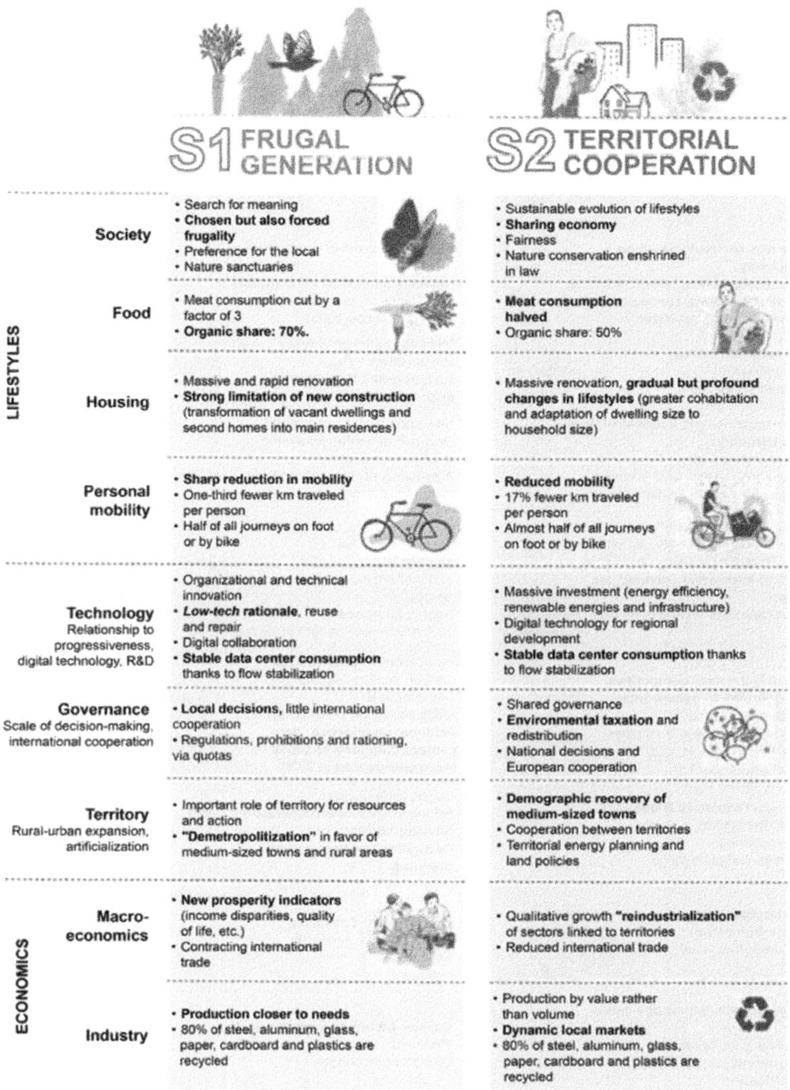

Figure 1.2. *Summary of ADEME's S1 and S2 transition 2050 scenarios [ADE 21]*[8]

8 See: https://transitions2050.ADEME.fr/.

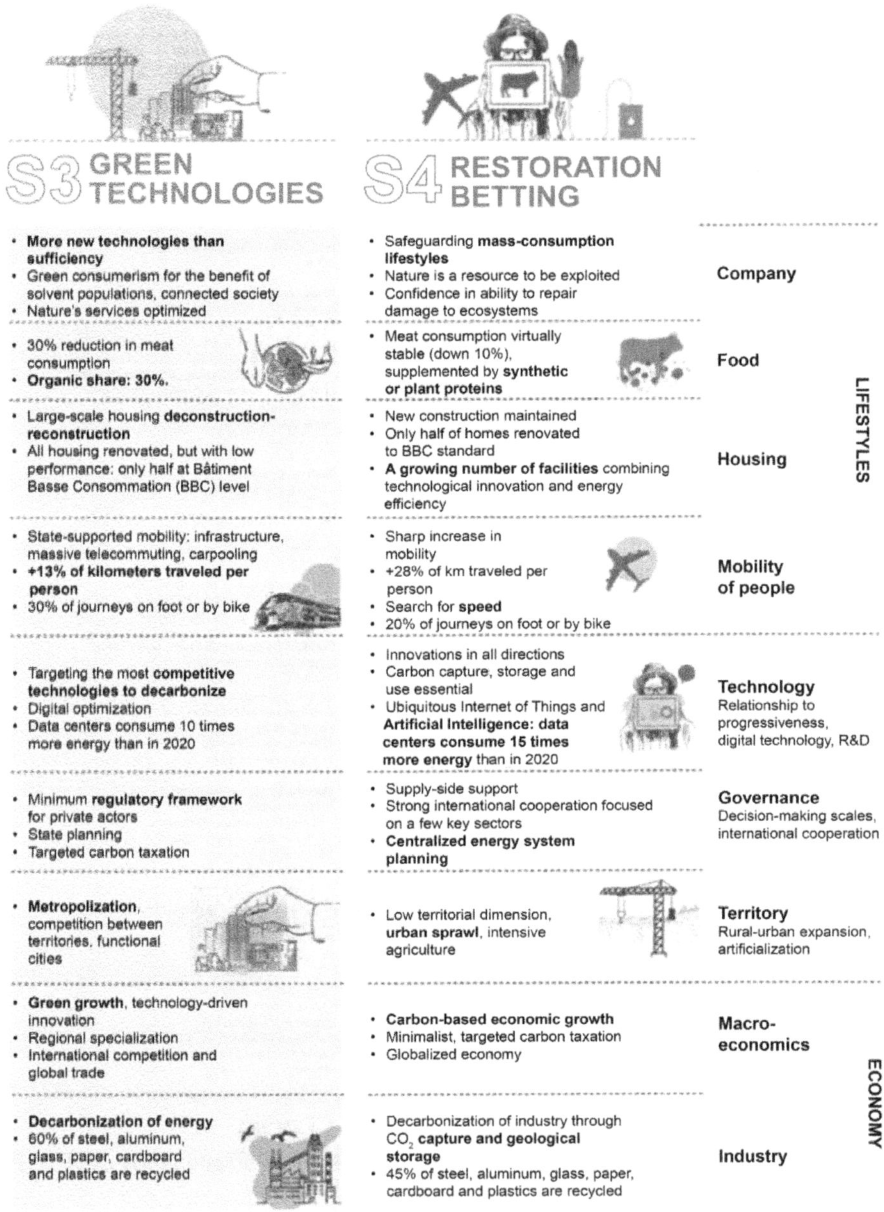

Figure 1.3. *Summary of ADEME's transition 2050 S3 and S4 scenarios [ADE 21]*[9]

9 See: https://transitions2050.ADEME.fr/.

This scenario is based on a search for meaning, leading to a consensual frugality (but also partly constrained by legislation), a preference for local products and a sanctuary for nature. It advocates a threefold reduction in meat consumption; a sharp reduction in mobility, stipulating that half of all journeys be made on foot or cycle; and the renovation of 80% of the building stock to be converted into low-energy buildings. Governance favors local decision-making, and the introduction of new economic indicators (based on income disparities, quality of life, etc.), the development of low-tech production (with a low carbon footprint) as close as possible to needs and a circular economy (with a high material recycling rate). To achieve these objectives and make this scenario realistic, we sense the need for strong governance, a reinvention of democracy (a strong democracy with a mandate to change lifestyles?) and the existence of a truly ecological political force that believes in its legitimacy. To make this scenario desirable and sought after, we need to show the link between human sufficiency and the prosperity of the planet and its biodiversity and the link between sufficiency and a quest for meaning and spirituality. A major obstacle to this scenario is the greed that has become the moral and economic model of the 20th century, but which is also inherent to human nature [BHO 19]. These issues will be discussed later in this chapter and touched upon throughout this book.

Jeremy Rifkin, who was the promoter of the Third Industrial Revolution [RIF 12], which was implemented in particular in the Hauts-de-France region [ROB 23], gives his vision of the future, updated for the world of today:

> The change in consciousness is real, it is resilience versus efficiency. Be more sufficient? Okay, we can make an effort, consume less, etc. But that is still staying within the same framework. It means staying within the same framework, in a more cramped environment. That is not enough. We need to move to a completely new system of economic governance, from growth to fulfillment, from financial capital to ecological capital, from productivity to recycling, from GDP to quality-of-life indices, from hyper-consumption to cyclical economics, from vertically integrated scales to a lateral economy, from giant conglomerates to networks of connected SMEs, from intellectual property to open sources, from geopolitics to bio-management or biopolitics [RIF 22, authors' translation].

In [ADE 21], ADEME compares the different scenarios applied to France, according to a number of criteria. By way of illustration, Figure 1.4 compares the energy consumption of the four scenarios in 2050 with that of France in 2015, specifying the share of renewable energies (REs). In scenario S1, energy consumption in 2050 is 44.6% of that in 2015, while in scenario S4 it is still 77.6%.

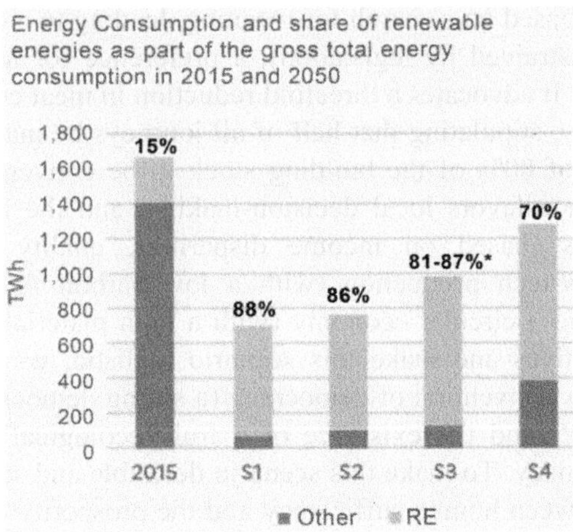

Figure 1.4. *Comparison of the energy consumption of the four scenarios in 2050 with that of France in 2015, specifying the share of renewable energy (*values dependent on industrial policy choices to develop floating wind farms or nuclear power) [ADE 21]. For a color version of this figure, see www.iste.co.uk/robyns/smartgrids.zip*

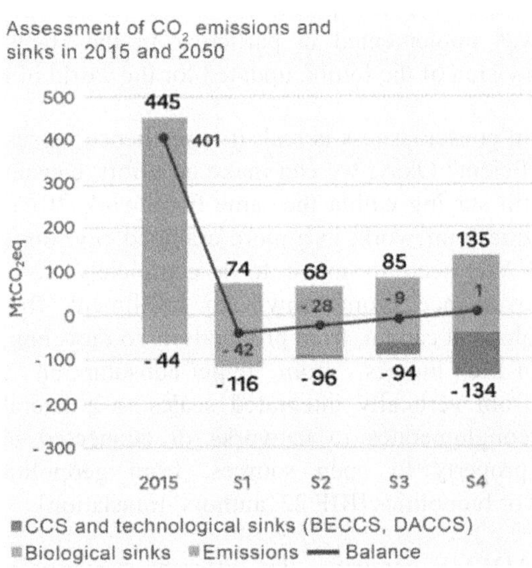

Figure 1.5. *Carbon emissions and sinks for the four scenarios in 2050, compared with the 2015 assessment for France [ADE 21]. For a color version of this figure, see www.iste.co.uk/robyns/smartgrids.zip*

Figure 1.5 shows the emissions and carbon sinks for the four scenarios for France in 2050, compared with the 2015 figures. Natural carbon sinks are shown in green, and the carbon that would be recovered by technological solutions is shown in gray. The importance of technological CO_2 capture in 2050 in scenario S4, which frankly is a bold bet on the future, makes this scenario less credible than scenarios S1, S2 and S3.

1.4. Sufficiency or prosperity

1.4.1. *Personal, shared and organizational sufficiency*

Sufficiency is therefore one of the major areas to be developed. It takes on several dimensions, partly conceptualized by the négaWatt Association[10] [VIL 18, WIK 20a, ROB 23], in a spirit of equity and sharing:

– dimensional sufficiency: choosing equipment that is adapted to needs, when making purchases or investments (e.g. preferring a small surface area for your home, using a vehicle that is adapted to the load and number of passengers, etc.);

– cooperative sufficiency: pooling uses, whether of space, goods, etc. (e.g. carpooling, car-sharing, sharing, lending equipment to neighbors, etc.);

– sufficiency of use: reasonably managing the use of appliances and goods (e.g. eco-driving, precautions to limit breakage and premature wear of goods, heating regulation, etc.);

– organizational sufficiency: structuring activities differently in terms of space and time (e.g. promoting telecommuting, regional planning, provision of public transport, etc.);

– material sufficiency: reducing the consumption of material goods and products (e.g. reducing the level of equipment, limiting packaging, etc.). These same goods indirectly require energy to design, assemble, transport, etc. This is known as indirect energy or gray energy.

Sufficiency can therefore be a personal commitment, shared and collaborative, but it also concerns organizations (companies, administrations, local authorities, associations, etc.). It is the commitment to sufficiency at all these levels that will contribute to achieving the 2050 objective of the energy and societal transition. While organizational sufficiency can be imposed, at least in part by the management

10 French non-governmental think tank for energy experts. Recommends solutions on energy demand and supply in the context of France's energy transition.

of these organizations or by the legislator, personal and shared sufficiency requires a strong commitment on the part of each individual.

1.4.2. *From smart sharing to wise sharing*

For businesses, and also for healthcare organizations and other sectors, Navi Radjou [RAD 19, RAD 21] proposes a business-to-business (B2B) sharing economy, which complements the peer-to-peer sharing economy (consumer-to-consumer sharing (C2C), strongly developed since the 2008 economic crisis) whose champions are now known to all: Uber, Airbnb, BlaBlacar, etc. For Radjou, the B2B sharing economy has potential benefits for social inclusion, the environment and health, particularly if we can move from smart sharing to wise sharing. This sharing can involve waste that can become raw materials for other companies, buildings and other equipment; joint purchasing; skills via human resources; customers; or even intellectual property.

But why would a company want to commit to B2B sharing? Companies need to be clear about their motivation and make clear their commitment to this path. Radjou's analysis highlights the following fact:

> If most companies answer the "why?" question with "we just want to save money or make more money", we shall see the rise of smart sharing, whose limited objective is simply to make the current capitalist economic model, with all its flaws, even more efficient. However, if most companies answer "we want to have a positive social and ecological impact", then they can enable wise sharing, whose lofty goal is to radically reinvent entire industries and accelerate the transition to a frugal economy in the post-Covid-19 world [RAD 21, authors' translation].

Radjou summarizes his analysis graphically, shown below in Figure 1.6.

One potential benefit of wise sharing for companies is to contribute more rapidly to the 17 SDGs to save the world, outlined by the United Nations (UN) and shown in Figure 1.7. These Sustainable Development Goals set out the path to a better, more sustainable future for all. They address global challenges such as poverty, inequality, climate change, environmental degradation, prosperity, peace and justice. These 17 objectives are broken down into 169 sub-objectives, to which the contribution of each actor (company, community, association, university, etc., as well as individuals) can be assessed by means of indicators.

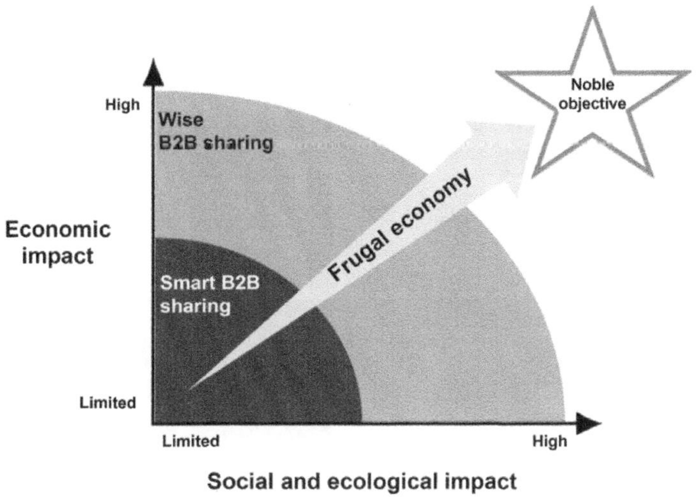

Figure 1.6. *Economic, societal and ecological impacts of smart sharing and wise sharing between companies (B2B) [ARD 21]. For a color version of this figure, see www.iste.co.uk/robyns/smartgrids.zip*

Radjou points out, however, that:

> Given the climate emergency, companies must go far beyond the concept of "sustainability" and rapidly adopt the principles of regeneration, a set of science-based practices that aim to reverse global warming, for example by sequestering carbon in products and processes. Whereas a sustainable business just aims to reduce its carbon footprint and waste (negative impact), a regenerative business consciously aims to expand its positive socio-ecological footprint – i.e. "doing more good" – by boosting the health and vitality of people, places (communities) and the planet [RAD 21].

Food companies, for example, can help reduce world hunger by sharing their intellectual property with others (in the same way that Open Science is developing within academic research by making scientific publications accessible to all). Medical care can be improved by sharing care equipment and pooling purchases (including medicaments), thus improving health and well-being worldwide.

Figure 1.7. *The 17 UN Sustainable Development Goals (SDGs)*

Radjou also cites a thought-provoking example of carpooling in France:

> 70% of French people drive to work and spend an average of €6,049 a year on car maintenance. 70% of these home-to-work journeys are made alone. Private cars contribute to almost 16% of France's greenhouse gas emissions. Public transport is not practical for all workers, given that 40% of French people live in areas poorly serviced by public transport. Inspired by the great success of BlaBlaCar, which facilitates carpooling between individuals, startups Klaxit and Karos have built carpooling networks dedicated to companies, offering their employees a sustainable, comfortable, user-friendly, and economical mobility solution. Klaxit and Karos alone facilitate more than three million home-to-work carpools a day. Karos estimates that, thanks to its carpooling services, its business users save an average of two full tanks of gas per month and 26 minutes per trip [RAD 21, authors' translation].

1.4.3. *From sufficiency to prosperity*

Sufficiency, however, is not the dream of the majority of the population, even though over the centuries, philosophers, hermits, monks, clerics, movements seeking simplicity or even a break with the dominant world, etc., have extolled its virtues or lived it. However, the greed advocated as a core value in the West throughout the 20th century is at odds with sufficiency. In his book *Le Bug Humain* [BHO 19], Sébastien Bohler argues that:

> In reality, the brain is a largely defective organ, bent on destruction and domination, pursuing only its own interests and unable to see beyond a few decades. We are swept along in a headlong rush of overconsumption, overproduction, overexploitation, overeating, over-indebtedness, and overheating, because part of our brain is automatically pushing us to it, without us currently having the means to curb it [BHO 19, authors' translation].

This controversial thesis needs to be qualified, however, by considering the influence of other biological mechanisms, the weight of socialization, education, norms, etc., and all the new contributions of the human and social sciences [WIK 22].

According to economist Gaël Giraud:

> [I] do not believe that over-consumption is deeply rooted in the human soul. Material abundance as such is not part of the Enlightenment project. Contemporary consumption is a matter of impulse, but desire is not impulse: the Lacanians will be happy to remind you of this. Billions are spent every year on marketing and communication to keep these artificial needs alive, this deadly idea that life is one big supermarket, and that happiness lies in the devouring impulse. This conditioning will evaporate in a few years as soon as we ban advertising for polluting goods and services, starting with SUVs. Grenoble has banned all advertising in public places. What a liberation for our brains! Imaginations can change very quickly [VIO 20, authors' translation].

Sociologists and philosophers Bruno Latour and Nikolaj Schultz [LAT 22] encourage the development of an ecological class based on the following analysis:

> Energies are easily mobilized when it comes to increasing production. Yet increased production, development and progress are aberrations associated with the destruction of the Earth's habitability. However, our mental structures, our moral systems, our organization, our administration, our laws, all our "equipment" is designed to manage increased production (authors' translation).

The ecological class must therefore work towards a:

> decisive change of tack [...] giving priority to maintaining the habitability conditions of the planet and not to developing production. [...] As the long history of the Earth shows, it's living things that have

> enabled the continuity of terrestrial life [...] To produce means to assemble and combine resources; it does not mean to engender, that is, to carefully induce the continuity of beings that the world's habitability depends on. Instead of the strange metaphor of development, it would be more helpful, if we want to capture this reversal, to talk about envelopment: all the issues of production are surrounded, wrapped up in the practices of engendering they depend on. We're used to seeing "growth" as the sole means of getting out of trouble, forgetting about the destruction it causes, whereas prosperity has always depended on engendering practices. It's not a matter of degrowing but finally prospering [LAT 22].

Prosperity, then, rather than sufficiency, in the knowledge that the prosperity of nature is generally conducive to the prosperity of human beings. But how can we reverse this worldview and return to the principles of growth and development of living things to be protected and imitated (biomimicry) and concern ourselves not only with the world in which we live, but also with the world from which we live [BOU 22, LAT 22]?

Let us try to interpret these different considerations by means of the synthetic diagram in Figure 1.8[11]. Human sufficiency contributes to the prosperity of a nature that is no longer systematically enslaved by humankind to satisfy its growing needs. As nature regains its freedom, it becomes abundant; observing it, contemplating it, meditating within it on our link with it, living a spiritual experience that connects us to humanity past, present and future, to nature, to the cosmos and beyond become sources of inspiration for a more sufficient life that goes to the essentials. Encouraging existential and civic commitments, including concern for otherness, the most vulnerable and the most disadvantaged, opens up intelligence in all its various forms.

Obviously, greed distances us from nature and is not conducive to its prosperity. Nevertheless, awareness of the benefits of a more sufficient life must be rooted in a world order experienced by each individual. This order can be religious, secular or cosmological. Ideally, it will link what we observe, contemplate, meditate on or experience spiritually, inscribing it in a narrative that feeds our imagination and positively affects our emotional world. This will enable us to stop living in cognitive dissonance, i.e. according to psychologists, in tension between our thoughts, beliefs and emotions, in the hope of leaving our grandchildren a pleasant planet to live in, and our attitudes, which are still largely consumerist and predatory towards nature.

---

11 Benoît Robyns thanks Anne-Laure Boursier, a doctoral student in philosophy at the ETHICS Unit of the Université Catholique de Lille, for her inspiring exchanges on this subject.

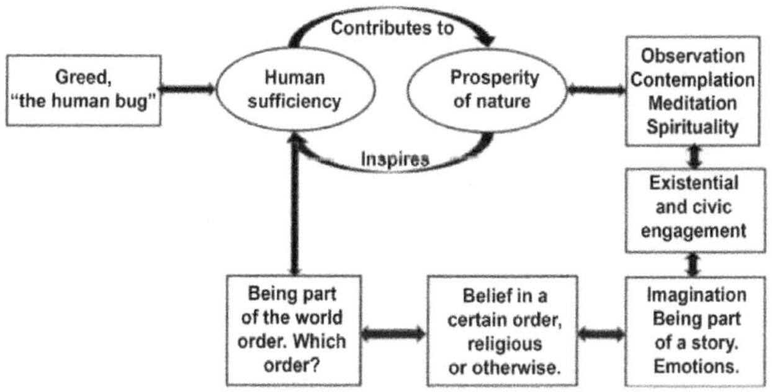

Figure 1.8. *The link between human sufficiency and nature's prosperity, what strengthens this link and what detracts from it*

1.4.4. *Spirituality and ecological transition*

In his encyclical *Laudato si* [POP 15], Pope Francis urges us to listen to the clamor of the Earth and the clamor of the poor! To become aware of a common origin, a mutual belonging and a future shared by all.

This awareness will be greatly helped by a spiritual experience that everyone can have, like the one described by André Comte-Sponville in his essay *L'esprit de l'athéisme*:

> We were walking […]. The starry sky above me, immense, unfathomable, luminous, and nothing in me but this sky, of which I was a part, nothing in me but this silence, this light like a happy vibration, like a joy without subject, without object […]. Nothing in me but the dark night, the dazzling presence of everything […] [COM 06, authors' translation].

Or as Spinoza wrote in *Ethics*: "We feel and experience that we are eternal" [SPI 11, authors' translation].

> Well, there you have it: I had felt and experienced it, indeed, and it was like a revelation to me, but without God. It was the most beautiful moment I've ever experienced, the most joyful, the most serene, and the most obviously spiritual. […]. Intellectually, I see no proof of anything: but I can't pretend it didn't happen either [COM 06, authors' translation].

Philosopher Glenn Albrecht speaks of the "Earth Emotions":

> What does a surfer feel when he catches a beautiful wave? How do we feel when we walk into a forest? When we walk beside the ocean? For much of our history, we have not needed to name these psychic states, to describe what moves and affects us when we come into contact with the Earth. However, as a result of deforestation, land artificialization, global warming and the sixth mass extinction, we are losing these emotions, along with landscapes, species, sounds and the fine-tuning of the senses that brought them to life within us. At the same time, we are increasingly mediatizing our relationship with nature: held back by our telephones, obsessed by our screens, we find it harder and harder to literally feel the beauty of our environment. We live entrenched in technological bubbles, monopolized by the pulses of work and consumption [VIO 20, authors' translation].

Thierry Magnin analyzes the following fact:

> Faced with the great challenges of ecological transition, for example, human beings realize that without an inner drive, shared with others, no real change is possible. Lifestyles cannot be transformed without transformation of the heart, without "conversion" in the etymological sense of turning around. Does not the return of spirituality in our Western societies correspond to a lack that may have dried up man's heart and enclosed him in a dehumanizing materialism? [MAG 22, authors' translation].

According to André Comte-Sponville, "the spiritual [...] is that which appeals to man's interiority, makes him reject the inhuman, invites him to fulfill himself in a search for transcendence and to give meaning to his actions, puts him in touch with others and leads him to give, exchange and receive" [COM 06, author's translation].

In an essay, Thierry Magnin analyzes the spiritual dimension of the human being, linking its body-psyche-spirit dimensions [MAG 21] from non-religious and religious angles. He stresses the difference and complementarity of psychological and spiritual experiences. Medical experience highlights four dimensions of an individual's spiritual life:

– meaning (that which guides life, what gives a profound reason for living);

– values (what characterizes what is good and true for a person);

– transcendence (relationship to the ultimate, elements that transcend a person, in relation to that which a person names his or her existential relationship);

– identity (singular coherence of the person, uniqueness of the subject, whether at the level of his or her thoughts, emotions or relationships) [MON 10].

More recently, the life sciences have highlighted the relationship between the biological, the psychic and the spiritual. Biology shows that the environment influences the evolution of living organisms, right down to the expression of their genes (epigenetics). Epigenetics shows that certain genes are inhibited, while others are strongly expressed, depending not only on the biological environment, but also on the behavior of living beings themselves. In the case of humans, for example, it has been shown that nutrition, exercise, stress management, pleasure and social networks can all influence epigenetic mechanisms, demonstrating that the biological and psychic realms are in a permanent reciprocal relationship. Epigenetics opens up new horizons, suggesting that what humans pass on to their descendants could be the fruit, in part, of their behavior [MAG 17, MAG 21, MAG 22]!

Thierry Magnin notes that in so-called "mindfulness" meditation (presently very popular), we initially focus on the level of the mind (the mental, in terms of the psyche), favoring a possible opening to the universal and to the spirit (and therefore potentially to awareness of nature and its prosperity, and of our interdependence with other humans and with nature) [MAG 21].

Teilhard de Chardin, a forerunner of the "everything is connected" concept, emphasized the collective aspect of human interaction, of creative union at the level of humanity, under the influence of the Omega point of convergence in the universe, the source of the "amorization"[12] that irrigates all creation (the Noosphere nourished by Love, by Spirit). Love is thus the summit of the spiritual life. A love that personalizes within a linked humanity: a link between two humans "from center to center", says Teilhard, a link between humans and the cosmos, all of which enables a dynamic of Socialization and Personalization [DEC 76, MAG 21]. For Teilhard, Personalization is that internal deepening of consciousness upon itself that characterizes the particular destiny of the element (individual) that has become fully self through reflection. Teilhard extends this notion to the collective future, to the evolutionary equilibrium of the Noosphere [DEC 55].

12 "Amorization" refers to the act of being primed, strengthened or invigorated by divine love. "The priming of the Universe [...] The Universe "amorising" and personalizing itself in the very dynamism of its evolution [...]: provided sufficient attention is paid to its extraordinary and significant power of "pan-amorization"" (Pierre Teilhard de Chardin, *Le Christique*, 1955) (French Wiktionary definition, authors' translation).

For Luc Ferry, the question of meaning and that of the sacred – why it would not be foolish to consider sacrifice – are inseparable [FER 96]. The sacred can be religious, it can be our children, grandchildren and great-grandchildren, it can be nature, etc. Ferry highlights a secular spirituality in *La révolution de l'amour* [FER 10], which does not deny a secular transcendence (in immanence, which assumes that there is something that comes from outside, of which we are the discoverers and not the creators), and which can be a source of powerful commitment in the ecological transition.

Spirituality is therefore a human dimension linking humankind's perception of themselves to their relationship with the world, and with transcendence – that which is greater than us and passes through us – whether religious (the attribute of God, "the transcendent" par excellence) or secular (that which is beyond what we commonly admit being humanly capable of thinking). It is an experience that appears to be a powerful motor for committing to the ecological transition by transforming our habits and lifestyles over the long term. This spiritual dimension broadens our horizons, encouraging sufficiency and frugality, gaining freedom, and finding peace by uniting body, mind and spirit, divided between the gravity of the planetary situation and the absence of decisive changes in our behavior. Drawing on transcendence provides us with external resources and powerful energy, whereas human beings, however militant they may be, who draw solely on their internal energy, run the risk of burning out more quickly.

1.5. Ethical and political issues

1.5.1. *Ethical issues in the ecological transition*

In [ROB 23], Malik Bozzo-Rey summarizes some of the ethical issues raised by the energy and societal transition. To sum up, climate issues and the necessary transition raise immense challenges such as:

– the confrontation between the interests of people living now and those who will live in the future (with the difficulty of acting in the interests of people we will never know, such as our great-great-grandchildren);

– the prospect of a less prosperous future and/or one with living conditions different from today's, forcing us to reassess what we mean by a "good life" or a "fulfilled life" (or a "more sufficient life");

– if the basic needs of the majority of the world's population cannot be met in the future, then we will have to face tragic situations where the choice between life and death will no longer be simply hypothetical.

Bozzo-Rey stresses the need to rethink and question three presuppositions that have structured modern ethical thinking:

– the interests of present generations coincide with those of future generations;

– future people will be better off than we are;

– favorable living conditions will continue indefinitely.

He concludes that in order to identify the best ways of influencing behavior and lifestyles, the energy and societal transition requires the unification of private (individual) and public (institutional) ethics, something which presents a major challenge for our democracies today: to combine ethics and politics in a way that respects individuals and protects their interests, as well as those of future generations [ROB 23].

In *The Imperative of Responsibility*, Hans Jonas stresses that "we are responsible for the future, all the more so insofar as we know neither certain consequences (of our technological choices, our lifestyles), nor their probabilities of occurrence (of the harmful effects induced by our choices), yet these may have an "infinite" destructive power (or at any rate be sufficiently great to compromise human survival on a large scale)". More than 40 years ago, Hans Jonas was already advocating the Responsibility Principle, which reads as follows: "Act in such a way that the effects of your action are compatible with the permanence of authentically human life on Earth" [JON 77, authors' translation].

In his encyclical *Laudato si* [POP 15], supplemented by the encyclical *Fratelli tutti* [POP 20] on fraternity and social friendship, Pope Francis insists that "it is not enough to balance, in the medium term, the protection of nature with financial gain, or the preservation of the environment with progress [...] it is a matter of redefining our notion of progress [...] A technological and economic development which does not leave in its wake a better world and an integrally higher quality of life cannot be considered progress" (LS 194).

1.5.2. *Questions of governance or the need to reinvent democracies*

ADEME's various scenarios are based on technological developments and changes in individual behavior, as well as on modes of governance that may differ considerably from one scenario to the next. Scenario S1, "Frugal Generation", assumes a strong state presence, but which kind of state? Scenario S2, "Territorial Cooperation", assumes a decentralization of power to local authorities that cooperate with each other. Scenario S3 "Green technology" assumes a regulatory state. Scenario S4 "Restoration Betting", on the contrary, is part and parcel of today's globalization, a globalization that has been undermined by the Covid-19 crisis and the geopolitical situation brought about by the war in Ukraine (to name but one example). Nevertheless, ultimately, does not the ecological emergency call for new planning "from the top down"? According to Patrick Jolivet of ADEME:

> To achieve carbon neutrality by 2050, we need to effectively implement structuring choices now. However, planning does not exclude participation. Just as participation does not induce paralysis. Didn't the 150 members of the Citizens' Climate Convention[13] manage to come to an agreement on a measure as divisive as limiting freeway speeds, even though it was spontaneously rejected by a large majority of French people? This means that with a well-informed process of collective deliberation, we can make choices that go beyond our immediate interests, in the name of a democratically defined common good [ADE 22].

In 1972, the *Meadows Report* by Dennis Meadows, Donella Meadows and Jørgen Randers, entitled *The Limits to Growth*, was published, setting out for the first time the dramatic consequences of exponential growth in a finite world: resource depletion, oil crisis, serial shortages, etc. In an interview with Usbek and Rica in April 2022 [MAY 22], Jørgen Randers deplores the fact that this prescient report has concretely affected barely 5% of Westerners, while bringing about some changes in the daily lives of ordinary people, voters and workers. According to Randers' analysis:

> We cannot grow our population, our use of natural resources, our CO_2 emissions, or our land use forever. Our global footprint must be reduced. On the other hand, we are in no way saying that economic activities should stop. GDP is measured in dollars, not in tons of CO_2 emissions: theoretically, GDP can grow as long as these two curves remain decoupled [although, in [ROB 23], Chapter 1, a link between GDP and CO_2 emissions is highlighted via the economist Kaya's equation]. The problem is that this distinction very quickly disappeared in favor of a longsuffering discussion on "sustainable development" […] contrasted with degrowth [note that the UN SDGs presented in Figure 1.7 do not assume economic degrowth] [MAY 22, author's translation].

Although Randers agrees with the IPCC that fossil fuels should be phased out, enabling a rapid reduction of 70–80% in CO_2 emissions, he notes that humanity can

13 The Citizens' Climate Convention, an unprecedented democratic experiment in France, was designed to give citizens a say in accelerating the fight against climate change. It was held in 2019–2020. See: https://www.conventioncitoyennepourleclimat.fr/.

continue to increase greenhouse gas emissions until at least 2050, without depleting natural and food resources. In this case, he predicts that "a global collapse is not the most likely outcome – at least not in the next fifty years. What is more likely is that we will not be able to bring down the curve of greenhouse gas emissions fast enough. We will probably exceed +2.5°C, which will lead to social and localized collapses" [MAY 22, authors' translation].

By "collapses", Randers considers that:

> Inequality levels will continue to rise, as will unemployment rates. People will become increasingly dissatisfied with the unpredictability of weather events and the inaction of their governments. Public services will be outdated. Politicians will no longer have a clear majority for their decisions to be sufficiently consensual. This will create very strong social tensions [...]. There will be a fall in the level of well-being, but GDP will continue to rise – because it represents the total value of wealth produced, even if it is unbalanced [MAY 22, authors' translation].

To limit the consequences of these potential collapses, the question of governance arises. According to researcher and political scientist François Gemenne, it is quite simply "impossible to convince a majority of people to vote for a program compatible with the Paris Agreement"[14] (authors' translation), at least in unindustrialized countries. Randers points out:

> The weakness and limitations of our democratic systems since the 1990s [...]. The short-term interests that parasitize our institutions are very powerful. Similarly, at the individual level, no ordinary worker will willingly agree to pay the additional taxes needed to implement policies adapted to the challenges of the future. The energy transition, for example, implies the closure of all jobs linked to the fossil fuel industries and the opening up of new jobs in renewable energies. The problem is that, in the meantime, many jobs are disappearing. That is why, in my view, these policies can only be led by strong states, able to provide the necessary financial compensation to the employees concerned. Private companies will never be able to take on this task alone, as their interest will always remain short-term profit [MAY 22, authors' translation].

14 The Paris Climate Agreement is an international treaty on global warming adopted in 2015. It covers climate change mitigation, adaptation and financing. The agreement was negotiated by 196 parties at the 2015 Paris climate change conference in France (Wikipedia).

However, is it a strong democratic state or a strong authoritarian state?

Randers continues:

> In the West, it seems that neoliberalism has been pushed to the limit. We have now reached a point where it is impossible not to reintroduce a strong state to provide for the needs of the population. The difficulty is that, from a political point of view, the divisions into two or three blocs – as in the United States with the Republican bloc and the Democratic bloc, or as in France with the far-right bloc, the centrist bloc and the leftist bloc – are likely to deepen in the coming decades. To overcome this difficulty, I have been advocating for some years now the creation of a small assembly in each country, entirely dedicated to the long term. It would be made up of around fifteen people elected for life, rather like the American Supreme Court. It would have veto power over all public decisions impacting on greenhouse gas emissions. The only other alternative is the Chinese model: a central party that enlists the most educated. They become the masters of all decisions because they are considered to have the most experience. Conversely, it is a far less democratic system. It would take us back several centuries [MAY 22, authors' translation].

Nevertheless, as economist Gaël Giraud points out:

> At the heart of the fossil fuel world, private banks are the main obstacle to ecological reconstruction. Many of their assets are linked to oil, coal, or gas. This is hardly surprising: their balance sheets reflect past economic activities, which have been massively dependent on sub-surface energy sources. If tomorrow, by some extraordinary occurrence, these investments were to lose all value and become stranded assets, most banks would have to declare bankruptcy. This was recognized in 2015 by the head of the Bank of England, in a speech that became famous under the title Tragedy of the Horizons [CAR 15]. These institutions have all the more reason to slow the greening of the economy as British Petroleum has already begun to devalue its assets [VIO 20, authors' translation].

Giraud points out that "despite this, in 2018, an Oxfam report showed that for every €10 invested by French banks, seven still go to fossil fuels, and two to so-called renewable energies [POI 18] […]. The banks are back at it again, talking about "sustainable finance" and "green bonds" to boost their image. Unfortunately, for the moment, this is nothing but greenwashing [EKE 19]. Collectively, we need to decide on their future. Should we nationalize them and create a bad bank, as we did in 2008, to rid them of stranded fossil assets? If so, who will foot the bill?" [VIO 20, authors' translation]. In the end, public authorities have a decisive role to play, one that will certainly require a great deal of courage!

Radjou also stresses the importance of governance to accelerate B2B sharing between companies and organizations and to move from smart sharing to wise sharing. Public authorities must be urged to accelerate the movement, and governments must set up a new regulatory framework and introduce tax incentives. Some regulatory constraints need to be eased in critical areas such as legal liability, taxation, intellectual property protection and workers' rights [ARD 21].

However, despite the urgency of climate change, the weakness of ecological political parties in most countries is questionable. According to Latour and Schultz's analysis:

> On the contrary, ecology seems to propose a step backwards, questioning the idea of progress and wanting to limit our freedoms. It therefore generates negative emotions. To mobilize people, we need to find ways to get them excited about ecology. However, the ecological class has not yet sufficiently engaged in the cultural battle, and it sorely lacks an aesthetic capable of fueling political passions [LAT 22, authors' translation].

A priori, sufficiency does not inspire dreams, it is not really a spontaneous practice, and lifestyle changes are not encouraged by a natural resistance to change. Latour and Schultz ask "how to bring about the emergence of an 'ecological class, conscious and proud of itself', which would contribute to giving ecology a central political place, as has been the case in the past for liberalism, socialism or neo-liberalism" [LAT 22, authors' translation]. In 76 propositions, they present an action plan designed to make ecology the new political paradigm.

In section 1.4.3, we have already introduced the notions of "engendering" and "prosperity", concepts that Latour and Schultz develop to provide a new vision, a new positive narrative for ecology. The authors propose using the concept of "prosperity" to designate wealth that does not imply increased production, which destroys ecosystems or material accumulation. In fact, they call for an effort to express ecological issues in enthusiastic terms, whereas today ecology is often perceived as punitive. Indeed, to prosper is a word with a positive connotation that can be interpreted, in line with its etymology, as conveying hope: to prosper is to enjoy healthy air, clean water, quality food and to be surrounded by natural ecosystems that are not degraded and that are aesthetically pleasing and that also prosper. It is also the possibility of being free from the frenzy of consumerism, while still having enough material goods, without destitution or excessive clutter. The idea of prosperity conveys a utopian vision of ecology. Taking care of the world in this way allows us to "maintain the conditions of habitability". Prosperity is the consequence. Indeed, "prosperity has always depended on begetting practices. It is not a question of decreasing, but of finally prospering" [BOU 22, LAT 22, authors' translation]. It is about re-enchanting the imaginary world around ecology and allowing nature to prosper once again, to the delight of humanity and the planet's biodiversity!

Finally, let us make mention of François Gemenne's very pragmatic vision:

> Of course, it is very difficult to anticipate the effects of global warming: we can reduce uncertainties, but we shall never achieve perfect model transparency. That is why we need to work on developing the resilience of populations and territories. Among the "vulnerability" factors that need to be taken into account are the configuration of space (how dangerous is submergence), the sensitivity of the environment (the degradation of sand dunes, for example), the cohesion of society (social redistribution, the feeling of solidarity), the diversity of economic and food-producing activities, the political and institutional organization and the living conditions of the population (income). [...] These models take us away from the purely economic model: the poorest are not necessarily the most vulnerable, even if economic resources play an important role. What is more, this approach frees us from the dichotomy, which I believe to be futile, between ecology and capitalism. Rather than waiting for the big night or a utopian upheaval in our behavior, it is preferable to redirect investments towards technologies and services that emit less carbon. It is not capitalism we should be aiming for, but short-termism: for the moment, the expected rate of return on capital is directing investment towards the "fossil fuel" economy. Adaptation saves us from long-term costs, which is inevitably less attractive, but states can regulate finance and markets [VIO 20, authors' translation].

The European Union (EU) has presented the Green Deal for 2019. This is a set of measures designed to start the EU on the path of ecological transition, with the goal of achieving climate neutrality by 2050. It promotes the transformation of the EU into a fair and prosperous society with a modern and competitive economy. The Green Pact has been defined by Commission President Ursula von der Leyen as the EU's "new growth strategy" to reduce greenhouse gas emissions "while creating jobs and improving our quality of life"[15]. The EU's target of reducing greenhouse

15 See: www.consilium.europa.eu/fr/policies/green-deal/.

gas emissions by 55% between 1990 and 2030 (Fit for 55) was revised at the end of 2022 to achieve a more ambitious reduction of 57%.

1.5.3. *Eco-anxiety*

Parts of the population, mainly young people, feel abandoned by those in power who are deemed unaware of the climate emergency; in particular, by politicians who are often powerless in a democratic system yet to be reinvented, as explained in section 1.4.2, or caught up in power games.

The concept of "eco-anxiousness" has been around since the late 1990s. The "eco-anxious" develops anxiety in the face of predicted ecological disasters: climate disruption, loss of biodiversity, pollution, deforestation, etc. This anticipatory anxiety is not a pathology, but a healthy natural reaction to awareness of the short- and long-term impacts of climate change, which are already observable today. A study published in The Lancet [MAR 21], based on a survey of 10,000 young people aged between 16 and 25 carried out in 10 different countries, reveals that, for 74% of the young French people questioned, the future is frightening, and that for 77% of them, we have failed to take care of the planet. Many young people stressed that their anxiety was linked to political decisions perceived as ineffective and compared to a form of betrayal on the part of the generations that preceded them [FOU 21, OGO 22].

Not all classes of the population are affected in the same way, as illustrated by Figure 1.9 showing greenhouse gas emissions by income group in France in 2021. Average greenhouse gas emissions are 8.5 tonnes per French person per year in 2021. However, this average conceals a wide disparity: French people with the lowest incomes emit far less greenhouse gases.

"Eco-anxiety" is to be distinguished from solastalgia. For Glenn Albrecht, "solastalgia is the pain or illness caused by the loss or lack of comfort and sense of isolation associated with the current state of one's home and territory" [ALB 05, authors' translation]. In this case, it is not a question of anticipation, but of living in a deteriorating habitat and environment, of feeling homesick.

These new concepts point to a malaise linked to the environment, highlighting a cognitive dissonance between our obsolete categories and practices, and the forms of human life to be invented to make the planet habitable in the future.

It should be noted that this phenomenon of "eco-anxiety" is part of a fundamental framework in which we already have difficulty integrating our vulnerability and finitude [GRE 22], vulnerability and finitude further exacerbated by the rapidity of climate change and its consequences. Thierry Magnin points out that:

> The feeling of being misunderstood, even despised by the powers that be, is transformed into a demand for personal freedom against a backdrop of injustice, discrimination, and uncertainty about the future. In a world plagued by the Covid-19 pandemic, psychological insecurity is superimposed on social and economic insecurity, fueled by new fears linked to global concerns about the effects of global warming, terrorism, growing immigration and uncontrollable geopolitics. And many are wondering what might happen to them in these uncertain times. The hardships of life are all the more keenly felt [MAG 22, authors' translation].

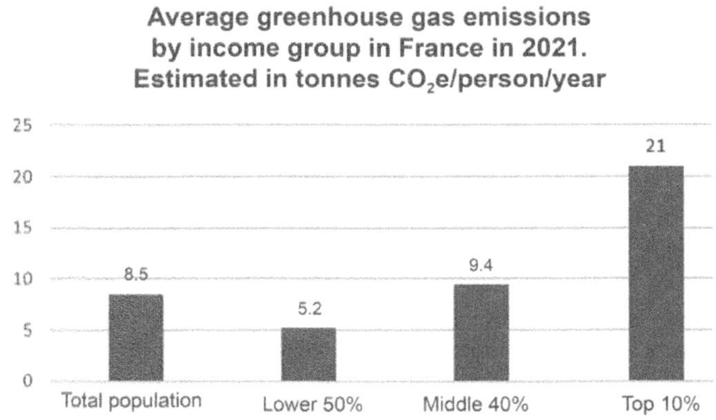

Figure 1.9. *Greenhouse gas emissions by income group in France in 2021 [OGO 22] (Source: Lucas Chancel)*

Against this backdrop and given that our societies are finding it difficult to make the ecological transition, we need to learn to live with our eco-anxiety. Individually, as physician Alice Desbiolles [DES 20] explains, solutions do exist. "Reconjugating our lives to the present" by not constantly anticipating the future, "avoiding hyper-responsibility and hyper-guilt", "learning to let go without giving up", "getting rid of anxiety-provoking observations *to move towards solutions and courses of action*" are just some of the possible keys. Being part of a collective is also an excellent remedy [OSO 22, authors' translation].

1.6. Bifurcating research

Implementing the energy and societal transition will require full-scale experimentation with a wide range of technological, economic and local societal solutions, while accepting trial, error and adjustment. The aim is to validate the viability of technologies, their effectiveness and impact on the environment, their acceptance and social appropriation, which are by no means automatic, to find new economic models and to design new legal frameworks. It is also a question of convincing, to limit resistance to change. It is about being examples, being exemplary [ROB 23].

Science and technology are generally developed in laboratories under controlled test conditions[16]. This generally leads to the development of prototypes for innovations. The complexity and rapidity of the transition make it necessary to test emerging solutions more systematically under real-life conditions, where not everything is under control, particularly climatic conditions and human behaviors. For example, smart building technologies not only need to be tested, but also ideally designed, with the users of a building under normal operation.

Confronting the realities of the field with what is happening requires us to engage with a wide range of disciplines, not just electrical and communication technologies (for smart grids, for example), but also sociological ones (to assess the acceptability and involvement of stakeholders), economic ones (to find viable new models), legal ones, ethical ones, etc. Interdisciplinarity, and even transdisciplinarity, will be essential to the success of the transition.

The legislator needs to define a framework that allows, and even encourages, the deployment of demonstrations, by delimiting "playgrounds" with "sandboxes", by allowing for the right to experiment over several years.

To make sure that energy and societal transition is a success, we need to get the various scientific and humanities disciplines to work together, in order to develop the interdisciplinary approaches between the humanities, social sciences and engineering sciences that are essential to a successful transition that belongs to everyone. Each field of science has developed its own methods and language, which are shared by an often-restricted community. This makes dialogue between the sciences difficult. Researchers need to simplify their discourse to make it more accessible to non-experts in their discipline. Exchanges between disciplines on vocabulary and methods are essential to identify more precisely the possibilities for dialogue. Interdisciplinary research, which crosses disciplines, and transdisciplinary

16 The growing use of AI, however, may call this presupposition into question. This topic is discussed in section 1.11.

research, which makes equal use of the expertise of at least two disciplines, must be encouraged and recognized by the authorities who guide and evaluate research, because this effort to cross disciplines is an adventure that takes time to be productive; and up to now, evaluations of research activities have been essentially mono-disciplinary. Time is needed to build trust and respect for different disciplines on an ethical basis.

Some innovative and promising technologies have not been as successful as expected, because potential users have not accepted them or appropriated them. These technologies were developed primarily with a technical logic in mind, without considering the interests and acceptability of users from the outset. Examples include Google Glass, a pair of augmented-reality glasses that can film people without their knowledge, and France's smart electricity meter (Linky), which, based on a fine-grained measurement of a home's consumption, would provide insight into the habits of its inhabitants. It cannot be decreed that this or that technological solution will be good for the planet without integrating the sociological, psychological, ethical, economic and legal dimensions as far upstream as possible in their design, in addition to assessing the impact or even the gains, for the environment. In other words, we need to develop interdisciplinary approaches and support the changes brought about by transition.

Interdisciplinary research with a desire to develop full-scale, manned demonstrations can fuel "action research", which can be defined as research in which there is a deliberate action to transform reality, research with a dual objective: to transform reality and to produce knowledge concerning these transformations [WIK 20b].

In an article published in Le Monde, "Science et Médecine", David Larousserie notes that "more and more researchers are deciding to modify their way of working, even to change their subject of study, to branch off, to get out of inertia in the face of global warming, to maintain coherence between their values and their practices or quite simply to "make humanity possible tomorrow", as Jean-Michel Hupé, one of these "bifurcators", simply sums it up" [LAR 22, authors' translation]. At the end of 2020, this CNRS researcher left his neuroscience laboratory in Toulouse for another team in the human and social sciences. Gone will be the study of synesthesia and vision, replaced by work in political ecology. "I was increasingly uncomfortable with the idea that nothing was happening, while climatologists were making clear observations about the climate situation", he recalls (authors' translation). By the end of 2018, he had set up Atécopol[17] in Toulouse, a collective made up of around 160 members designed to "share knowledge and make it operative in the face of ecological devastation".

17 An amalgamation of "Atelier d'écologie politique": Political Ecology Workshop.

The year 2019 saw the creation of "Labos 1point5" [LAB 22], which is committed to carrying out greenhouse gas assessments and helping labs make the transition. In [HUM 21], scientists recount their intimate encounters with the Anthropocene.

While it is important to assess the environmental impact of research (laboratories, publications, conferences, travel, etc.), it is becoming necessary to assess the societal impact of the results of a research activity. This assessment can be based on the UN's Millennium Development Goals (Figure 1.7) and its 169 sub-goals. It is no longer just a question of scientific evaluation based on the impact of scientific publications in specialized journals, but of assessing whether the results and innovations of the theme are accessible, appropriable and have concrete spin-offs for society (i.e. a wide audience) through popularization actions (conferences, publications, mass media, social networks, etc.), training, dissemination of technical or software applications or services or new economic models (circular economies), etc. Open science, which makes scientific publications freely available on the Internet, is a step in the direction of accelerating the transition, but these publications are generally not understandable and appropriate for a wide audience.

By way of example, Figure 1.10 illustrates the sub-objectives of the UN's SDGs, research themes that Junia[18] builds on and contributes towards, following a pre-analysis of the outputs of these themes.

The results presented in this book come from research work that is generally interdisciplinary, often carried out in partnership with companies or local authorities and in many cases tested by means of demonstrations. They will thus be positioned in relation to the UN's SDGs and ADEME's scenarios.

1.7. Smarter energy networks

1.7.1. *From 100% renewable energy to a combination of solutions*

Energy is a fundamental issue for a successful energy and social transition. We need to stop consuming fossil fuels, and to do so, we need to consume renewable energy, either directly in thermal form or in electrical form. As the latter is a very practical energy vector for many uses, electrical energy consumption will tend to increase, even though, overall, energy consumption will have to decrease; for the planet, the best energy is the one we do not consume!

In 2021, the French electricity transmission system operator published its six energy evolution scenarios for France up to 2050 [RTE 21]. The focus is on

18 The Junia Engineering Graduate School in Lille, France.

electricity, but this is a key aspect of all the scenarios that set out our possible futures, because while overall energy consumption is set to fall, electricity consumption will increase as it replaces fossil fuels in many sectors: transport, including electric vehicles, industry, hydrogen production from renewable sources, etc. This increase would be 35% in 2050 compared with 2020.

Figure 1.10. *Sub-objectives of the UN's SDGs, research themes that Junia builds on and contributes towards (Source: Muriel Billamboz and Franck Chauvin)*

However, the report insists on the need to reduce overall energy consumption by increasing the energy efficiency of our energy systems and through energy sufficiency, which some see as the first response to the environmental crisis, while others reject the principle in the name of individual freedoms and the maintenance of a subjective form of "comfort" [RTE 21]. This reduction is estimated at around 40% by 2050 compared to 2020, but new European targets aim to accelerate this reduction to 55% by 2030. The various ADEME scenarios forecast reductions of between 22.4% and 55.4%, depending on the type of scenario (Figure 1.4).

A scenario based on 100% renewable solutions, for electricity generation only, has been proposed by ADEME [ADE 18] and RTE [RTE 21]. Given the fluctuating nature of many renewable sources – wind, photovoltaic, small-scale hydro and marine energies [ROB 21] – it will be essential to develop electrical energy storage [ROB 15, ROB 16, ROB 19] and load flexibility, i.e. to modulate electricity consumption over time, in particular by adapting to the variability of renewable sources, and also to investigate and adapt electricity networks. RTE imagines different variants of this scenario, depending on the level of development of onshore and offshore wind power, as well as photovoltaics. One scenario predicts that one out of every two houses in France will be equipped with photovoltaic panels for partial self-consumption [ROB 19].

However, RTE also foresees scenarios in which part of France's electricity would still be generated by nuclear power, by extending existing plants and building new ones. According to RTE, these scenarios would be less costly in terms of investment than a 100% renewable scenario, except that uncertainties over the extension of certain power plants and delays in the construction of new plants will probably lead to these scenarios reaching their objectives in 2060 rather than 2050. In this case, to achieve the 2050 targets, in spite of everything, the economic balances of the different scenarios would be similar.

Scenarios based on combined energy solutions, with renewable and nuclear energies, include in some cases the capture of CO_2 emitted by fossil fuels that are still in use, even though they are declining significantly, and by certain industrial processes [ROB 23]. These scenarios are proposed, for example, by ADEME [ADE 21] and the IPCC [INT 18].

However, by 2022, the financial assessments of these scenarios had to be revised, due to the impact of the recovery from the Covid-19 crisis, the war in Ukraine, and the shutdown of many French nuclear reactors for maintenance, as well as corrosion problems in these reactors (32 out of 56 by mid-2022 and another 15 or so shut down by the end of 2022). In its July 19, 2022 edition, the French newspaper Les Echos ran the headline "Wind Power: Surprise $8 billion gain for the State"[19] [WAJ 22]. This expected gain for 2022 and 2023 is explained by the supplement remuneration system. Under this system, the state supplements operators'

19 Proposed translation for: *Éolien : un gain surprise de 8 milliards pour l'État.*

remuneration when the market price is lower than the tariff granted to renewable energy projects (EUR 80 to 120 per MWh). Conversely, when the market price is higher than the agreed feed-in tariff for renewable electricity, operators pay back the excess to the state (by mid-2022, MWh was fluctuating between EUR 300 and 500 or even more). This compares with EUR 30 to 50 per MWh in January 2020. Forecasters predict that this cost could reach EUR 1,000/MWh by early 2023.

The same applies to photovoltaic solar power, which is tending towards financial equilibrium. These renewable energies are clearly becoming competitive, all the more so given that we can expect electricity costs to continue to rise (but hopefully to levels below EUR 1,000/MWh) and gas costs to rise (by a factor of 10 between early and mid-2022).

According to Jeremy Rifkin:

> Solar and wind power are the cheapest and most competitive energies, and they also consume the least water. In the United States, 41% of freshwater reserves are used to cool nuclear power plants. There are 12,000 nuclear and thermoelectric power plants worldwide. In their place, renewable energies would save 95% of the world's freshwater. Why continue with nuclear and fossil fuels when solar or wind power are so much cheaper and less predatory on natural resources? [RIF 22, authors' translation].

1.7.2. *Towards the decentralization of electricity grids*

The historical organization of an electricity grid is based on centralized management of the transmission network, to which traditional nuclear, thermal and hydro generation units are connected. At the time, the distribution network only served consumers, and only carried power flows from the higher voltage levels, from the interconnection points with the transmission network, to the lower voltage points. At the distribution level, however, control is limited, and system services (voltage and frequency control) are therefore provided at the transmission network level by the generating units connected to it [ROB 15, ROB 19, ROB 21].

The development of decentralized generation, generally low-power, unplanned and not centrally controlled, is changing the picture considerably; it is often connected to the distribution grid and dispersed over a territory, unlike conventional high-power generation, which is concentrated in a limited number of well-identified sites. Today, the impact of the integration of this generation – mainly wind and photovoltaic, and also run-of-river hydro and marine resources [ROB 21] – is becoming increasingly significant and is creating new constraints. The variable and forecast error-prone nature of these renewable sources obviously greatly complicates grid management.

The liberalization of the electricity market, which has been underway in the European Union since the beginning of the 21st century, has led to the separate management of electricity generation, which is subject to competition, from the management of transmission and distribution networks, as the latter's infrastructure cannot be multiplied. In France, the CRE[20] [COM 19] is responsible for ensuring that the competition mechanisms that are being put in place are respected, while regulating this competition to ensure that it is not unfavorable to consumers and does not jeopardize an infrastructure that is vital to the country's economy and security. This liberalization calls for new approaches to managing the power system, and new market mechanisms that take into account the characteristics of the new decentralized sources. Starting from the premise that electricity grids cannot be rebuilt, three types of evolution will be necessary:

– at source level: by using the functionalities offered by power electronics to develop new control and supervision strategies to participate in system services, notably by associating energy storage, and by developing multisource systems (integrating intermittent renewable generation, predictable conventional sources and storage) with integrated and optimized energy management [ROB 15];

– at grid level: by deploying smart grids, developing new grid architectures such as microgrids to increase the efficiency, security and availability of electricity grids and increasing energy storage capacities, whether centralized or dispersed throughout these networks [ROB 15, ROB 19];

– at consumer level: in industrial processes, tertiary and residential buildings, electric and plug-in hybrid vehicles, as well as guided transport systems (trains, metros, streetcars) [ROB 17, ROB 19], in order to modulate energy demand according to demand, available renewable production and electricity grid constraints.

Actions between these different system components require a certain degree of coordination, which raises the question of the degree of decentralization that is desirable or acceptable, and the need for a system of communication between these components. These issues are not purely technological, but integrate economic and sociological aspects and will require new developments in the legal framework, energy consumption patterns and the link between production and consumption, in turn raising ethical questions.

20 Commission de régulation de l'électricité: an independent body that regulates the French electricity and gas markets. It is a member of the European Union organisation ACER and the all-European CEER.

1.7.3. *Smart grids, self-generation and self-consumption*

This calls for the implementation and use of new communication technologies, combined with advanced management tools. System and network intelligence has two facets. The first corresponds to the intensified deployment on transport and distribution networks of a telecommunication network, devices and equipment enabling a higher degree of telecontrol and automation of network operation. The second involves advanced management of generation (centralized and decentralized) and load, resulting in the development of new products and services by producers and suppliers for network operators to increase the degrees of freedom in network control, and for end consumers, who could benefit from service offerings and pricing systems enabling them to adopt ambitious behaviors with regard to the control of instantaneous electricity demand and the integration of renewable energies [ROB 19].

Smart grids will really unfold their innovative potential by fostering interactions between the various actors in the electricity system (producer, consumer, storage operator, network manager), themselves becoming "electrically" smarter actors. These actors have very different consumption and/or production profiles and very different economic and societal objectives and/or constraints. New sets of actors may therefore emerge, enabling us to target new business models, but also to meet energy and climate challenges, by promoting the development of renewable energies. One of the challenges of this evolution is to ensure that all actors win, not forgetting those in fuel poverty [ROB 19].

According to Erik Orsena:

> The time has come for consumers to take revenge on pro-producers. The most striking example is energy. The ancient mode of production was in the hands of monopolies. No other source was available. The market opened wide to many new actors [ORS 18].

The development of renewable energy sources therefore encourages local production close to the loads to be supplied and therefore local consumption. This is particularly true for energies where the primary source varies independently of demand, where there is no natural storage (e.g. in the form of water or fuel) and where production forecasts are subject to error, as in the case of wind power and photovoltaics. In this context, we define self-generation and self-consumption of renewable energy, based on Figure 1.11, as follows [ROB 19]:

– Self-generation is the share of total consumption provided by local renewable energies, on one day for example; i.e. area C/area B (generally calculated over one year).

– Self-consumption is the share of renewable production consumed in real time, on a given day for example; i.e. area C/area A (generally calculated over a year).

On the basis of Figure 1.11, we can also specify definitions for zero- or positive-energy buildings by extending the reasoning over one year:

– A building is zero-energy if surface area B is equivalent to surface area A over one year.

– A building is energy-positive if surface area B is less than surface area A over one year.

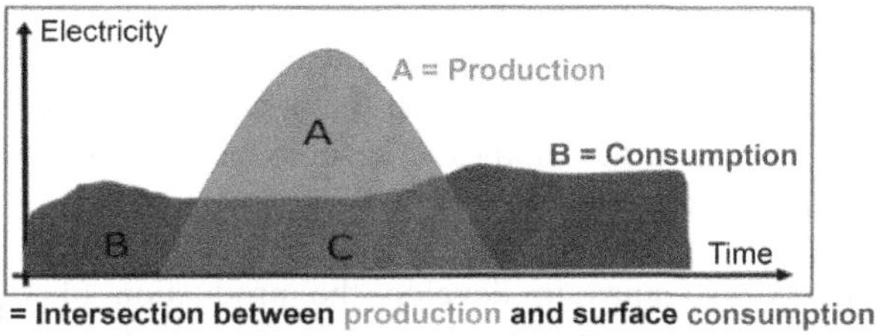

Figure 1.11. *Self-generation and self-consumption over a 24 hour day. For a color version of this figure, see www.iste.co.uk/robyns/smartgrids.zip*

A local network will only be able to operate autonomously, or be islanded from the main electricity grid, if production A is at all times, in the absence of energy storage devices, equal to or greater than consumption C.

Self-generation and self-consumption of energy will encourage the development of renewable energy communities encouraged by the European Union and a return to short circuits and local production [ROB 23]. It should be noted that self-production and self-consumption tend to make users more aware of the origin of the energy they consume, thus encouraging them to seek greater efficiency in their consumption, control it better and be more sufficient.

1.7.4. *An increasingly miraculous electricity fairy – yes but?*

The development of self-consumed locally generated energy raises fundamental ethical questions, since energy, particularly electrical energy, has become indispensable in industrialized societies to ensure our lifestyles in terms of comfort, health, safety, education, etc. Indeed, self-consumption calls into question today's highly centralized electricity grids in terms of production and management; in fact, it constitutes an energy revolution that could be compared for collective self-consumption, in an extreme situation, i.e. if the public distribution grid operator no longer played the role of trusted third party, to the "uberization" (exchange of services between private individuals without going through large established companies thanks to Internet applications) of urban car transport services or short-stay accommodation. Except that the availability of electricity has become crucial to the functioning of our societies, which have become highly dependent on it[21], and that self-consumption could call into question the French principles of energy solidarity and equal access to energy (in terms of cost). These last points raise new ethical questions, particularly in view of the risk of increasing energy insecurity or even the emergence of energy communitarianism.

Is it not conceivable that self-consumption could enrich consumers who are already in a favorable position, such as well-off households who are able to install solar panels on the roofs of their homes? These considerations echo smart sharing and wise sharing (see section 1.4.2).

Self-consumption also aims to develop the production of "new" renewable energy sources (mainly solar, and also wind and small hydro), which are by their very nature variable and weather-dependent (with significant diurnal and seasonal effects). Ongoing climate change means that there is added uncertainty as to how these new technological approaches will behave in the future.

These considerations should lead us to bear in mind an ethical rule proposed in [GIO 18]:

> You will not entrust your children with the task of solving problems that you have voluntarily created, that are vital for your descendants and for which you are not certain of the existence of a realistic solution, now or in the future. On the other hand, the advances generated by scientific discovery and or technical development under consideration will aim to strengthen the common good and promote the restoration of the original ecosystems, if they provide balance and harmony, wherever possible [GIO 18, authors' translation].

21 One novel masterfully illustrates the rapid collapse of our societies in the event of a widespread and sustained power cut. It is *Black-out* by Austrian author Marc Elsberg (Penguin Books, 2016). A novel can undoubtedly do more to open up imaginations than a scientific work.

The aim is not to limit research on the development of smart grids and self-consumption, but to ensure that this research is regularly challenged by the ethical issues highlighted above (even though the risks appear to be of a lesser magnitude and of a different nature than those induced by the development of nuclear energy). This will require interdisciplinary approaches to these issues, combining science and technology with sociology, economics, ethics and even legal issues. Lawmakers will play an important role in providing an "ethical counterweight" [GIO 18] to new methods of energy production and consumption.

These new modes of production and consumption, combined with better control of consumption with a view to reducing it and making it flexible, to adapt as much as possible to the times of availability of renewable energies, but also to contribute to reducing the constraints to which electricity grids are subject in this new decentralized context, will make increasing use of communication and data management resources. These abundant data, or big data, will enable AI (see also section 1.11) algorithms to assist users and actors in buildings, energy networks and cities to make them more economical and sustainable, to turn them into smart buildings, smart grids and smart cities[22].

However, we need to be careful since digital technology and AI are energy hungry. AI requires machines with rare materials for batteries, electrical power for operations and lots of energy to cool and air-condition hardware. Solutions are emerging, however, to recover the heat released by data centers and use it for other purposes. We also need to think about what data are *really* useful, how often they need to be acquired and what operations need to be carried out wisely [ROB 23] or even whether we need to implement low-tech solutions (approaches to the design and evolution of products, services, processes or systems that aim to maximize their social utility while controlling their environmental impact [ADE 22]).

In 2018, the digital sector accounted for 3.7% of global greenhouse gas emissions, and 4.2% of global primary energy consumption; 44% of this footprint was due to the manufacture of terminals, data centers and networks and 56% to their use [FRE 20]. Currently, the digital sector emits as much carbon as the air transport sector. In 2019, the digital sector accounted for 2% of France's carbon emissions. Growth of up to 6.7% in 2040 is predicted if nothing is done to reduce the carbon

22 Once again, a novel illustrates very well the potential excesses of the all-digital world through the global network that is the Internet, combined with ever more powerful AI (thanks to Deep Learning), which some imagine or dream could make machines conscious, or even become a planetary brain, drawing parallels with Teilhard de Chardin's Noosphere. The book in question is *Immortal* by Portuguese author J.R. Dos Santos (Hervé Chopin, 2020, for the French edition). Enough to open up the imagination!

impact of digital technology [FRE 20]. ADEME's S3 scenario predicts that data centers will consume 10 times more energy in 2050 than in 2020, while the S4 scenario predicts a 15-fold increase (Figure 1.3 and [ADE 21]).

In their book *La servitude électrique* [DUB 21], sociologists Gérard Dubey and Alain Gras draw our attention to the invisible or hidden sides of the electricity fairy and the risk of enslavement to digital technology:

> Contemporary digital society is indeed the product of an electrical history in which polluting generators and waste are relegated to the background. The machines of our e-world seem to be borne by the enchantment of a fairy to whom the algorithm owes its social existence. The image of green transition is an old story [DUB 21, authors' translation].

These authors note that "even though coal's share of the energy generation is declining in wealthy countries, including the United States, on a global scale, coal is once again the leading energy source, accounting for almost 40% of the total due to demand for electricity" [DUB 21, authors' translation]. In fact, growth in electricity demand is outstripping the development of renewable energies, so the trend is to meet this demand by exploiting energy that is still abundant and available just about everywhere (particularly in Asia, led by China and India, and in Africa). This is illustrated by Figure 1.12, which shows emissions from fossil fuels (and cement used by the construction sector) from 1990 to 2022; after a decline in 2020 due to Covid-19, the rise continues and is particularly high for coal, the biggest CO_2 emitter. As Dubey and Gras put it:

> Make no mistake, […] a large part of the relative sufficiency of the rich is due to globalization and the outsourcing of their energy consumption. The world's various workshops work for the West and deliver their goods with energy that is counted as a debit. Here we have another layer exposing the conditions for the existence of the electrical phenomenon, but in the second degree [DUB 21, authors' translation].

Figure 1.13 illustrates the evolution of greenhouse gas emissions according to the economic situation of countries: emerging, developed, less advanced. Only in developed countries will emissions fall significantly. Total emissions for 2022 are estimated to have risen to 40.6 gigatons of CO_2, compared with around 23 gigatons of CO_2 in 1990.

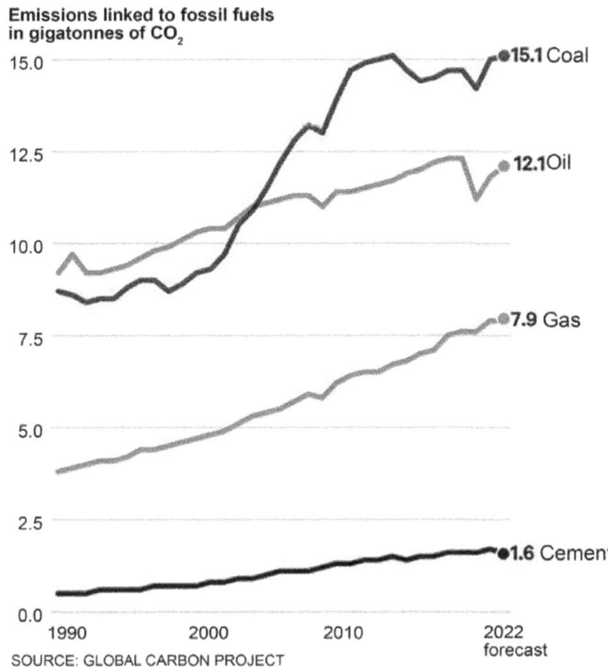

Figure 1.12. *Emissions linked to fossil fuels (and cement used for construction) from 1990 to 2022 [JAC 22]*

In 2020 and 2021, the Covid-19 crisis clearly highlighted the shortcomings of globalization for many countries, particularly in the West, making them overly dependent on a small number of (mainly Asian) countries manufacturing drugs and medical devices (both of which require energy and therefore generate CO_2). In 2022, the war in Ukraine highlighted the overdependence of European countries on Russia for energy (especially gas and oil). This energy crisis is encouraging some countries to make greater use of coal (e.g. Germany, Poland); even France has re-commissioned a coal-fired power station, despite its substantial but aging nuclear fleet and significant renewable energy potential. These successive crises highlight the need to return to a more local production of goods and energy, thus favoring short circuits that reduce carbon emissions. For a country like France, the only locally available energy sources are renewables (uranium for nuclear power plants is non-renewable and imported).

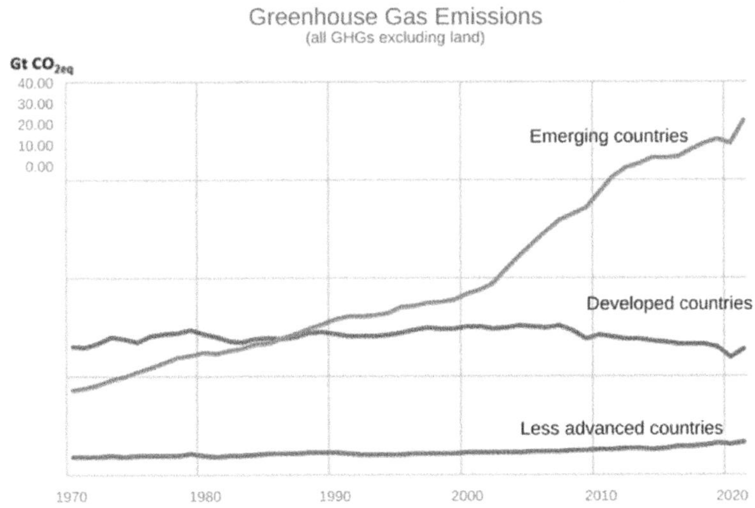

Figure 1.13. *Trends in greenhouse gas emissions according to the economic situation of countries: emerging, developed, less advanced. A significant decline in emissions is only effective for developed countries [PER 22]*

According to Dubey and Gras:

> The intermittency of renewable energies, which we hope today to be able to tame by perfecting existing techniques and systems, is in fact an invitation to change times and social organization. It is a far more conflict-ridden and revolutionary process than the euphemisms of "energy transition" and "digital transition" suggest. Not just because powerful positions and interests are at stake, which those who benefit from them have no intention of abandoning. But because the irregularity, variability and unpredictability of renewable energies remain irreducible to the temporal regime of digital capitalism and its mode of apprehending reality. To welcome intermittence, then, is to welcome disorder in all its creative and subversive implications, by dispelling the confused fear it inspires [DUB 21, authors' translation].

The development of renewable energies is clearly in line with ADEME's S3 "Green Technologies" scenario. Self-consumption, particularly collective and in local energy communities, is in line with scenario S2 "Territorial Cooperation", and if we move from smart energy sharing to wise energy sharing, these modes of local energy production and consumption are in line with scenario S1 "Frugal Generation". In this way, they contribute to UN Sustainable Development Goal 7,

“Clean and affordable energy” (Figure 1.7). This SDG emphasizes energy efficiency, takes into account the need to reduce fuel poverty and focuses on developing regions with little or no access to electricity (Africa, Asia, etc.). Nearly a billion people in the world still have no access to electricity, a figure that is slowly declining (in addition to the 1.6 billion who had no access to drinking water at the start of the 21st century, and the 2.5 billion who had no access to proper sanitation).

In addition to the need to reduce fuel poverty, we need to remain vigilant to the fact that the development of digital technology associated with the “smartization” of our environment can marginalize part of the population. According to INSEE[23], 20% of French people will have difficulty using the Internet in 2019.

According to Dubey and Gras:

> This “illectronism” or digital divide refers not only to the under-equipment of some and the lack of computer literacy of others, but to a major anthropological fact: some people just can’t adapt to the norms of a touchless world. Although increasingly marginalized, this “deviant” fringe of the population puts its finger on what is incommensurable about digital technology, and at the same time raises a crucial anthropological question that should, in any case, be of interest to all those concerned about the future of our societies. Undoubtedly impressed by the prevailing techno-legalistic fatalism, few dare to tackle the issue. And yet, as we all know, it is always at the frontiers and margins that the important things happen [DUB 21, authors’ translation].

We must realize that not all users will be smart users.

1.8. Smarter buildings in a desirable habitat

1.8.1. *Buildings and living spaces*

The buildings in which we live, work and shop, as well as those used for leisure and sport, are the main CO_2 emitters. In France, in 2016, residential and tertiary buildings accounted for 45% of total final energy consumption. The French PACTE law, passed in 2019, aims to reduce building consumption by 40% by 2030 and 60% by 2050. In 2019, these residential and tertiary sectors accounted for just over 20% of greenhouse gas emissions, with a downward trend thanks to energy efficiency and

23 Institut national de la statistique et des études économiques: the National Statistics Bureau of France.

renovation efforts [GAS 22]. The technologies used to construct and operate these buildings play an important role in the scale of these carbon emissions, but the ways in which they are lived in and used also contribute to these emissions. These buildings are equipped with devices that enable them to better understand and control energy. They can become energy producers and store energy, thanks to the new materials used to construct these buildings, and also increasingly thanks to new energy and digital technologies, while ensuring the comfort of their occupants. They are becoming smart buildings.

And yet technology alone is not enough to optimize a building's operation and reduce its carbon footprint. Building users also play an important role, with activities and comfort needs varying from one individual to the next, to the point where we might well ask whether, for smart buildings to achieve their low-carbon footprint objective, should they not be used and inhabited by smart users who are adapted or integrated into the building's intelligence? In Chapter 5 of [ROB 23], sociologist Hervé Barry develops a socio-technical analysis of buildings becoming smart, based on feedback from two smart building demonstrations at the Université Catholique de Lille.

1.8.2. *Building trends in 2050*

The different ADEME scenarios, presented in Figures 1.2 and 1.3, assume a reduction in energy consumption and greenhouse gas emissions in the building sector. Compared to 2015, when energy consumption in buildings in France stood at 725 TWh of total energy, scenario S1 assumes a decline of almost 50% in energy consumption, while scenario S4 assumes a decline of only 16%. The decline in GHG emissions induced by direct energy consumption (Scope 1 and 2, Figure 2.1 in [ROB 23]) is even more drastic: in scenario S1, the decline is 96% compared to 2015 emissions (85 MtCO2eq), while in scenario S4, the decline is 80% [GAS 22].

The ADEME scenarios show other trends, such as the surface area of dwellings per person, which evolves from 49 m^2/person in 2050 to 48 m^2/person in S1 and 54 m^2/person in S4. The temperature set point of air conditioners also rises from 22°C in 2015 to 26°C in 2050 in S1; however, it remains unchanged in S4 [GAB 22].

In [GAS 22], several scenarios for the evolution of buildings up to 2050 are presented, based on the four ADEME scenarios. These scenarios are based on the availability of resources (materials, labor, finance, etc.), the speed at which actors (companies, owners, users, and also the legislator) evolve, and the form this evolution will take. If the actors evolve little and/or if certain resources experience shortages, the scenarios that emerge, entitled "Difficult to do everything" and "Shortage", do not herald a very desirable future. On the other hand, if the actors evolve strongly, two more positive scenarios emerge, entitled "Building as a service" and "Rebalancing". The "Building as a service" scenario is marked by the rapid adaptation of building and real estate actors to the challenges of the ecological transition and demographic growth, and more specifically to two key issues: avoiding urban sprawl and renovating the building stock. The relationship between buildings and property undergoes profound change. In the city, buildings increasingly become services made available to their occupants by institutional actors. This allows for greater flexibility in space management, aided by digital solutions [GAS 22]. This scenario is in line with ADEME's S3 "Green Technologies" scenario (Figure 1.3). The "Rebalancing" scenario is marked by a territorial rebalancing in favor of peri-urban areas, medium-sized towns and rural areas. Preserving the environment and anticipating systemic crises are becoming high priorities for society. We are witnessing the emergence of dynamics based on greater frugality, the use of local resources and reinvestment in housing and local services [GAS 22]. This scenario is in line with ADEME's S1 and S2 scenarios "Frugal Generation" and "Territorial Cooperation" (Figure 1.2). It focuses on renovation rather than construction, which can reduce GHG emissions by 30% compared to the most planet-friendly new buildings, and on a far-reaching change in lifestyles towards frugality, particularly in the use of everyday and digital equipment.

1.8.3. *Smart buildings*

Buildings will therefore have to considerably improve their energy performance. They will also play a fundamental role in the development of smart energy grids, energy microgrids, eco-neighborhoods and smart cities. Figure 1.14 illustrates the different characteristics of a smart building, which integrates local energy production and storage, controllable loads whose consumption can be modulated (lighting, heating, electric vehicles, etc.); this building can be connected to the electrical distribution network and to external sources or operate in isolation, i.e. not connected to the electrical distribution network [ROB 19, ROB 23].

An intelligent building integrates a range of technical equipment interfacing between:

– all or part of the weather conditions (heat, light, air quality, wind);

– all or part of the energy-consuming equipment that accompanies use of the building (heating, air conditioning, ventilation, lighting, electric vehicles), local energy production and storage;

– all or part of the spaces and times of use for the building, whether for residential or business use.

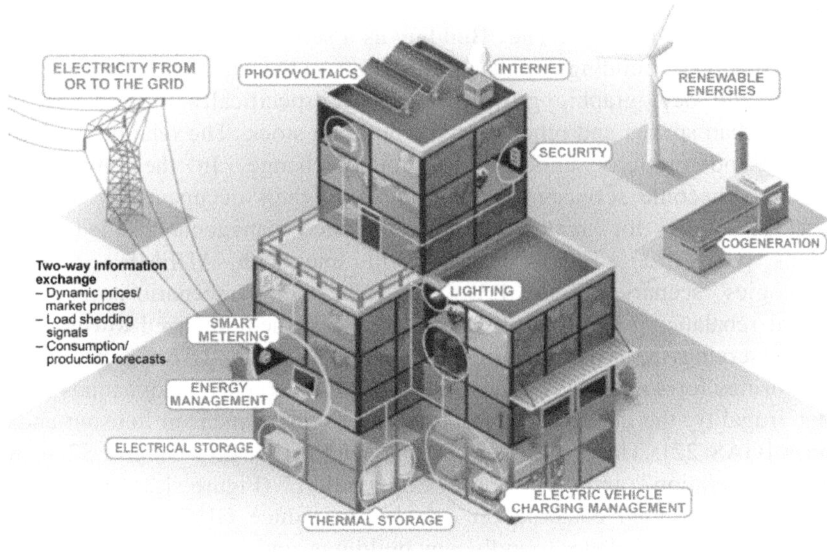

Figure 1.14. *Smart building connected to a smart grid*[24]

A building's level of intelligence is not standard. It varies according to the range of services provided by the technical equipment. Several dimensions are involved in the graduation of this intelligence, developed with the dual aim (ideally) of maximizing energy efficiency and maximizing comfort of use. There are four such dimensions [ROB 19, ROB 23]:

– *Automation*: according to a typical scheme, in response to a measurement or time setting, a technical tool (e.g. a sensor) triggers an adjustment action (more or less heat, more or less light, more or less ventilation). For example, brightness sensors can control the switching on of lights or the activation of blinds according to sunshine and time. The use of such automated systems in the home is usually referred to as home automation. Actions can be triggered by a strict dialogue between devices, with or without human intervention.

– *Centralization*: all or part of the device is controlled by a single system, such as a thermostat for home heating. There are, however, frequent exceptions to this centralization rule, such as thermostatic valves on radiators.

– *Regulation*: this generic term covers all actions, technical or otherwise, automatic or otherwise, designed to adjust energy demand and/or user comfort levels

24 See: http://www.objetconnecte.com/batiments-intelligents-marche-iot/.

according to their needs. For example, a window that can be opened is a regulation device in its own right. In the same way, regulations, information on consumption or awareness-raising through eco-gestures are all ways of regulating (or modifying, to put it another way) the socio-technical conditions of energy use.

– *Information*: consumption needs to be tracked as closely as possible, i.e. over the year, month, week, day, time, equipment, etc. The challenge is to establish the link between technical devices, consumer appliances, usage periods and actual consumption. This information is not enough to know consumption practices, but simply the times and sometimes the intensity of use (e.g. with an oven set more or less hot).

Smart buildings are becoming increasingly digital and are therefore processing a growing amount of data. According to the ADEME scenarios (see Figures 1.2 and 1.3), the rate of growth in the volume of data processed in small data centers varies greatly from one scenario to the next. Considering the extremes, for S1, there is an increase of 50%/year between 2020 and 2023, then a decrease of −10%/year between 2024 and 2032, followed by a rate of 0% until 2050. In scenario S4, this rate increases perpetually by 50%/year between 2020 and 2025, by 60%/year between 2026 and 2028, and then by 25%/year until 2050 [GAB 22].

The low-tech approach makes a lot of sense in the building sector, even though it is still not very widespread. Low-tech obviously concerns energy renovation, with a set of simple techniques and know-how that take into account what is already there, mobilize local, accessible and renewable resources in materials and energy and call on transferable know-how for low-cost implementation, maintenance and use, accessible to all. Some key aspects of a low-tech approach include: definition of needs, user involvement, short circuits, reuse, etc., with constant questioning of energy, carbon, resources and biodiversity, while dealing with program, operating and financing constraints; in short, it is techno-discernment [ADE 22b].

Low-tech approaches are at the heart of ADEME's S1 and S2 scenarios (Figure 1.2), while a high-tech smart building approach is more in line with scenario S3 (Figure 1.3). Smart buildings can also be approached from a more socio-technical angle, which is not incompatible with scenario S2 or even ADEME's scenario S1 (this subject is extensively covered in Chapter 5 of [ROB 23]).

The evolution of buildings will have to integrate into their environments, which are part of the habitat in which we live; the aim is to preserve and even contribute to the development of biodiversity, and more generally to bring nature back into the city [ROB 23]. With this in mind, buildings undergoing renovation or construction must respond to the UN Sustainable Development Goal 11 "Sustainable cities and communities" by becoming part of sustainable urbanization, with a reduced or, better still, positive environmental impact and contributing to the resilience of territories in the face of climate change.

1.9. Smart buildings as nodes of smart grids

As nodes in a smart energy network, smart buildings will become increasingly active participants in the energy ecosystem, going beyond simple service to the smart grid to become an integral part of it. This means developing a true smart building vision, considering buildings as a network (buildings of different kinds, energy-coherent blocks, neighborhoods, etc.), strongly integrating all the technological dimensions and interests of users, operators and owners and positioning them as intelligent actors in energy networks, integrating all the potential of digital platforms.

Demand for electricity continues to grow worldwide. Traditional approaches to meeting this demand would require significant resources for additional generation.

Making the electricity grid "smarter" through information technology can offset some of the growth in demand by more finely optimizing grid management, but this is only half the solution. The other half of the solution must be to make buildings smarter, and to better integrate them into an extended electricity grid. Smart buildings, with their intelligent interaction in networks of buildings, islands of buildings, districts, etc., will enable the implementation of innovative control technologies to save energy, reduce energy costs and reduce the constraints induced on electrical networks by the development of intermittent renewable energies and electric mobility by modulating consumption/production as seen from these networks (by bringing flexibility to the networks). They integrate the needs of users and residents, ensuring their comfort and involvement.

When properly connected to the electricity grid, a smart building or building island can be configured as a virtual back-up station or virtual power plant (behaving like a conventional power plant [ROB 15, ROB 21]) for electricity grid regulation and balance.

Each building block is potentially a node in the electrical grid and can be likened to a microgrid providing multiple services to its owners and occupants and to the energy networks [ROB 19, ROB 23]. An increasing number of these microgrids will integrate energy production and storage services. The operators of these "micro-companies", both owners and businesses, will be able to optimize services and energy consumption at each node in conjunction with energy network managers and potential energy and service aggregators [MEM 17].

Figure 1.15 illustrates the smart buildings as nodes of smart grids concept by combining the smart grid (SG) vision, i.e. the multiple benefits for the power grid in this case, and the smart building (SB) vision, integrating the benefits for buildings and building clusters in terms of reduced CO_2 emissions, and also quality of life for users, financial savings and even business. This concept will be a key component of future smart cities. Nodes are characterized by transformer substations and/or meters that can communicate.

An aggregator will probably be needed to aggregate interactions with energy networks [ROB 19].

At present, few tertiary, commercial and residential buildings have the infrastructure to participate in such a transformation. However, the deployment of smart meters will help to accelerate this transformation. Several aspects of this smart buildings as nodes of smarts grids concept are discussed in Chapters 4–8.

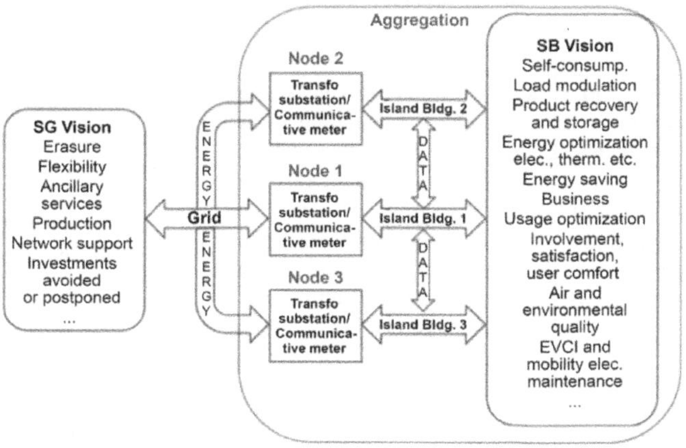

Figure 1.15. *Smart buildings as nodes of smart grids (electric vehicle charging infrastructure (EVCI,))*

Once again, this approach can respond to ADEME's S3 "Green Technologies" scenario with a highly technological vision. However, if we integrate smart sharing or wise sharing visions, the smart buildings as nodes of smart grids approach can respond to the S2 "Territorial Cooperation" or S1 "Frugal Generation" scenarios. In terms of contributions to the UN's SDGs, the aim is to develop an integrative approach to SDGs 7 and 11.

1.10. Methodological contributions

Chapter – Theme	Objectives	Methods, tools and disciplines used to build an energy management strategy
2. The "smart" imaginary	To show that so-called smart techniques are based on an imaginary growing distance between technical systems and their users This raises the question of how to bring together technology and the user? The rest of the book aims to provide some answers to this question	Sociology and anthropology
3. Production and consumption forecasting	Forecast local photovoltaic production and building consumption to enable self-consumption of local production and flexibility for the public distribution grid and reduce CO_2 emissions	Neural networks Satellite data Camera data (from a camera)
4. Taking into account actors in energy management strategies	Identification of stakeholder profiles with a view to building an energy management system	Sociology Microeconomics
5. Energy supervision of a local residential smart grid integrating stakeholder involvement	Smooth out the district's electrical load drawn from the distribution grid Increase consumer satisfaction by considering costs and renewable energy consumption Respect acceptation by each actor	Game theory Dynamic programming Fuzzy logic
6. Auto-Consumption within a local renewable energy community of tertiary buildings	Encourage self-consumption of local photovoltaic production by taking into account actor profiles Assessing the environmental impact of blockchain	Digital twin Game theory ADMM algorithm Blockchain technology
7. Potential of smart buildings	Smart building services Temperature control taking user comfort into account Indoor air quality Environmental impact	Digital twin thanks to BIM Multiple linear regression models Genetic algorithm
8. Demonstration smart grids and buildings	Demonstration of energy flexibility Acceptability of flexibility solutions	Digital twin Full-scale demonstration Sociology

Table 1.1. *The different contexts considered, the objectives pursued, the tools and disciplines, as well as the methods used to build supervision systems*

In the remainder of this book, various contributions of smart grids and smart buildings to energy and societal transition are developed and applied to real-world case studies. Several methodologies are developed and implemented, based on interdisciplinary approaches between the engineering sciences, the humanities and the social sciences. these methodologies call on artificial intelligence tools such as neural networks, fuzzy logic and genetic algorithms, game theory, optimization algorithms such as alternating direction method of multipliers (AMM) typically used to minimize the sum of two functions dependent on different variables coupled by a constraint, blockchain technology, etc., while integrating sociological and economic considerations. The performance of the solutions developed is tested by means of numerical simulations on models (or digital twins), case studies and demonstrations at real scale and in real operating conditions. Table 1.1 summarizes the different contexts considered and the objectives, tools and disciplines implemented, as well as the methods used to build energy, electrical or thermal supervision systems.

1.11. The question of artificial intelligence

Various AI techniques are used throughout this book. These tools are highly effective when it comes to dealing with complex systems that are difficult to model precisely or that require a great deal of computing time to simulate. AI is thus very useful for making forecasts, modeling a system, supervising a system, etc. AI seeks to copy a number of living mechanisms (reasoning, neuronal functioning of the brain, genetics, social behaviors, etc.) that have proved their worth in the deployment and functioning of life. In addition to this attempt to copy the living, AI makes use of a lot of data, and if there are a lot of data, we talk of Big Data. A priori, the more data there are, the better the AI's performance. A distinction is made between weak AI, whose operating stages are mastered and whose results can be physically interpreted – even though the AI does not use a model, the model may exist – and strong AI, such as Deep Learning, for which no model exists.

Pierre Giogini and Thierry Magnin warn of the dangers of strong AI [GIO 21]. Thierry Magnin summarizes the problem as follows:

> The problem comes when you do not have the basic mathematical equations, you do not know the phenomena, and you use AI and massive data processing by "deep learning" to predict possible evolutions [of a system]. By going digital without knowing the mathematical equations of the phenomena under study, the search for correlations will take precedence over the search for causes. Prediction changes status, as it is not based on statistics linked to physical phenomena. This is a promising approach, provided we can

conduct "critical experiments" based on the "possible futures" provided by AI. However, this is not always possible. This represents a major break for scientists accustomed to subjecting their hypotheses and theories to the sieve of critical experimentation. This does not call into question digital tools as such, but it does alert us to the framework in which they are used. The transition from "data-assisted science" to "data-driven science" is thus problematic in terms of "scientific truth" in the sense of the usual experimental sciences. The hypothesis of man losing his grip is no longer a fiction. It is already here, be it in a form that is still latent [MAG 22, authors' translation].

1.12. References

[ADE 18] ADEME, Trajectoires d'évolution du mix électrique 2020-2060, Report, ADEME, 2018.

[ADE 21] ADEME, Transition(s) 2050, choisir maintenant, agir pour le climat, Report, ADEME, available at: https://transitions2050.ADEME.fr/, 2021.

[ADE 22] ADEME, "Low-tech : une nouvelle façon d'innover", *ADEME*, no. 154, 2022.

[ALB 05] ALBRECHT G., ""Solastalgia", A new concept in health and identity", *PAN*, vol. 3, pp. 44–59, 2005.

[BHO 19] BHOLER S., *Le bug humain*, Robert Laffond, Paris, 2019.

[BOU 22] BOURSIER A.-L., Recension sur l'ouvrage de Bruno Latour et Nikolaj Schultz, Mémo sur la nouvelle classe écologique, available at: https://culturecom-larevue.com/bruno-latour-et-nikolaj-schultz-memo-sur-la-nouvelle-classe-ecologique-paris-la-decouverte-les-empecheurs-de-penser-en-rond-2022-isbn-9782359252187-94-p/, 2022.

[CAR 15] CARNEY M., Breaking the tragedy of the horizon – climate change and financial stability, Speech, 2015.

[CAS 20] CASSORET B., *Transition énergétique. Ces vérités qui dérangent*, De Boeck Supérieur, Louvain-la-Neuve, 2020.

[COM 06] COMTE-SPONVILLE A., *L'esprit de l'athéisme*, Albin-Michel, Paris, 2006.

[COM 19] COMMISSION DE REGULATION DE L'ENERGIE, Smart grid, available at: www.smartgrid-cre.fr, 2019.

[DEC 55] DE CHARDIN T., *Le phénomène humain*, Le Seuil, Paris, 1955.

[DEC 76] DE CHARDIN T., *Oeuvres complètes*, Le Seuil, Paris, 1976.

[DES 20] DESBIOLLES A., *L'éco-anxiété. Vivre sereinement dans un monde abîmé*, Fayard, Paris, 2020.

[DUB 21] DUBEY G., GRAS A., *La servitude électrique*, Le Seuil, Paris, 2021.

[EKE 19] EKELAND I., LEFOURNIER J., L'obligation verte : homéopathie ou incantation ?, Working paper, available at: http://www.chair-energy-prosperity.org/wp-content/uploads/2019/06/publication2019_obligation-verte_ekeland-fournier.pdf, 2019.

[FER 96] FERRY L., *L'homme-Dieu ou le sens de la vie*, Grasset & Fasquelle, Paris, 1996.

[FER 10] FERRY L., *La révolution de l'amour. Pour une spiritualité laïque*, Plon, Paris, 2010.

[FOU 21] FOUGIER E., *Eco-anxiété : analyse d'une angoisse contemporaine*, Fondation Jean Jaurès, Paris, 2021.

[FRE 20] FRENCH SENATE, Pour une transition numérique écologique, Report, 2020.

[GAS 22] GASPARD A., VISIER J.-C., "Le bâtiment à l'horizon 2050", *Futuribles*, no. 449, pp. 25–41, 2022.

[GIO 18] GIORGINI P., DEPREZ S., WALD LASOWSKI A., *La tentation d'Eugénie. L'humanité face à son destin*, Bayard, Paris, 2018.

[GIO 21] GIORGINI P., MAGNIN T., *Vers une civilisation de l'algorithme ? Un regard chrétien sur un défi éthique*, Bayard, Paris, 2021.

[GRA 16] GRANDJEAN A., MARTINI M., *Financer la transition energétique*, Les Éditions de l'Atelier, Ivry-sur-Seine, 2016.

[GRE 13] GREENPEACE, Scénario de transition énergétique, Report, 2013.

[GRE 22] GREGOIRE-DELORY V., MAGNIN T., "L'humain troublé par sa vulnérabilité : exemple en sciences et technologies du vivant", *MSR*, pp. 41–54, 2022.

[HUM 21] HUMEAU J., VERCHERE C., ESTEVEZ-TORRES A. et al. *Rencontres intimes avec l'anthropocène* : *récits personnels de scientifiques*, La Chopienerie, Lyon, 2021.

[INT 18] INTERGOVERNMENTAL PANEL ON CLIMATE CHANGE, Global warming of 1, 5°C, Summary for Policymakers, 2018.

[JAC 22] JACQUE M., "Climat : les émissions mondiales de CO_2 se maintiennent à des niveaux record", *Les Echos*, 2022.

[JON 77] JONAS N., *Le principe responsabilité*, Champs essais, Paris, 1977.

[LAB 22] LABOS POINT 5, Réduire l'empreinte de nos activités de recherche sur l'environnement, available at: https://labos1point5.org/, 2022.

[LAR 22] LAROUSSERIE D., "La recherché mène sa transition écologique", *Cahier du Mondes*, no. 24098, 2022.

[LAT 22] LATOUR B., NIKOLAJ S., *Mémo sur la nouvelle classe écologique*, La Découverte, Paris, 2022.

[MAG 17] MAGNIN T., *Penser l'humain au temps de l'homme augmenté*, Albin-Michel, Paris, 2017.

[MAG 21] MAGNIN T., "Pour une anthropologie spirituelle", *MSR*, pp. 83–94, 2021.

[MAG 22] MAGNIN T., *Foi et neurosciences. Dialogue sur l'homme vivant*, Salvator, Paris, 2022.

[MAI 22] MAILLE P., "Il n'y aura pas d'effondrement global mais des effondrements sociaux et localisés", *Usbek & Rica*, 2022.

[MAR 21] MARKS E., HICKMAN C., PIHKALA P. et al. "Young people's voices on climate anxiety, government betrayal and moral injury: A global phenomenon", *The Lancet*, 2021.

[MEM 17] MEMOORI, Smart buildings can be 'the nodes' of the smart grid, available at: www.memoori.com/smart-buildings-can-nodes-smart-grid [Accessed 13 January 2017], 2017.

[MON 10] MONOD S., ROCHAT E., BÜLA C.J. et al. "The spiritual needs model: Spirituality assessment in the geriatric hospital setting", *Journal of Religion, Spirituality and Aging*, vol. 4, no. 22, p. 271, 2010.

[NÉG 21] NÉGAWATT, Négawatt, available at: https://negawatt.org/Scenario-negaWatt-2022, [Accessed 21 December 2021], 2021.

[OGO 22] O'GOMES I., "Qu'est-ce que l'éco-anxiété et sœur jumelle la solastalgie ?", *Sciences et avenir*, available at: https://www.sciencesetavenir.fr/nature-environnement/developpement-durable/qu-est-ce-que-l-eco-anxiete-et-soeur-jumelle-la-solastalgie_163637, 2022.

[ORS 18] ORSENNA E., QUERE S., *La fabrique du neuf*, Le cherche midi, Paris, 2018.

[PER 22] PERTHUIS C., "Solidarité Nord-Sud, financements, débats sur le 1,5 °C, méthane : ce qu'il faut retenir de la COP27", *The Conversation*, 2022.

[POI 18] POIDATZ G., Banques françaises, les fossiles raflent la mise, Report, Oxfam, 2018.

[POP 15] POPE FRANCIS, *Loué sois-tu*, Bayard, Paris, 2015.

[POP 20] POPE FRANCIS, *Fratelli tutti (Tous frères)*, Bayard, Paris, 2020.

[RAD 19] RADJOU N., PRABHU J., *Le guide de l'innovation frugale – Les 6 principes clés pour faire mieux avec moins*, Diateino, Paris, 2019.

[RAD 21] RADJOU N., "La révolution du partage B2B", *Terra Nova*, 2021.

[RIF 12] RIFKIN J., *La troisième révolution industrielle*, Éditions les liens qui libèrent, Paris, 2012.

[RIF 22] RIFKIN J., BOUCHER Y., "La terre est beaucoup plus puissante qu'on l'imaginait, l'espèce humaine est beaucoup plus insignifiante que prévu", *L'atlas éco de la Voix du Nord*, pp. 1–11, 2022.

[ROB 15] ROBYNS B., FRANCOIS B., DELILLE G. et al., *Energy Storage in Electric Power Grids*, ISTE Ltd, London and John Wiley & Sons, New York, 2015.

[ROB 16] ROBYNS B., SAUDEMONT C., HISSEL D. et al., *Electrical Energy Storage in Transportation Systems*, ISTE Ltd, London and John Wiley & Sons, New York, 2016.

[ROB 19] ROBYNS B., DAVIGNY A., BARRY H. et al., *Electrical Energy Storage for Buildings in Smart Grids*, ISTE Ltd, London and John Wiley & Sons, New York, 2019.

[ROB 21] ROBYNS B., DAVIGNY A., FRANCOIS B. et al., *Electric Power Generation from Renewable Sources – Revised and Updated Second Edition*, ISTE Ltd, London and John Wiley & Sons, New York, 2021.

[ROB 23] ROBYNS B., LENGLET C., BARRY H. et al., *Smart Users for Energy and Societal Transition*, ISTE Ltd, London and John Wiley & Sons, New York, 2023.

[RTE 21] RTE, Futur énergétique 2050, Report, 2021.

[SPI 11] SPINOZA B., *Éthique*, Le Livre de poche, Paris, 2011.

[THE 15] THE GLOBAL COMMISSION ON THE ECONOMY AND CLIMATE, *La nouvelle Économie climatique. Une meilleure croissance, un meilleur climat*, Les Petits Matins, Paris, 2015.

[VIL 18] VILLABLA B., SEMAL L., *Sobriété énergétique. Contraintes matérielles, équité sociale et perspectives institutionnelles*, Quae, Versailles, 2018.

[VIO 20] VIO-DURY P., NOYON R., *Face aux chocs écologiques*, Marabout, Paris, 2020.

[WAJ 22] WAJBRT S., "Energies renouvelables : vers un jackpot d'au moins 8,6 milliards d'euro pour l'Etat", *Les Echos*, 2022.

[WIK 20a] WIKIPEDIA, Sobriété énergétique et effet de serre, available at: https://fr.wikipedia.org/wiki/Sobri%C3%A9t%C3%A9_%C3%A9nerg%C3%A9tique [Accessed 6 May 2020], 2020.

[WIK 20b] WIKIPEDIA, Recherche-Action, available at: https://fr.wikipedia.org/wiki/Recherche-action [Accessed 19 May 2020], 2020.

[WIK 22] WIKIPEDIA, Le bug humain, available at: https://fr.wikipedia.org/wiki/Le_Bug_humain [Accessed 25 July 2022], 2022.

2

Smart City, Smart Building, Smart User: The Imaginaries of Smart and its Dead Ends

2.1. Introduction

Achieving climate targets for greenhouse gas reduction requires a profound rethinking of our lifestyles and energy systems. Many forward-looking models, such as the négaWatt scenario, agree on three stages and objectives for achieving this energy transition: sufficiency, energy efficiency and renewable energies. Sufficiency is based on changes in practices, while the development of efficiency and renewable energies (REs) is based on technical changes (devices and systems). The energy transition thus crystallizes two main issues: on the one hand, the need to reduce energy consumption, and on the other hand, the need to change energy production by using decarbonized energy sources with low environmental impact.

Substituting RE for fossil fuels alone will not achieve climate targets. It must be accompanied by a drastic reduction in energy consumption. This can be achieved through energy efficiency measures (such as insulating buildings) and/or by changing practices (such as lowering the temperature at which buildings are heated).

The aim of this chapter is to discuss the limits of both an approach focused on technical change (energy efficiency) and one focused solely on changing practices. Demonstrating the intrinsic link between technical systems and practices implies thinking of technologies and their uses together, so that the appropriation of technology by the user plays a central role in determining uses, and thus in achieving energy-saving practices.

This raises the question of the "effectiveness" of smart technologies (smart city, smart grid, smart building) in meeting environmental challenges, since all these technical developments are based on the same imaginary of an ever-increasing distance between technical systems and their users.

2.2. Reducing energy consumption: changing technologies or changing practices?

2.2.1. *Limits to energy efficiency*

Numerous studies point to the limits of energy efficiency measures. The aim is to achieve the same level of service, but with lower energy consumption. The limits of energy efficiency are linked in particular to the "rebound effect" [JEV 65], which leads either to a transfer of the energy saved to another item of energy consumption (e.g. saving on heating will enable a trip to be made), or to an intensification of use which cancels out the savings made, or even to higher energy consumption: a better-insulated house will be heated to a higher temperature than previously, a more energy-efficient car will be used more often and so on. Numerous empirical studies in the human and social sciences have investigated the rebound effect in household consumption [HER 08, NEM 12, GAL 15]. Recent empirical studies show that, in the case of energy-efficient building renovations, energy gains can be cancelled out by the persistence of energy-intensive practices, combined with potential rebound effects [RAU 20].

Energy efficiency measures alone – in other words, technical solutions – are therefore not enough to meet the challenge of reducing consumption. Increased energy efficiency can also be accompanied by new social demands, which often cancel out the energy gains. For example, a 1955 Volkswagen Beetle (730 kg and 30 hp) consumed 7.5 l/100 km, while a 2005 New Beetle (1,200 kg and 75 hp) consumes 7.1 l/100 km. The energy gains linked to the car's increased efficiency have been cancelled out by the increase in its weight (safety issues) and power.

Another contemporary example of this phenomenon is the SUV craze, which has led to the replacement of the existing car fleet with cars that consume more energy than their predecessors. The very notion of efficiency is therefore questionable from the point of view of technical change or "evolution" [GRA 93, GRA 03, GUC 05].

Finally, the central criticism that can be leveled at the notion of efficiency is that, by aiming to achieve an equivalent service, the pursuit of energy efficiency not only fails to call into question our lifestyles (which are ever more energy-intensive) but also rules out the very possibility of doing so. As energy sociologist Elizabeth Shove demonstrates: "[Efficiency] diverts attention away from the project of developing new, non-modern, configurations of nature and society, and of material culture and practice" [SHO 18, p. 9].

The danger of energy efficiency therefore lies less in its "inefficiency" – or failure to reduce final energy consumption – than in the techno-solutionism imaginary that accompanies it, which conveys the idea that it would be possible to maintain identical lifestyles, the same relationship to nature and technology, while respecting environmental limits. On the contrary, a radical change is needed.

2.2.2. *Limits of an approach focused (solely) on practices*

Faced with the limits of energy efficiency, the impetus for a reduction in consumption would therefore be a change in practices.

Some pioneering work has studied energy consumption using the tools of the sociology of consumption [BOV 87], such as life-cycle analysis [DES 96]. However, within the currents of transition or sustainable consumption studies, it is mainly from the angle of the theory of social practices [SCH 96, REC 02, SCH 02, SHO 02] and more recently the energy culture framework [BAR 13, STE 10, STE 15] that energy practices have been studied.

The theory of practices, which emerged and was formalized at the turn of the 2000s, proposes to study the routine practices of everyday life, which nonetheless consume large amounts of natural resources, as an arrangement of the various elements that make them up, that is, materials, skills, social norms and conventions [SHO 02, DUB 13]. Practices are seen as routine, sedimented, and the originality of this approach lies in the fact that there is little (or no) interest in the individual who implements them, but in the practice itself, whether in terms of its constituent elements, its dissemination or its stabilization. This "practice turn" [SCH 01] shifts the focus from individuals to the social dimensions of their practices.

By refuting the idea of an autonomous consumer who has a choice and constructs their identity through their consumption practices [WAR 05, SHO 10], this current is part of a critique of a symbolic approach to consumption [HAL 11, DUB 13].

It developed in response to the emergence of public environmental policies aimed at changing behavior, as well as to psychological and economic work proposing models for redirecting consumer behavior by playing on instrumental rationality, via economic or psychological incentives (such as nudges, for example). This work is based on the hypothesis of the rational individual, much favored by public policies in Western countries (Anglo-Saxon, and also French). However, they have been widely called into question, whether by the work of Elizabeth Shove (2010) and the theory of practices or, in the wake of Foucault's work, by the work of Sophie Dubuisson-Quellier and the notion of governmentality of conduct [DUB 16] or that of the "ideal client" in [CHA 11].

According to E. Shove, the social practice theory highlights the role of collective dimensions in change, which are neglected or marginalized in approaches based on individual choice, which she categorizes with the letters ABC (Attitude, Behavior, Choice).

Nudges, labels (the Nutri-Score, energy labels, etc.), awareness-raising campaigns, etc., are all based on the idea of guiding individuals (who are assumed to be rational) to make the "right" choices. This governmentality of practices (getting an individual to do what is expected of them) raises ethical questions: who determines what is desirable and what is not? According to what legitimacy? Do the ends (saving the planet) justify the means (manipulating individuals without their knowledge)?

It also raises social and political questions: by making individuals responsible for the environmental crisis (it is because individuals have "bad" practices that there is an environmental crisis, and by changing these practices, such as the famous "eco-gestures", we could solve the crisis), this approach completely denies the political and collective dimension of environmental issues. And yet, not only are individual practices not exclusively individual, but they are also social and collective, as they are linked to social norms and imaginaries, as well as to collective material infrastructures (technical networks, for example). For example, the social norms of dress codes according to profession (shirt, jacket and tie or others that require the wearing of light clothing) are not the free choice of the worker but are professional social norms to which he or she must submit, and which may require offices to be air conditioned or, on the contrary, heated. The individual alone has no power over this tangle of social norms and technical systems, however well-intentioned and rational they may be.

Individual practices are not solely responsible for the environmental crisis; there are collective and therefore political choices underlying lifestyles. However, by individualizing environmental issues, through public policies that target individuals alone, governments are absolving themselves of their responsibility for public action on a collective scale and for technical infrastructure (development of public transport, cycle paths, renewable energy production, etc.).

However, these approaches account for the overwhelming majority of public environmental policies. In fact, a survey of over 1,000 household energy initiatives in 30 European countries[1], whether promoted by public authorities, municipalities, associations, companies, etc., showed that almost 75% of these initiatives promoted a change in individual behavior (48.2%) or a technical change (26.4%) [JEN 18]. In Switzerland, for example, the majority of energy initiatives are based on this vision of change – or government of conduct [DUB 16] – aimed at guiding individuals towards "better choices" via rationality, evaluation or awareness-raising [SAH 19].

The contribution of practice theory is precisely to demonstrate the specificity of consumption practices, in order to go beyond public policy measures aimed at energy efficiency (i.e. technical change) and changing individuals' behavior [LAB 18].

Another contribution is the emphasis placed on the routinization of practices, which makes particular sense for energy practices, showing that they are grounded by several constraints (time, skills, norms, material resources, etc.) and thus stem from several rationalities. These, however, are not very reflexive, as routines lighten the mental load: the best option from a range of choices is not constantly recalculated [DUB 13, DUB 16]. However, technical infrastructures play a role in this dynamic of stabilization or change.

The limitation of practice theory, however, is that it fails to take sufficient account of the technologies and the technical system in these dynamics of change and stabilization. As Gordon Walker explains, in social practice theory:

> You don't start from the technology; you start with the activity or practice and see how multiple technologies and materialities are integrated into practice entities and in practice performances [SOV 17, p. 713].

Technologies, or material devices, are indeed taken into consideration, but as this quotation underlines, technologies are not the input; they will only be taken into account as a constitutive element of the practice, and not as a determinant in their own right. This limitation is echoed in the energy culture framework (ECF), which appeared a little later (2010).

The ECF does include the material dimension of energy consumption, but the focus and scale of analysis remain consumption (or behavior), and primarily domestic consumption, irrespective of production structure: the question of whether practices and representations (and therefore consumption) vary according to whether we use an autonomous wood-burning stove, fueled by our own wood resources, or

1 Online database: http://www.energise-project.eu/projects.

electric heating supplied by a nuclear power plant via a large technical system (LTS) is not questioned. Yet the relationship to the world in which both uses are situated is incommensurable. Nor is the relationship to energy outside of consumption (such as representations or imaginaries) questioned.

What is more, in both theories, the analysis often takes into account only those technologies that are directly linked to energy practices and with which the user is in contact (boilers and insulation within the domestic space, for example, when studying heating practices), rather than the technical system that underpins the existence of these technical objects. Yet the technical system influences our perception of energy, our awareness of how it is used on a daily basis and our representations of energy. And all this ultimately has an impact on (quantitative) energy consumption [DOB 16].

While the technical dimension is taken into account in the social practice theory and energy culture framework approaches, this is only partly the case. The fact that energy is produced through a large technical system and its impact on practices are not really taken into account.

2.2.3. *Usage dependence on the technology used*

If it is necessary to consider technologies and practices together, it is because the technologies used have an influence on practices: the proximity of the system and technical infrastructures, their understanding and appropriation play a central role in reducing energy consumption and in energy sufficiency. This was highlighted by two empirical studies on the use of renewable energies, in partial or total energy autonomy, one on an individual (residential) scale and the other on a collective (municipal) scale.

On a residential scale, the proximity of energy consumption and production sites (solar panels on the roof, wind turbines in the garden, etc.) enables a direct understanding of the technical system and the attribution of a symbolic value to energy, leading to energy saving [DOB 08].

On a collective scale, a comparative analysis of five rural communities in Germany, Austria and France that are energy self-sufficient through the use of renewable energy shows that proximity to production within a decentralized energy system and an understanding of the issues and difficulties involved in energy production influence the representation of energy and its value and consequently contribute to moving towards more sufficient practices [DOB 16].

It is therefore a question of shifting from individual practices to the collective dimension of technology and its impact on practices. As pointed out above, studies based on practice theories are highly critical of the emphasis placed on individual responsibility for environmental problems in public policy (and more broadly, in the political paradigms of environmental issues[2]) and emphasize the social dimension of energy practices. However, in so doing, they often overlook the collective and social dimension of technology.

How can social norms and social imaginaries change in favor of less energy-intensive and less resource-intensive practices, if the technology used – which acts as an intermediary between individuals and nature, as well as between members of a group – remains identical? In other words, how can social norms change without technical change?

Changing our relationship with the world, our imaginaries, our values and our conception of well-being and comfort is necessarily accompanied by technical change, insofar as technology is the expression of a particular relationship with the world. There is in fact a correlative link between technical change and social change, as well as between technical (or energy) systems and social organization [DOB 08]: if technical changes are part of social changes (imaginaries, values, aspirations, policies, etc.), once appropriated by the social body, these technologies modify it in return. Through them, they bring about a relationship with the world that until then could only be situated in the imaginaries and conceptions of the world.

Therefore, the *modern* relationship to the world, although pre-existing as a conception of the world to modern technologies, will only come into being with the use of modern technologies (e.g. the railway), a use that will in turn bring about social changes: new relationships to time (such as punctuality) and space (travel, proximity to certain places), new social relationships, etc. [GRA 03]. However, this special relationship with the world, made possible by modern technology, runs counter to the very idea of sufficiency. Consuming "less" energy, goods, etc., implies a very different way of looking at our social and political relationships, our relationships with nature, space, time and so on. And such a relationship to the world necessarily implies the use of technologies *different* from those we currently use.

The challenge of the energy transition therefore also rests on the implementation of technologies that are compatible with a peaceful, less predatory relationship with the environment, in terms of the consumption and destruction of natural resources.

---

2 Michael Maniates [MAN 01] is a good example of the paradigm of individual responsibility in environmental issues in the United States. For France, consider the work of Sophie Dubuisson-Quellier [DUB 16].

Since there is no such thing as technology neutrality, the use of certain technologies (and technical systems) would be more likely than others to lead to an ecological and energy transition.

This is suggested by Gibson's concept of "affordance" [GIB 79], and more specifically that of technological affordance, developed by Hutchby [HUT 01]. With this concept, Hutchby aims to go beyond both constructivist radicalism and technical determinism, by postulating that artifacts "enable" (or not) certain uses and practices by virtue of their properties. Affordances are therefore the possibilities for action that the artifact offers:

> The affordances of an artefact are not things which impose themselves upon humans' actions with, around, or via that artefact. But they do set limits on what it is possible to do with, around, or via the artefact. By the same token, there is not one but a variety of ways of responding to the range of affordances for action and interaction that a technology presents. We can analyse the development of those responses empirically, but in order to do so we have to accept that technological artefacts do not amount simply to what their users make of them; what is made of them is accomplished in the interface between human aims and the artefact's affordances [HUT 01, p. 450].

This concept opens up interesting theoretical and empirical perspectives. In the case of energy, for example, we need to consider the cultural dimension of affordance and its place in the environment: is affordance territorialized in the case of energy?

However, the assumption of the non-neutrality of technology and its potential "affordances" should not be seen as an assumption on the determinism of technology. A wide variety of practices and variations in (quantitative) consumption can be observed, for example, on identical energy networks between different countries, regions or households [COU 10]. This is because norms, values, representations, socio-economic and cultural contexts, etc., influence energy consumption and explain these variations in use within the same technical system. Technology is only one determinant among others, but one that must be taken seriously.

Decentralizing and relocating energy production make the constraints of production (aesthetic, technical, environmental, etc.) tangible, and thus make the challenges of energy consumption visible.

The proximity and appropriation of energy production, self-consumption, energy sources (e.g. RE) and technologies used all have an impact on the way energy is represented and used, contributing to a move towards less-intensive energy practices.

2.3. The smart imaginary and its dead ends

The "smart" imaginary (smart city, smart networks, smart buildings, etc.) is based on the paradigm of remote, automated management of technology and networks, without user intervention. The vision of the user here is, on the one hand, that of a disrupter of the technical system, and on the other hand, of an agent definitively reluctant to adopt the necessary energy-saving and efficiency practices. To get around this "behavioral" problem, the technology will be automated and managed remotely, without the end-user's intervention. This technician imaginary of network optimization and efficiency is embodied, for example, in the vision of a smart grid which, via smart meters, would enable households to remotely switch off their electric heaters for ten minutes an hour during periods of peak consumption: this cut-off would not be perceived thermally by the user, but would enable consumption on the power grid to be smoothed out on a large scale, optimizing it and thus generating significant energy savings. In this vision, the possibility of users deciding to turn down their heating and put on a pullover is excluded.

This imaginary is problematic, to say the least: on the one hand, it is based on a vision of the citizen as incapable of grasping environmental issues and changing their practices of their own volition; on the other hand, it is based on a technical imaginary in which users no longer have any possibility of acting on the technology that surrounds them; they find themselves completely dispossessed of it, through centralized management.

In the case of the two empirical studies on the use of renewable energies cited above, in contrast, it is the proximity of the technology, the understanding of the technical system and its appropriation that enable low-energy use [DOB 08, DOB 16]. Indeed, how can we become aware of our energy consumption when energy production and the technical system that deploys it are totally invisible to the user, whether by veiling the materiality of technical infrastructures such as networks or by making all the devices and artifacts through which the user can interact with the technical system (e.g. radiators for the heating system) disappear?

This process of making technical systems invisible is incompatible with the objective of reducing consumption, which requires, on the contrary, making visible the energy used and consumed daily.

2.3.1. *Technical distancing as a common denominator*

However, this distancing from the user can be found in all the deployments of smart technologies, whether in terms of buildings, networks, cities and so on. This distance is both technical and operational (the user is unable to control the technology) and symbolic [GAR 14, CHE 17]. Paradoxically, in the smart imaginaries, the user is surrounded by technology, and at the same time isolated from it. Their environment is invaded by technologies that generate and analyze data, but over which they have no possible means of action, as these are automated and managed by a control center. In the end, they find themselves a spectator of the technical system and its functioning, with no possibility of action [REI 23].

2.3.2. *The smart city or the imaginary of a city without inhabitants*

Therefore, whether in representations of the smart city or the smart building, the inhabitant, placed at a distance, is conspicuous by their absence. According to philosopher Tyler Reigeluth, the depopulated smart city presents an image of "a deserted city, a city with almost no life left in it, a city abandoned to its technical workings" [REI 23, pp. 12–13, authors' translation]. If we try to draw up the ideal type of smart city, he continues to say:

> First and foremost, we find the structural idea of a city that exists for the sake of "technology itself": a city that would function all by itself, that would be as automated as possible, that would involve only a small amount of human intervention and would be fluid and ergonomic to use. A city that runs smoothly, does not break down, runs in the background, and silently watches over the best of all possible worlds. In this world, the inhabitant is an afterthought, a sub-optimal variable that always interferes with the ideal functioning of the perfectly oiled machine, the unfortunate blemish on the glittering dashboard. This fetish of automatic intelligence, which would consist in a kind of technologically calibrated *laissez-faire*, is only possible because its material conditions are constantly disdained, euphemized, distanced [REI 23, p. 19, authors' translation].

This representation of the inhabitant as a "parasite" is recurrent in the "smart" imaginaries, whether in the smart city or the smart building [DOR 24]. We therefore need to keep the inhabitants at as great a distance as possible, until they disappear completely.

Putting the user at a distance also means making the technical system that enables this intelligence invisible. This intelligence is itself located "elsewhere", outside the city. The data, produced by the user without their knowledge, is stored, managed and analyzed remotely, via the cloud and control centers. This production, management and use of data are not without legal implications [POU 23].

Therefore, "in line with one of the precepts of seamless design, a "good" smart city would present itself as a set of continuous interfaces of which users would ideally not even be aware, even though they interact with them" [REI 23, p. 20, authors' translation].

This invisibility of the technical system is precisely what prevents it from being appropriated, and thus, ultimately, from being used sparingly. Yet, paradoxically, or ambivalently, this logic of rendering technical systems invisible in the deployment of the smart city goes hand-in-hand with a logic of transparency:

> In many respects, the rhetoric and projects of smart cities reflect the logical culmination of a technologically transparent world. The smart city, with all its sensors, data and predictions, would be one where there are no more grey areas, no more secrets, where everything is immediately visible and accessible [REI 23, p. 12, authors' translation].

Here, we see the paradox of striving for visibility, transparency and exposure of the city and its data, while at the same time, the technical systems surrounding the inhabitant are rendered invisible and operate in an almost magical way for the latter, since no externality is visible or perceptible. Here again, the problem lies in achieving sufficiency: the invisibility of the technology, services and energy used, which prevents us from being aware of the impact of our lifestyles and consumption patterns.

The deployment of the smart city therefore raises issues of inequality (between favored neighborhoods where this intelligence is deployed to the detriment of disadvantaged neighborhoods, ditto between urban and rural areas, etc.), vulnerability (malfunctions, hacking and failures) [REI 2023] and inclusion (intelligence serving which user?).

The ideal of citizen participation in the smart city, thanks to information and communication technologies (ICT) that would enable greater participation [ROB 23] and the smart management of data produced by AI and algorithms, has not been realized.

The bibliometric analysis carried out by Corsini et al. [COR 19] shows the gap between this ideal of citizen participation in the smart city imaginary, and its actual implementation, particularly in academic research, where citizens are rarely taken into account:

> The bibliometric analysis of 74.932 research publications shows that public participation and community involvement was a minority interest in the planning, design, construction and operation of energy projects in urban contexts. Results presented show an almost overwhelming interest with the technology of energy generation or efficiency of consumption rather than the social context that might motivate change to a more sustainable energy model for urban conurbations [COR 19, p. 330].

The low priority given to residents in the smart city is not, however, a question of scale: it applies just as much to the building itself.

2.3.3. *The smart building or the imaginary of a building parasitized by its users*

This raises the question of how the user is included – and what kind of user – in the implementation of smart technical devices. On a building scale, or that of a smart building, the building is designed for the user, but without the user.

A study on the imaginary world of efficient buildings also reveals a technical approach that sidelines the user in order to achieve transition objectives. As Cherry et al. [CHE 17] observe in the case of passive or smart homes:

> Whilst portrayed as divergent futures, both visions aimed to "design out" the role of occupants, achieving emissions reductions through changes to the built environment and maintaining current lifestyles; a position that was reinforced by an imagined public that was unable or unwilling to accept the need for lifestyle change [CHE 17, p.36].

Therefore, in this notion of smart, the user, considered incapable of changing his or her practices, is materially and symbolically distanced.

However, empirical studies have clearly shown the limits of a purely technical approach to reducing residential consumption, whose effects are cancelled out by the continuity of pre-existing practices or by the rebound effect (notably [RAU 20]).

As Isabelle Garabuau-Moussaoui points out, the technical distance is compounded by a symbolic distance: the building, especially if it is a demonstration site, is given pride of place, to the detriment of the users:

> They [users] can also feel that they are left by the wayside, compared to the building itself, which is valorised through external communication – for example, inaugurated with great ceremony, without them (showcase building); yet at the same time they are expected to make commitments and to participate in the "building" project [GAR 14, p.10].

In the smart building, there is thus an ambivalent expectation of a user who is absent from the building, as least disruptive as possible, and at the same time engaged and proactive in the "life" of the building.

Indeed, smart building designers and technicians draw an analogy between buildings and living organisms: smart buildings "breathe", "live" and even "survive", thanks to technical systems such as automated ventilation. Its performance and "survival" therefore depend on the smooth operation of technical systems and the assistance of technicians to maintain and optimize them. There is thus a binary relationship between technicians and building, which the user disrupts. They parasitize this balance, constantly threatening the building's performance and smooth operation [DOR 24]. The smart building, a complex organism, finds itself parasitized by its users.

It is this ambivalent expectation, through the figure of the smart user, that we will examine in conclusion.

2.4. Conclusion: in search of the smart user?

The figure of the smart user is thus divided between, on the one hand, the representation of the least disruptive possible user, ideally absent, and on the other hand, a proactive user, familiar with the subtleties of the technical system and able to adapt to it: in short, the engineer.

We thus find both the idea that users must adapt their behavior to energy and "play an active role in any future-oriented energy concept" [ROB 23, p. 137] and that "a smart user is a user who does not need to take care of anything. When they enter a room, the technology adapts to them, to their profile. For example, when a teacher enters a class with their students, the luminosity and the ventilation adapt. The teacher can thus concentrate on their tasks without worrying about their environment" [ROB 23, p. 187].

We may well wonder to what intelligence we are referring to in this last representation, that of the passive user. The user's intelligence seems to be limited to not interfering with the technical system: basically, the intelligence of "doing nothing" and "touching nothing".

The intelligence of the smart building will then be to anticipate the diverse needs of users, that is, to differentiate between uses within the same building: whether some rooms are heated/illuminated and others not, some more or some less, in a variable way according to time slots, etc.

Conversely, in the representation of a proactive user, the smart user's intelligence lies in conforming and adapting (as it changes) to the building's performance requirements, such as lowering energy consumption when the building is producing little and consuming a lot and postponing it until the situation reverses. This can be achieved through applications that warn the user, knowledge of the technical system or user "training". This super-user, or proactive user, is ultimately the user who has acquired as much knowledge and skill as the smart building engineer or technician. But in this case, is the smart user still a user?

Whether the smart user is a completely passive user or, on the contrary, an engineer, it is the very figure of the user that is being erased from the "smart" imaginary.

Paradoxically, smart buildings are, for their designers, first and foremost about user comfort. However, comfort is seen here as a stable, factual given, not as a social norm that varies over time and between societies, and which, like any social construct, can be modified, (re)negotiated and redefined. In this representation, since performance objectives are non-negotiable (environmental objectives) and user comfort is also non-negotiable (stable data), it is up to technical systems to reconcile the irreconcilable and to obtain the same level of comfort while consuming less. This performance, or technical efficiency, requires the user to comply with the technical requirements on which it is based (e.g. not to counteract the building's efficiency by opening the windows).

Consider the social dimension of comfort, that is, the fact that uses change according to social norms, that is, a specific collectively shared representation, which varies according to cultures and times, defining what well-being and comfort are. For example, while recommended temperatures for heating homes have varied between 19 and 20°C since the 1970s, they varied widely in the 20th century alone, depending on the period and circumstances, from 17.5°C in 1906 to 15°C in 1918, rising to 18°C in 1934 and 16.5°C in 1948 [VIG 20]. The "ideal temperature" is a social construct which, as such, varies according to social context. Today's comfort standards can just as easily be deconstructed and redefined.

On the contrary, the equally social dimension of technology, that is, the impact of the technical system used and its appropriation on usage, demonstrates the need for a new approach to the issue of low consumption and the means to achieve it. That is why the "smart" imaginary, based on the idea that environmental performance depends on the efficiency of a technical system that increasingly distances the user, is a dead end.

2.5. References

[BAR 13] BARTON B., BLACKWELL S., CARRINGTON G. et al., Energy cultures: Implication for policymakers, Report, University of Otago, Centre for Sustainability, 2013.

[BOV 87] BOVAY C., ROLAND J., CAMPICHE F. et al., *L'Énergie au quotidien : aspects sociologiques et éthiques de la consommation d'énergie*, Labor et Fides, Geneva, 1987.

[CHA 11] CHAVANON O., JOLY, O., RAYMOND R. et al., "Le scénario facteur 4 : les rhétoriques institutionnelles au regard des conduites ordinaires en matière de consommation d'énergie", *Développement durable et territoires*, vol. 2, no. 1, 2011.

[CHE 17] CHERRY C., HOPFE C., MACGILLIVRAY B. et al., "Homes as machines: Exploring expert and public imaginaries of low carbon housing futures in the United Kingdom", *Energy Research & Social Science*, no. 23, pp. 36–45, doi: 10.1016/j.erss.2016.10.011, 2017.

[COR 19] CORSINI F., CERTOMA C., DYER M. et al., "Participatory energy: Research, imaginaries and practices on people' contribute to energy systems in the smart city", *Technological Forecasting and Social Change*, no. 142, pp. 322–332, doi: 10.1016/j.techfore.2018.07.028, 2019.

[COU 10] COUTARD O., "Services urbains : la fin des grands réseaux ?", in COUTARD O. and LEVY J.-P. (eds), *Écologies urbaines*, Economica, Paris, pp. 102–129, 2010.

[DES 96] DESJEUX D., BERTHIER C., JARRAFFOUX S. et al., *Anthropologie de l'électricité. Les objets électriques dans la vie quotidienne en France*, L'Harmattan, Paris, 1996.

[DOB 16] DOBIGNY L., Quand l'énergie change de mains. Socio-anthropologie de l'autonomie énergétique locale au moyen d'énergies renouvelables en Allemagne, Autriche et France, PhD thesis, Paris 1 Panthéon-Sorbonne University, Paris, 2016.

[DOB 18] DOBIGNY L., "Changement énergétique et rapport au monde", in MENOZZI M.-J., FLIPO F., PECAUD D. (eds), *Énergie et société. Sciences, gouvernances et usages*, Édisud, Paris, pp. 215–224, 2018.

[DOR 24] DORTHE G., DOBIGNY L., "From smart buildings to smart users? Energy transition to the test of parasitic humans", in GUGGANIG M., BRONSON K., MIRZA V. (eds), *The Smartification of Everything*, University of Toronto Press, Toronto, 2024.

[DUB 13] DUBUISSON-QUELLIER S., Plessz M., "La théorie des pratiques", *Sociologie*, vol. 4, no. 4, 2013.

[DUB 16] DUBUISSON-QUELLIER S., *Gouverner les conduites*, Presses de Sciences Po, Paris, 2016.

[GAL 15] GALVIN R., "The rebound effect, gender and social justice: A case study in Germany", *Energy Policy*, vol. 86, pp. 759–769, 2015.

[GAR 14] GARABUAU-MOUSSAOUI I., "How people work and live in energy efficient workplaces. Logics of actions, social tensions and real issues for occupants in energy efficient office buildings", *BEHAVE Conference Proceedings*, 2014.

[GIB 79] GIBSON J.J., *The Ecological Approach to Visual Perception: Classic Edition*, Houghton Mifflin, London, 1979.

[GRA 93] GRAS A., *Grandeur et dépendance*, Presses Universitaires de France, Paris, 1993.

[GRA 03] GRAS A., *Fragilité de la puissance*, Fayard, Paris, 2003.

[GUC 05] GUCHET X., *Les sens de l'évolution technique*, Editions Léo Scheer, Paris, 2005.

[HAL 11] HALKIER B., KATZ-GERRO T., MARTENS L., "Applying practice theory to the study of consumption: Theoretical and methodological considerations", *Journal of Consumer Culture*, vol. 11, no. 1, pp. 3–13, 2011.

[HER 08] HERTWICH E., "Consumption and the rebound effect: An industrial ecology perspective", *Journal of Industrial Ecology*, vol. 9, no. 1, pp. 85–98, 2008.

[HUT 01] HUTCHBY I., "Technologies, texts and affordances", *Sociology*, vol. 35, no. 2, pp. 441–456, doi: 10.1177/S0038038501000219, 2001.

[JEN 18] JENSEN C., AUDLEY G., EDINA V., Publication of reports of results for subsequent WP6 and WP7, ENERGISE – European Network for Research, Good Practice and Innovation for Sustainable Energy Grant Agreement No.727642, Deliverable 2.6, 2018.

[JEV 65] JEVONS W.S., *The Coal Question; An Inquiry Concerning the Progress of the Nation, and the Probable Exhaustion of Our Coal Mines*, Macmillan & Co, London, 1865.

[LAB 18] LABANCA N., BERTOLDI P., "Beyond energy efficiency and individual behaviours: Policy insights from social practice theories", *Energy Policy*, no. 115, pp. 494–502, doi: 10.1016/j.enpol.2018.01.027, 2018.

[MAN 01] MANIATES M.F., "Individualization: Plant a tree, buy a bike, save the world?", *Global Environmental Politics*, vol. 3, no. 1, pp. 31–52, doi: 10.1162/152638001316881395, 2001.

[NÉM 12] NEMOZ S., WALLENBORN G., "Comment comprendre les effets rebonds dans la consommation domestique d'énergie ? Pour une socio-anthropologie des pratiques en transition", in POIROT-DELPECH S. and RAINEAU L. (eds), *Pour une socio-anthropologie de l'environnement*, L'Harmattan, Paris, 2012.

[POU 22] POULLET Y., "Introduction à quelques aspects juridiques du smart building", *Revue Lamy droit de l'immatériel*, no. 198, pp. 43–50, 2022.

[RAU 20] RAU H., MORAN P., MANTON R. et al., "Changing energy cultures? Household energy use before and after a building energy efficiency retrofit", *Sustainable Cities and Society*, vol. 101983, no. 54, doi: 10.1016/j.scs.2019.101983, 2020.

[REC 02] RECKWITZ A., "Toward a theory of social practices: A development in culturalist theorizing", *European Journal of Social Theory*, vol. 5, no. 2, pp. 243–263, 2002.

[REI 23] REIGELUTH T., *L'intelligence des villes. Critique d'une transparence sans fin*, Éditions Météores, Brussels, 2023.

[ROB 23] ROBYNS B., LENGLET C., BARRY H. et al., *Smart Users for Energy and Societal Transition*, ISTE Ltd, London and John Wiley & Sons, New York, 2023.

[SAH 19] SAHAKIAN M., DOBIGNY L., "From governing behaviour to transformative change: A typology of household energy initiatives in Switzerland", *Energy Policy*, no. 129, pp. 1261–1270, doi: 10.1016/j.enpol.2019.03.027, 2019.

[SCH 96] SCHATZKI T.-R., *Social Practices – A Wittgensteinian Approach to Human Activity and the Social*, Cambridge University Press, Cambridge, 1996.

[SCH 01] SCHATZKI T.-R. (ed.), *The Practice Turn in Contemporary Theory*, Routledge, London, 2001.

[SCH 02] SCHATZKI T.-R., *The Site of the Social: A Philosophical Account of the Constitution of Social Life and Change*, Pennsylvania State University Press, University Park, 2002.

[SHO 02] SHOVE E., WARDE A., "Inconspicuous consumption: The sociology of consumption, lifestyles and environment", in DUNLAP R.E., BUTTEL F.H., DICKENS P. et al. (eds), *Sociological Theory and the Environment*, Rowman & Littlefield, Lanham, pp. 230–251, 2002.

[SHO 10] SHOVE E., "Beyond the ABC: Climate change policy and theories of social change", *Environment and Planning A: Economy and Space*, vol. 6, no. 42, pp. 1273–1285, doi: 10.1068/a42282, 2010.

[SHO 18] SHOVE E., "What is wrong with energy efficiency?" *Building Research & Information*, vol. 7, no. 46, pp. 779–789, 2018.

[SOV 17] SOVACOOL B., DAVID K., HESS J., "Ordering theories: Typologies and conceptual frameworks for sociotechnical change", *Social Studies of Science*, vol. 5, no. 47, pp. 703–750, doi: 10.1177/0306312717709363, 2017.

[STE 10] STEPHENSON J., BARTON B., CARRINGTON G. et al., "Energy cultures: A framework for understanding energy behaviours", *Energy Policy*, no. 11, 2010.

[STE 15] STEPHENSON J., BARTON B., CARRINGTON G. et al., "The energy cultures framework: Exploring the role of norms, practices and material culture in shaping energy behaviour in New Zealand", *Energy Research & Social Science*, no. 7, pp. 117–123, doi: 10.1016/j.erss.2015.03.005, 2015.

[VIG 20] VIGUE R., "Une courbe pour bien se chauffer. Les températures de chauffage préconisées au XXe siècle", *Flux*, vol. 121, pp. 102–107, 2020.

[WAR 05] WARDE A., "Consumption and theories of practice", *Journal of Consumer Culture*, vol. 5, no. 2, pp. 131–153, 2005.

3

Forecasting the Production and Consumption of Electrical Energy

3.1. Introduction

As explained in Chapter 1 (see section 1.7), managing both power generation and consumption has become a necessity to make better use of energy, minimize losses and ensure user satisfaction. However, most renewable resources are known to be intermittent and unstable; the energy they produce is totally dependent on weather conditions. Moreover, consumption is not regular and depends not only on weather conditions but also on user behavior. Forecasting both energy production and consumption has therefore become essential to ensure better energy management and optimum user satisfaction.

In this chapter, forecasting methods for predicting photovoltaic production (with satellite images, camera and artificial neural networks) and electricity consumption are presented. The importance of forecasting in electrical systems is highlighted, followed by a presentation of the different forecasting methods for each variable, namely camera, satellite and neural network prediction. The chapter then focuses on forecasted consumption in commercial buildings using artificial neural networks. These methods were tested on the Université Catholique de Lille smart grid demonstrator. Results and feedback are described and discussed. Forecasting intermittent renewable energies contributes to SDG 7 by promoting and optimizing the use of clean energy. Forecasting building consumption contributes to SDG 11, “Sustainable cities and communities”, by enabling better control of this consumption and promoting the use of renewable energy to satisfy it.

The development of renewable energies is an objective of ADEME's S3 "Green Technologies" scenario; however, a better forecast of these intermittent energies, combined with consumption forecasting, contributes to the S2 "Territorial Cooperation" scenario, the life of renewable energy communities and the S1 "Frugal Generation" scenario, since better knowledge of consumption also makes it possible to act on reducing it.

3.2. Variability in production and consumption

In an electrical system (microgrid or power grid), production and consumption vary with time. This may be due to operational causes, such as variations in equipment consumption, charging or discharging of storage systems, charging of electric vehicles and the generation of renewable energies such as wind power or photovoltaics. It can also be caused by a fault in the network infrastructure or by line outages. We focus here solely on the variability of renewable production sources, in particular solar photovoltaics, and on load demand (with or without electric vehicles). Figures 3.1 and 3.2 show the variability of renewable photovoltaic and wind generation. This variability occurs on all time scales (from a few seconds to several years) and depends on location.

On the other hand, under partly cloudy skies, the power produced by a photovoltaic (PV) plant can easily fluctuate by three-quarters of its rated output in 10 seconds. This intermittency poses serious problems in terms of system stability, balance of production and consumption, compensation of reactive power, frequency response, etc., and can potentially lead to the oversizing of primary and secondary reserves. This limits the penetration rate of intermittent renewable energies (PV and wind) in power grids.

Demand is also subject to fluctuations, which can occur at different timescales (Figure 3.3). In addition, consumption peaks occur during extreme weather events, due to the thermo-sensitivity of electricity consumption. This can disrupt the supply–demand balance.

Consequently, forecasts of RE production and electricity demand can be useful for estimating energy reserves, planning power plants, managing congestion, coordinating storage with RE or trading on electricity markets. In this way, they have become an essential component of energy management systems. Accurate forecasts can help improve the quality of electrical energy supplied to the power grid, and thus reduce the ancillary costs associated with general volatility.

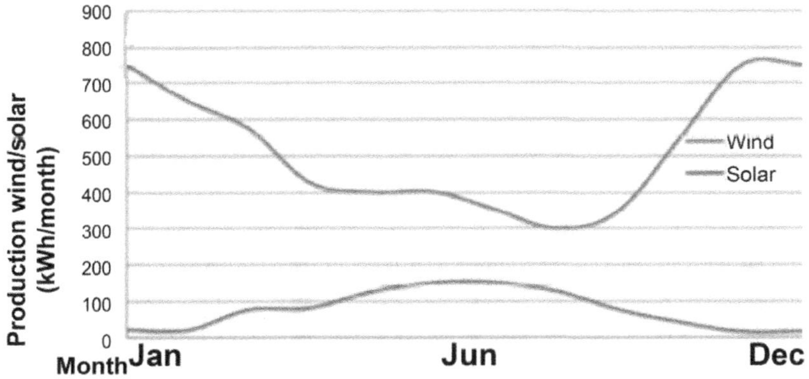

a)

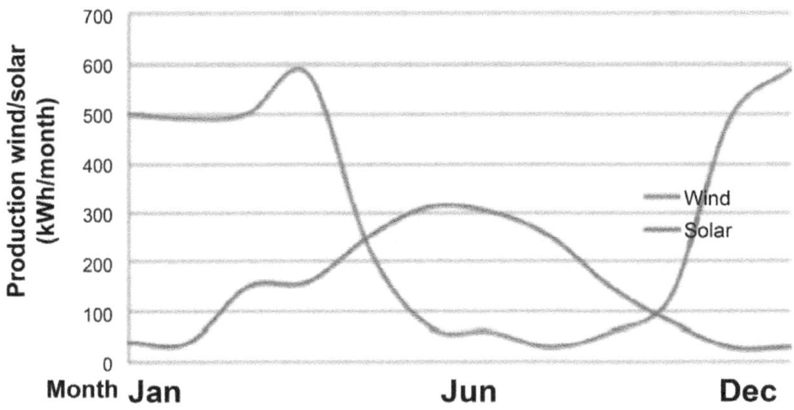

b)

Figure 3.1. *Variability of wind and photovoltaic production according to month and location: a) example of the distribution of annual production in Brest (windy with low sun) for the SKYSTREAM wind–photovoltaic hybrid system; b) example of the distribution of annual production in Nice (sunny with low wind) for the SKYSTREAM wind–photovoltaic hybrid system. For a color version of this figure, see www.iste.co.uk/robyns/smartgrids.zip*

In this chapter, the smart grid demonstrator presented in [ROB 23] and in Chapters 6 and 8 of this book is treated as a case study, with all forecasts used for its energy management. For example, photovoltaic production forecasts enable users to save on the cost of electric vehicle recharging by targeting the hours when the price of electricity is lowest.

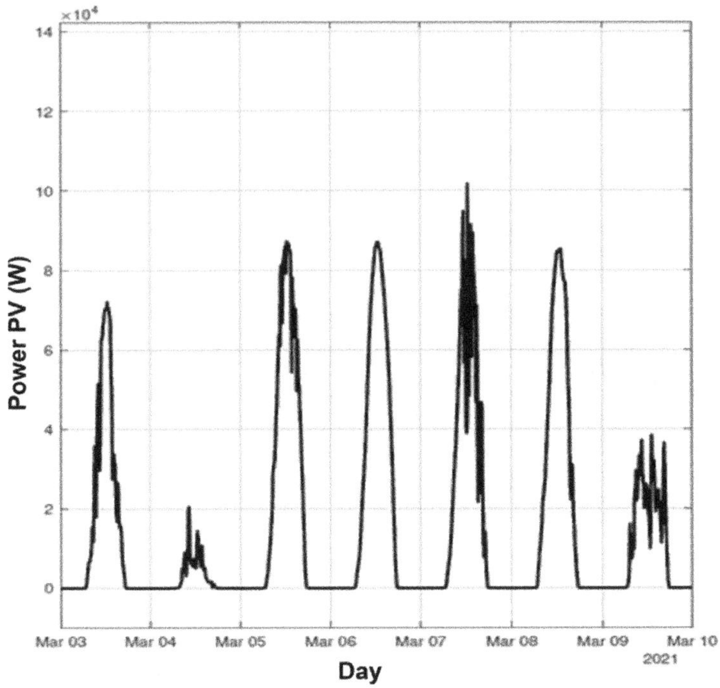

Figure 3.2. *Example from a real recording of a photovoltaic installation in northern France (Université Catholique de Lille smart grid demonstration)*

3.3. Photovoltaic production forecast

The increasing integration of renewable energy sources into the electrical system makes it more difficult to maintain its stability and reliability. It influences grid regulation, load-dependent generation, power scheduling and unit commitment. So, as an intermittent renewable energy source, solar power needs to be well forecasted and managed to ensure the smooth operation and efficiency of the electrical system. This forecasting depends on the use case. It may concern a single local PV system, or a large number of systems spread over a wide geographical area. The forecast may focus on the power output of the systems, or on their rate of change (also known as ramp rate). Forecasting methods also depend on the tools and information available, such as data from meteorological stations and satellites, PV system data and the results of numerical weather prediction (NWP) models.

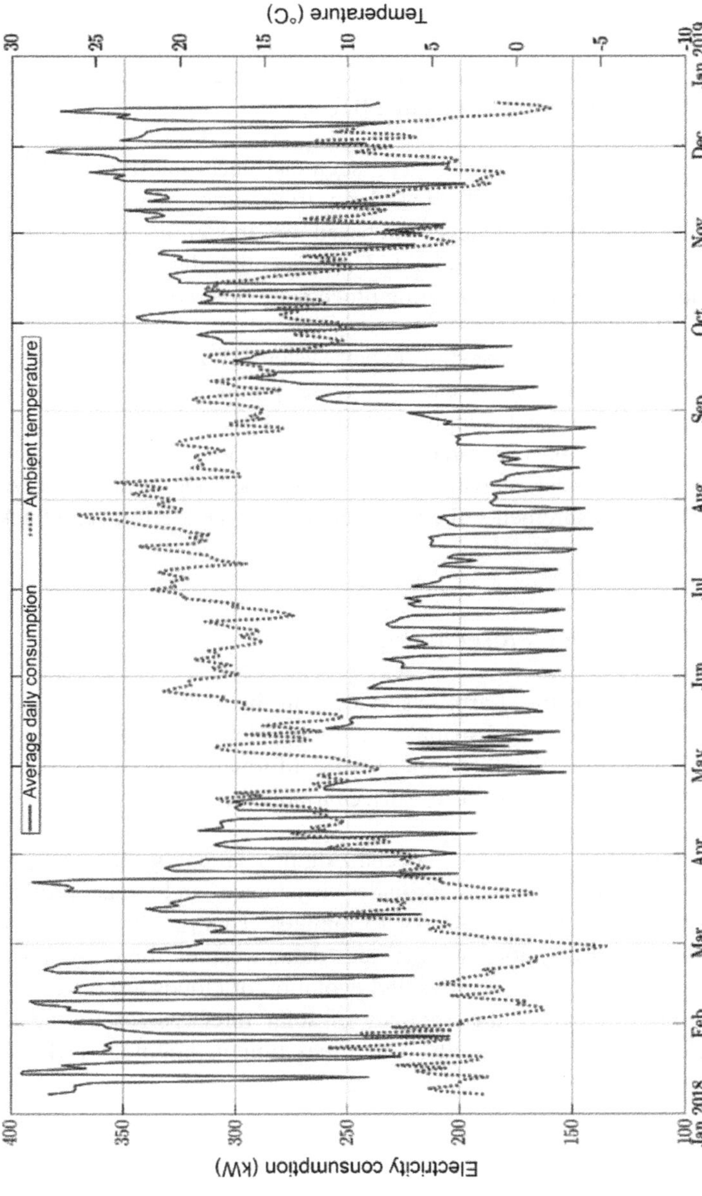

Figure 3.3. *Example of a real reading of a commercial building's consumption (in kW) and ambient temperature (in Celsius) in 2018 in northern France (Université Catholique de Lille smart grid demonstration). For a color version of this figure, see www.iste.co.uk/robyns/smartgrids.zip*

Consequently, various forecasting methods are used [ESP 10, PEL 13]. A first method consists of forecasting solar irradiation and then forecasting PV power using a mathematical model of the PV generator [LIU 15]. A second method proposes to forecast PV power directly from environmental data (irradiation, temperature, etc.) [BAC 09, YAN 14]. Indeed, energy production is strongly proportional to the overall irradiation and varies with temperature [HUA 10].

Statistical analysis tools are also used, such as linear/multilinear/nonlinear regression and autoregressive models based on time series regression analysis [KUD 09]. These forecasting models are based on modeling the relationship between influent inputs and power output. Consequently, the calibration of the mathematical model and the parameter fitting process are time-consuming.

At the same time, AI methods, such as expert systems, fuzzy logic and artificial neural networks (ANNs), are widely used to predict power generation in order to cope with uncertainties in load supply and demand [KER 22]. Among these methods, ANNs are the most commonly used technique. They offer several advantages over conventional statistical forecasting methods, such as ease of adaptation to online measurements, better tolerance to data errors and lack of superfluous information.

3.3.1. *Satellite image-based forecasting*

As solar irradiance of the Earth's surface can be deduced from satellite measurements at high temporal and spatial resolution, satellite images are used as a source of data for prediction. Satellite data provide information on cloud cover, the most important atmospheric parameter for surface irradiance.

For example, SteadySun, a company specializing in solar energy forecasting, has developed the SteadySat tool. This provides irradiance and solar production forecasts up to six hours in advance, every 15 minutes. This product combines the processing of satellite observations with in situ data, weather forecasts, physical models and AI techniques. The properties and evolution of clouds are observed and forecast in real time, improving the accuracy of solar forecasts for the next few hours, and anticipating risks of variability. The forecasting methodology is described in Figure 3.4.

Figure 3.4. *SteadySat methodology for predicting photovoltaics using satellite images*[1]

3.3.2. *Short-term forecast by camera*

SteadySun has developed a short-term solar forecasting solution called Steady Eye. Equipped with a 180° camera (Figure 3.5(a)), it can anticipate cloud movement over a 4 km^2 area with energy production forecasts for up to 30 minutes with a one-minute step using image-processing techniques. The camera takes hemispherical images of the sky at regular intervals (Figure 3.5(b)). Cloud and shadow maps are constructed by the cloud detector and motion field sensors using the method developed by SteadySun. These data are combined with data from the irradiance sensors and analyzed by the SteadySun algorithm to produce photovoltaic power forecasts. Analysis and forecasting can be carried out locally or remotely, and the tool can be used stand-alone or with an Internet connection.

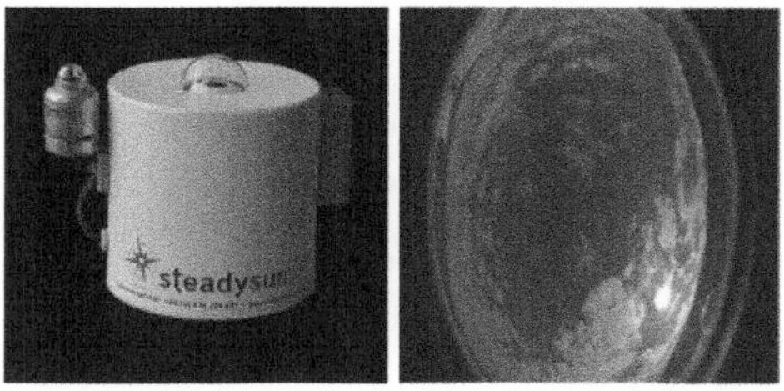

a) b)

Figure 3.5. *Sky image acquisition by Steady Eye camera for photovoltaics: a) Steady Eye camera; b) example of an image acquired by the Steady Eye camera*[2]

1 See: https://www.steady-sun.com/fr/steadysat-previsions-infra-journalieres/.

2 See: https://www.steady-sun.com/fr/steadyeye-previsions-solaires-pour-les-prochaines-minutes/.

The SteadySun server connects to the SCADA client control and data acquisition system using a WAN configured for a given area with an Internet connection at a download speed of 2 Mbps. Where the location is not equipped with a permanent Internet connection, an inter-process communication system is used to transmit data from the SteadySun server to the SCADA client. A set of components – including an X-series camera with smartwares (SW) lens, a pyranometer, a silicon reference cell and a SteadySun data processor – is used for data acquisition with SCADA, via an IP or Modbus connection. An overview of the tool's capabilities is outlined in Table 3.1.

Description	Capacity
Forecast zone	4 km^2
Temporal resolution	10 s
Spatial resolution	Local coverage
Forecast horizon	60 min
Components	Camera with SW lens (smartwares), silicon reference cell, pyranometer, data processor

Table 3.1. *Steady Eye solution features*

Figure 3.6 shows the block diagram of the Steady Eye solution for forecast generation.

Figure 3.6. *Steady Eye methodology for predicting photovoltaic production*[3]

3 See: https://www.steady-sun.com/fr/steadyeye-previsions-solaires-pour-les-prochaines-minutes/.

3.3.3. *Neural network prediction*

ANNs aim to mimic the capabilities of human neural networks. This multivariable, nonlinear and non-parametric mathematical tool has been used extensively to model complex nonlinear systems. A neural network can learn, store and model a system by processing external information for predictive and self-diagnostic purposes.

Today, neural networks have been successfully used in several electrical system problems, such as planning, control, analysis, protection, design, prediction, safety analysis and fault diagnosis [VAZ 14]. Compared with conventional statistical forecasting methods, ANNs offer additional advantages, including easy adaptability to online measurements, greater tolerance to data errors and the ability to handle superfluous information.

Figure 3.7 shows the basic model of an artificial neuron. Its components are the inputs x_i, the synaptic weights W_{ki}, the adder, the activation threshold b_k and the activation function φ.

There are several neural network architectures. The basic structure consists of an input layer including the input parameters required for prediction (e.g. irradiance, temperature, etc.), a hidden layer used to model the relationships and interactions between output parameters and inputs, and an output layer with the predicted variable(s), such as photovoltaic output (Figure 3.8).

The inputs and outputs of this structure are linked by the following equations:

$$v_j = f_{12}\,(\textstyle\sum_{i=1}^{n} x_i w_{ij})\; j = 1,2, \dots m \qquad [3.1]$$

$$y_k = f_{23}\,(\textstyle\sum_{j=1}^{m} v_j w_{jk})\; k = 1,2, \dots l \qquad [3.2]$$

where x_i are the neural input variables, v_j are the hidden layer units, y_k are the output variables, w_{ij} and w_{jk} are the interconnection weights and f_{12} and f_{23} are the transfer or activation functions. Several hidden layers are possible, depending on the complexity of the system.

Generally speaking, building a high-performance neural network involves several stages, as shown in Figure 3.9.

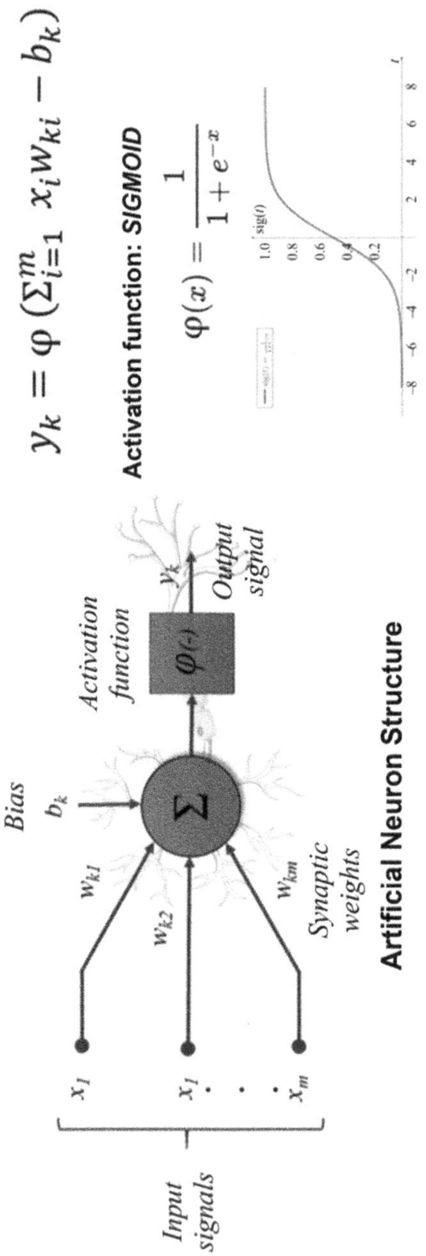

Figure 3.7. *Basic model of a neuron in an artificial neural network*

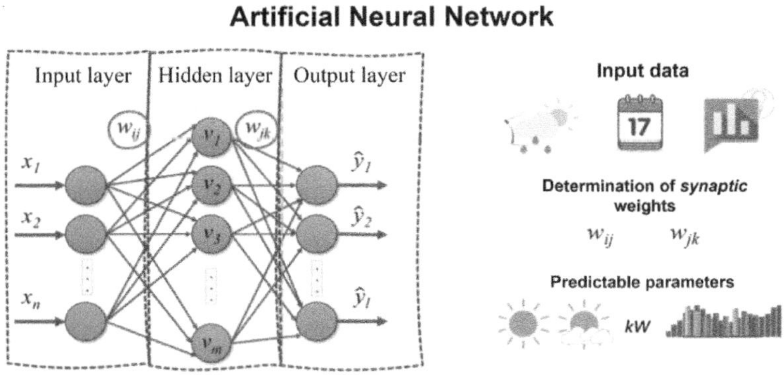

Figure 3.8. *Structure of a three-layer artificial neural network*

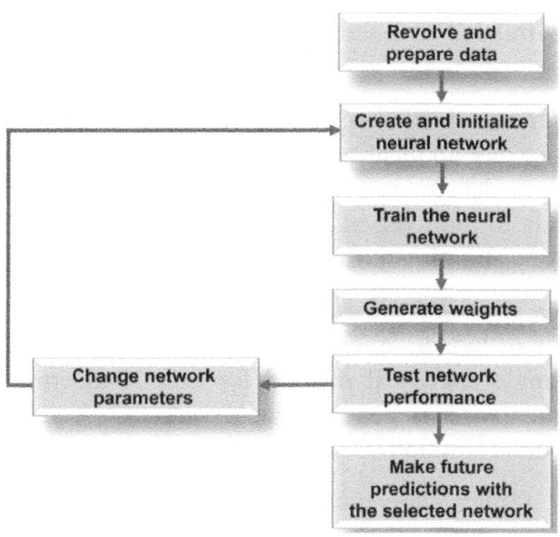

Figure 3.9. *How to set up a forecast using neural networks*

These can be summarized in three main stages:

– *Data preparation*: data preparation is one of the most important steps in solving forecasting problems, as it aims to eliminate, as far as possible, any noise that could affect the accuracy of the ANNs. The latter are highly sensitive to any inconsistencies in the training data, especially when their quantity is low. For this reason, meticulous and efficient data preparation is strongly recommended prior to the use of ANNs. Two key sub-steps are mandatory:

– a *correlation study* to determine which input factors have the greatest influence on the output, and which must be considered when training and testing ANNs. Specifically, the correlation coefficient between inputs X = {(xi) |1≤i≤N} and outputs Y = {(yi) |1≤i≤N} is calculated by the following equation:

$$\rho = \frac{\sum_{i=1}^{N}(x_i-\bar{x}).(y_i-\bar{y})}{\sqrt{\sum_{i=1}^{N}(x_i-\bar{x})^2}.\sqrt{\sum_{i=1}^{N}(y_i-\bar{y})^2}} \quad [3.3]$$

where:

$$\bar{x} = \frac{1}{N}\sum_{i=1}^{N} x_i \text{ and } \bar{y} = \frac{1}{N}\sum_{i=1}^{N} y_i$$

In general, the closer the correlation coefficient is to −1 or 1, the stronger the relationship between input and output variables. In this case, the output is sensitive to any change in the input variable. Consequently, it is important to take this into account when constructing the input vector of the ANN. For example, photovoltaic power is strongly correlated with the solar irradiation received by the panels and the local temperature;

– *Data manipulation*, aimed at correctly preparing the network input vector. This sub-step differs from one application to another. For example, in the case of time series inputs, the same data source and time scale must be respected. However, the normalization of data to the interval [−1,1] is common to all ANN applications, in order to avoid neuron saturation during the learning process [WAN 12].

– *Setting up the neural network*: at this stage, the network parameters are defined: the number of hidden layers and number of neurons per hidden layer. To date, there is no single analytical method for exactly determining these numbers [SHE 13]. Consequently, it is generally necessary to test several architectures in order to find the optimal solution.

– *Training and testing the network*: first, the network is trained using the backpropagation algorithm[4] (Figure 3.10). It is then tested with new data. Occasionally, the network adapts "too much" to the training data and fails to make reliable predictions – this is known as overfitting. In order to detect this problem, we need to evaluate the performance of the ANN during both the training and test phases.

4 In machine learning, gradient backpropagation is a method for training a neural network, consisting of updating the weights of each neuron from the last layer to the first. The aim is to correct errors according to the importance of each element's contribution to them. This correction is based on the calculation of the gradient using backpropagation. In the case of neural networks, synaptic weights that contribute more to an error will be modified more significantly than weights that cause a marginal error.

In this chapter, the code ("computer implementation of the algorithms") used for neural network prediction is taken from the course given by Andrew Ng, an expert researcher in machine learning[5]. This code brings into play several functions, numerical optimization and others, in order to set up a neural network capable of learning and rapidly generating forecasts with fairly high accuracy[6]. A few modifications have been made to adapt it to the application in question.

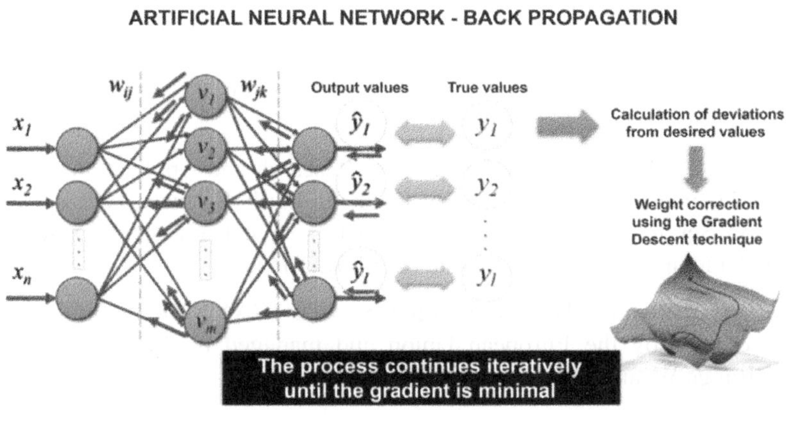

Figure 3.10. *Steps in the implementation of a neural network by backpropagation*

3.3.4. *Case study: 24-hour production forecast for the photovoltaic power plant at the Université Catholique de Lille*

This section concerns photovoltaic solar energy production in the RIZOMM building power plants of the Université Catholique de Lille smart grid demonstration (latitude: 50° 37' 59.5" north, longitude: 3° 02' 43.0" east, altitude: 24 meters). The data cover the period from September 17, 2018, to September 10, 2019, with a frequency of 10 minutes. For the training and validation phases, 300 days are reserved, 30 of which are randomly selected for validation. The remaining days are reserved for training. Finally, one day is dedicated to the test phase, to evaluate the model's error. This is generally the day before the day to be planned. It is recommended that the data extraction period cover at least one year, to take into account the impact of the seasons on variability.

5 His course is free online at coursera.com. See: https://fr.coursera.org/learn/machine-learning.

6 It is available at: https://github.com/aptiva0985/MachineLearning-Andrew-Ng-Coursera-/tree/master/ML004.

3.3.4.1. *Database for PV production prediction*

After studying the correlation of various data and carrying out several tests at different times, the input variables selected for building the database were: photovoltaic power measured in kW (data archive measured 300 days); global horizontal irradiation at the ground level (in W/m2); total cloud cover (in %); and temperature at 2 m above the ground level (in K).

Given the extreme sensitivity of PV production and electricity consumption to meteorology, it was essential to select reliable and accessible sources to implement the forecasting algorithms on the control platform. To this end, we chose two sources offering free access to weather archives:

– Modern-Era Retrospective Analysis for Research and Applications, Version 2 (MERRA-2), managed by the NASA Goddard Earth Sciences (GES) Data and Information Services Center (DISC). This data source has a spatial resolution of 0.625° and a temporal resolution of one hour [GLO 15].

– Copernicus Atmosphere Monitoring Service (CAMS) Radiation Service, a program created by the European Union and managed by European Centre for Medium-Range Weather Forecasts (ECMWF)[7].

In the historical database, temperature is extracted from MERRA-2, while global horizontal irradiance and total cloud cover are downloaded from CAMS.

Weather forecasts are automatically downloaded from the Global Forecast System (GFS) forecasts produced by the National Centers for Environmental Prediction (NCEP). This data system offers a spatial resolution of 0.25° (approx. 28 km) and a temporal resolution of three hours[8].

A summary of historical and forecast data sources is presented in Table 3.2.

<table>
<tr><th rowspan="2">Parameters</th><th colspan="3">Source</th></tr>
<tr><th colspan="2">Historical data</th><th>Forecast data</th></tr>
<tr><td>Temperature (K)</td><td>MERRA-2
(up to a month back)</td><td>GFS</td><td rowspan="3">GFS
(up to 16 days in advance)</td></tr>
<tr><td>Total cloud cover coverage (%)</td><td colspan="2" rowspan="2">CAMS
(up to J-2)</td></tr>
<tr><td>Global horizontal irradiation (Wh/m2)</td></tr>
</table>

Table 3.2. *Sources of historical and forecast meteorological data*

7 See : https://www.soda-pro.com/web-services/radiation/cams-radiation-service.

8 See : https://www.soda-pro.com/web-services/meteo-data/gfs-forecasts.

3.3.4.2. *Normalization*

The database must first be normalized between 0 and 1 as follows before being introduced into the prediction model:

$$f: \bar{x} \rightarrow x = \frac{x_i - x_{min}}{x_{max} - x_{min}} \qquad [3.4]$$

where x_i is the ith component of the original input data vector; and x_{max} and x_{min} are the maximum and minimum values of each set, respectively.

In the particular case of PV production, the plant's peak power is taken as the maximum production.

3.3.4.3. *Determining errors*

The effectiveness of the proposed prediction method was validated by analyzing the normalized root-mean-square error (nRMSE and the normalized mean absolute error) between predicted values $\tilde{y}_k$ and measured values $y_{k,}$ with n being the predicted samples. These performance indicators are described by equations [3.5] and [3.6]:

$$nRMSE = \sqrt{\frac{1}{n} \sum_{i=1}^{n} (\tilde{y}_k - y_k)^2} \qquad [3.5]$$

$$nMAE = \frac{1}{n} \sum_{i=1}^{n} |\tilde{y}_k - y_k| \qquad [3.6]$$

3.3.4.4. *Applying the algorithms*

3.3.4.4.1. ANN model for photovoltaic forecasting

A three-layer BP-ANN backpropagation model was developed to forecast photovoltaic energy production over 24 hours with a resolution of 10 half minutes. The input layer comprises solar irradiance, temperature and nebulosity (measured on the day or forecast for the following day). For the hidden layer, 200 neurons were determined using a trial-and-error method designed to minimize forecast errors. The output layer was made up of the PV power predicted for 24 hours in advance. Firstly, 90% of the measured data over a two-year period were used to train the ANN. Secondly, 10% of the measured data were used to create a set of validation models to assess the quality of the forecast. Figure 3.11 illustrates the prediction methodology implemented.

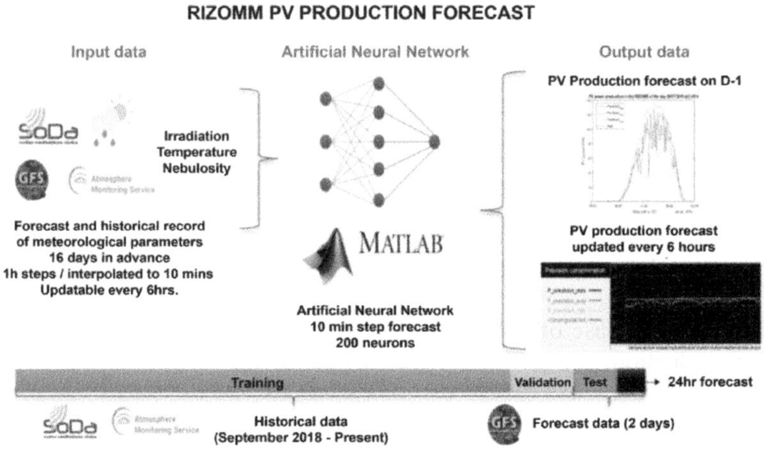

Figure 3.11. *Implementation of the D-1 photovoltaic power prediction method with an update every six hours. For a color version of this figure, see www.iste.co.uk/robyns/smartgrids.zip*

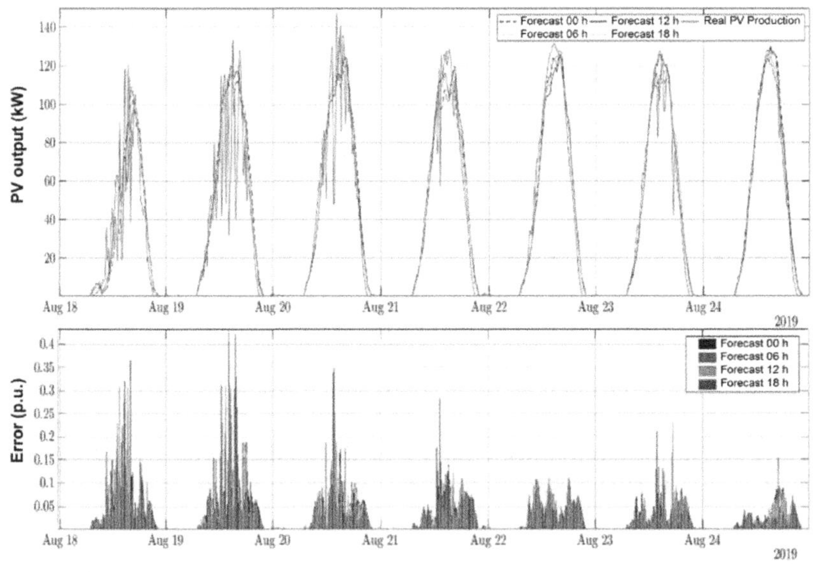

Figure 3.12. *Results of RIZOMM's three-week PV forecasts for the period from August 18th to September 7th. Forecasts (at 0 h, 6 h, 12 h, 18 h) are compared with real production. The error displayed is the absolute error. For a color version of this figure, see www.iste.co.uk/robyns/smartgrids.zip*

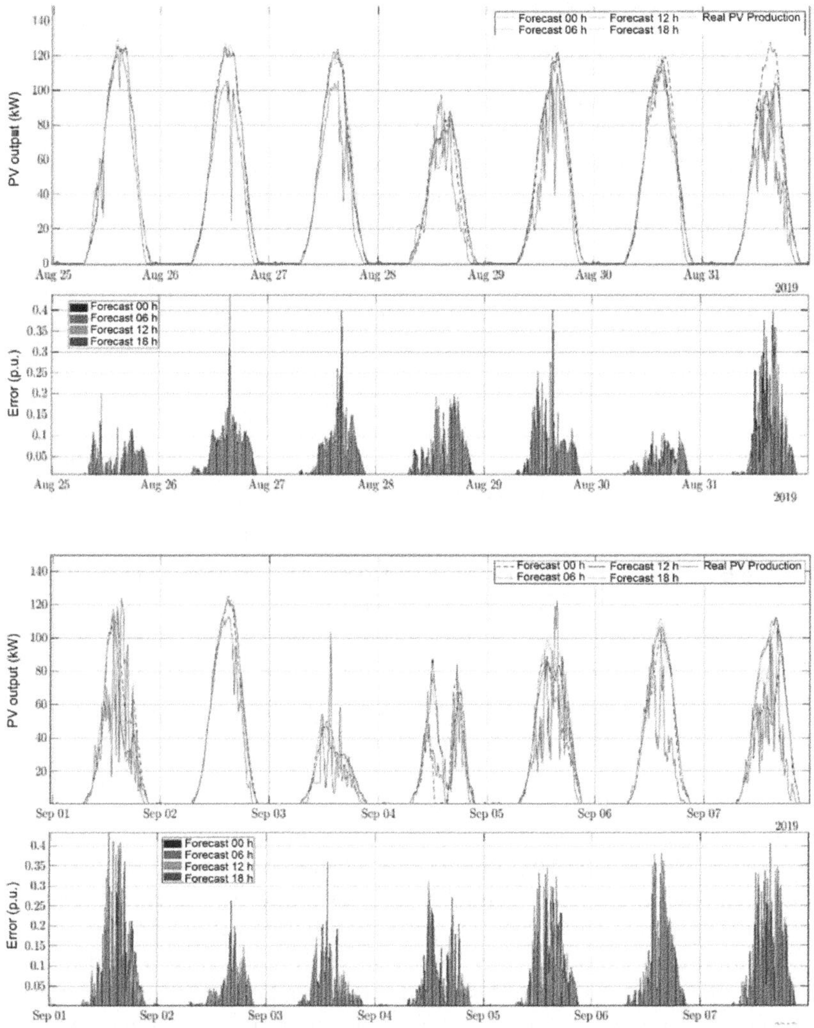

Figure 3.13. *Results of RIZOMM's three-week PV forecasts for the period from August 18th to September 7th. Forecasts (at 0 h, 6 h, 12 h, 18 h) are compared with real production. The error displayed is the absolute error. For a color version of this figure, see www.iste.co.uk/robyns/smartgrids.zip*

3.3.4.4.2. Forecasting results and performance of the ANN implemented

The algorithm implemented in the demonstration's control platform makes forecasts 24 hours in advance and is updated every six hours. In order to assess its performance, the PV production prediction algorithm ran continuously from August 14th to September 7th, 2019 (except on August 17th, when a preventive electrical shutdown of the block was carried out). Each iteration took about five minutes. Figures 3.12 and 3.13 show the forecasts and real production of the RIZOMM power plant during three weeks of testing. The absolute errors of each forecast are also shown per unit (p.u.). As expected, performance increased on sunny days compared to cloudy or partly cloudy days. This difference was mainly due to the sharp fluctuations in production caused by the passage of local clouds.

In this respect, daily updates at 6 a.m., 12 p.m. and 6 p.m. brought slight improvements in terms of performance. The nRMSE and nMAE are shown in Table 3.3.

Refresh rate	Errors			
	Totals		**Excluding night-time hours**	
	nRMSE	**nMAE**	**nRMSE**	**nMAE**
0 h	7.92%	4.22%	10.34%	7.11%
6 h	7.89%	4.18%	10.27%	7.07%
12 h	7.81%	4.03%	10.19%	7.03%
18 h	7.78%	4.12%	10.16%	6.92%

Table 3.3. *Forecast errors on D-1 with update at 6 h, 12 h and 18 h, for PV production with ANN*

To get a better idea of the accuracy of the PV prediction model developed, the forecast results were compared with those of SteadySat. Being an industrial module developed by the SteadySun company and powered by satellite imagery and advanced mathematical algorithms, its prediction horizon is six hours with updates every 15 minutes. Based on weather forecasts, satellite imagery is used to fine-tune cloud cover and production profiles for the coming hours. SteadySat offers three prediction curves: minimum forecast (Pmin), maximum forecast (Pmax) and average forecast (Pmoy). The latter was chosen to compare the forecasting results of our ANN with those of SteadySat. Figure 3.12 shows the comparison between the two models over the week from August 18th to August 24th.

According to Figure 3.14, the backpropagation ANN model used provides forecasts close to those of SteadySat, but with a longer prediction horizon. Error indicator values are shown in Table 3.4.

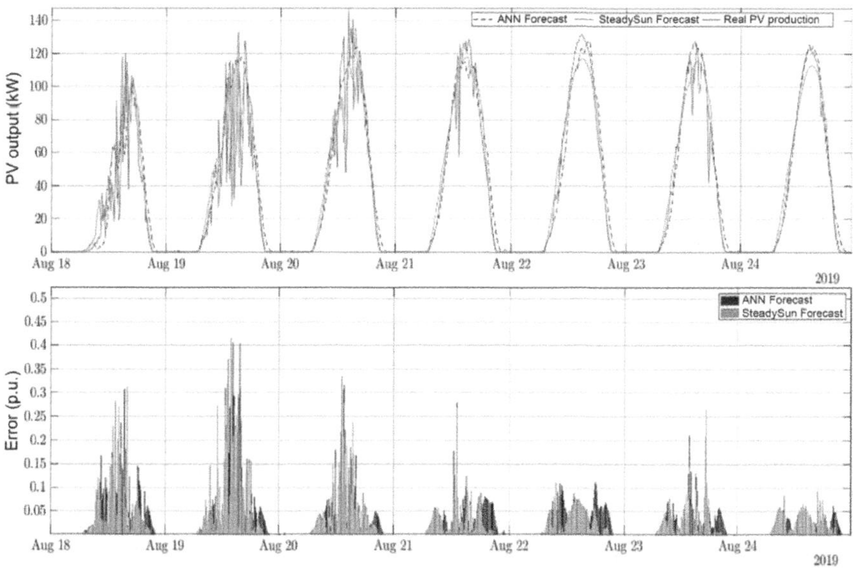

Figure 3.14. *Comparison of results between the ANN model and the SteadySat industrial solution, from August 18th to August 24th. The error shown is the absolute error. For a color version of this figure, see www.iste.co.uk/robyns/smartgrids.zip*

Type of forecast	Errors			
	Totals		**Excluding night-time hours**	
	nRMSE	**nMAE**	**nRMSE**	**nMAE**
ANN	7.81%	4.13%	10.19%	6.93%
SteadySat	6.49%	3.37%	8.46%	5.73%

Table 3.4. *PV production forecast errors with ANN and SteadySat*

These results show that the model developed can provide forecasts in line with the uncertainties associated with commercial solutions. In addition, it has a longer prediction horizon than these solutions. Finally, it is also important to note that the intrinsic error of weather forecasts is a contributing factor.

3.4. Forecasting electricity consumption

One of the major challenges for electrical system operators is to ensure the reliability of the electrical system, despite limitations in energy resources and variations in electricity demand. It is therefore essential for these operators to accurately forecast energy consumption so as to plan in advance the operation of generation at minimum cost. What is more, the de-centralization of generation plants, fluctuations and changes in weather conditions exert an even greater influence on energy and power demand. Depending on the different time horizons, load forecasts can be divided into three categories: long-term forecasts (over one year), medium-term forecasts (from one week to one year) and short-term forecasts (from one hour to several days). Long-term demand forecasts play an important role in planning, long-term maintenance and construction scheduling for the development of new production facilities. While underestimation leads to customer dissatisfaction, overestimation of load demand will also result in substantial investment. Unfortunately, it is not easy to accurately estimate load demand over a period of several years. Short-term load prediction (a few days in advance) is deemed necessary in the analysis of unit commitments, maintenance scheduling and economic dispatch diagnostics. It can also be useful to estimate load flows to avoid overloads. In this context, load prediction also concerns the prediction of hourly, daily, weekly, monthly and annual values of system load, system peak load and system energy demand.

3.4.1. *Important factors for forecasting electricity consumption*

The aim of electricity consumption forecasting is to improve accuracy in order to optimize electrical system planning and reduce costs. However, many variables directly or indirectly affect accuracy. Different prediction models apply to specific time horizons, and the factors to be taken into account are varied. Short-term forecasting, for example, generally depends on time factors, weather data and customer categories; medium- and long-term forecasts consider historical load and weather data, customer categories, economic data, demographics and other factors.

Temporary factors represent different periods of the year, day and time. For example, load demand may differ significantly between weekdays and weekends. In addition, energy consumption on summer days is different from that on winter days. Weather conditions have a major influence on forecasting: temperature and humidity are the most commonly used short-term load forecasting factors. In addition, France is particularly sensitive to temperature in terms of electricity consumption, with an estimated average sensitivity of 2400 MW per degree Celsius in winter, according to

RTE's 2017 energy balance. What is more, the different types of consumption (residential, commercial and industrial) lead to varying impacts on how electricity is used. Fortunately, load demand in a system can be modeled in a comprehensible and predictable way. This makes it possible to establish a basic production schedule, guarantee the security of the supply system and facilitate the dissemination of information.

In general, the factors influencing system load behavior can be grouped into four main categories [GRO 87]:

– *Economic factors*: the economy has a significant impact on the evolution of electricity consumption, particularly the demographics of service areas, levels of industrial activity and economic trends.

– *Temporal factors*: the main temporal factors influencing load patterns include seasonal effects, the weekly or daily cycle and vacations. Gradual changes in load patterns can be observed in response to seasonal temperature variations, such as summer and winter peaks. In addition, the weekly or daily periodicity of loads is linked to the active population zone, and vacations result in a different demand from normal working days.

– *Meteorological factors*: as most utilities have a temperature-sensitive load, such as air conditioning and space heating, temperature plays an important role in the configuration of load models. On the other hand, other factors such as humidity, wind speed, precipitation and cloud intensity also have a less significant impact on system load.

– *Random factors*: these include all unpredictable events, such as unexpected disturbances and sudden shutdowns of industrial plants.

3.4.2. *Electricity consumption prediction methods*

Over the last few decades, numerous methods and models have been developed to forecast long-term load. These methods are essential for the planning and development of future generation, transmission and distribution systems. According to the article [LAD 11], these methods can be divided into two categories: parametric methods and methods based on AI.

3.4.2.1. *Parametric methods*

Trend analysis focuses on past fluctuations in consumption to predict future demand. This method has the advantage of being simple, quick and inexpensive to implement. However, it does not provide detailed explanations of usage.

End-use models forecast load demand directly, based on detailed end-user information. They offer high predictive accuracy. However, they assume a constant relationship between supply and end-use, which is a long-term disadvantage as this relationship will evolve over time.

Econometric models make it possible to estimate the relationship between demand and the factors influencing it, by combining economic theory and statistical techniques. By calculating demand in different sectors, predicted data can be brought together with recent historical data. This provides information on the reasons and mechanisms by which various factors influence demand. However, it can be difficult to justify the assumption that electricity does not vary, which is essential for accurate forecasts.

3.4.2.2. *Methods based on artificial intelligence*

Traditional parametric methods cannot accurately predict the complex nonlinear relationships between demand and the factors influencing it. Consequently, methods based on AI, such as ANNs, wavelet networks, fuzzy logic models, etc., have been developed to predict load demand.

Detailed explanations, as well as the advantages and disadvantages of these methods, can be found in [LAD 11], while in [GRO 87, FEI 05, HUA 16], the various methods used for short-term load prediction are summarized, including the similar-day approach, various regression models, time series, neural networks, fuzzy logic and statistical learning algorithms. Long- and short-term load prediction plays a crucial role in formulating safe and reliable operating strategies for the electrical supply system. In this chapter, as in the case of photovoltaic production prediction, the method chosen for electricity consumption forecasting is backpropagation ANN prediction.

3.4.3. *Case study: 24-hour forecast of electricity consumption for a block of buildings at the Université Catholique de Lille*

3.4.3.1. *Database for electricity consumption forecasting*

The data used in this section correspond to observations of the total electrical load of the Université Catholique de Lille smart grid demonstration block at 10-minute intervals. They cover the period from January 1, 2018, to August 24, 2019.

The database used to forecast consumption was made up of meteorological, temporal and seasonal data, as well as historical demand data. The parameter selection methodology was the same as that used for the PV power forecast.

Given the thermo-sensitivity of the country's electricity demand, temperature plays an important role in predicting electricity consumption. Another important characteristic of this type of building is the periodicity of its activity regime. To take this into account, various temporary parameters were incorporated into the prediction model, such as the day of the week (whole number between 1 and 7), the month (whole number between 1 and 12) and the hour (in minutes). Historical consumption data are also included, such as demand on D-1 and D-7. In addition, due to the dual activity regime of the buildings studied, both academic and administrative, a difference in their behavior during public holidays and school vacations was noted. Binary parameters were therefore assigned to represent these two events.

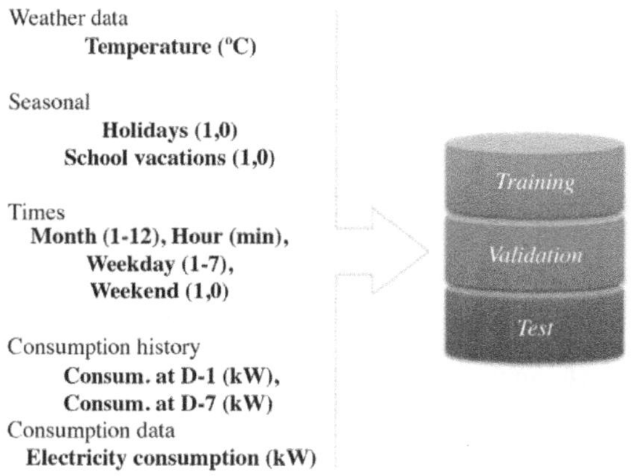

Figure 3.15. *Representation of the database used to forecast electricity consumption*

Like the PV production forecast model, 500 days are allocated to training and validation. Of these, 30 days are randomly selected for validation and the remainder for training. Finally, one day is reserved for the test phase, to measure model error. This is generally the day before the day to be forecast. Figure 3.15 shows the database used to forecast block electricity consumption.

3.4.3.2. *Applying the algorithms*

3.4.3.2.1. ANN model for electricity consumption prediction

In a similar way to photovoltaic forecasting, a three-layer backpropagation neu-drone network (BP-ANN) model was developed to predict 24-hour electricity consumption. This model includes:

– an input layer with corresponding input parameters;

– a hidden layer with 100 neurons determined by a trial-and-error method to minimize prediction errors;

– an output layer with electricity consumption predicted for 24 hours in advance.

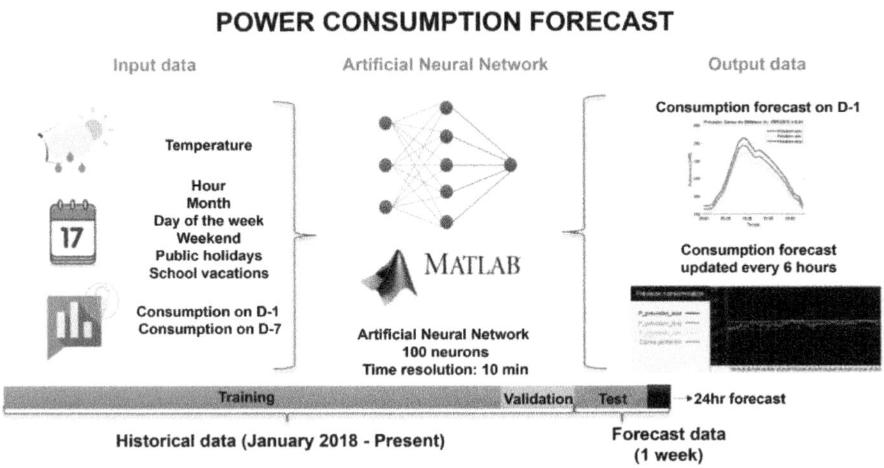

Figure 3.16. *Implementation of the D-1 electricity consumption prediction method with updates every six hours*

Input parameters were selected based on the analysis of historical consumption curves. A total of 90% of previously measured data (from a two-year database dataset) were used to train the BP-ANN, and 10% of measured data were used to create a set of validation models to assess forecast quality. Figure 3.16 illustrates the prediction methodology implemented.

3.4.3.2.2. Forecasting results and performance of the ANN implemented

The algorithm implemented in the demonstrator's control platform enables 24-hour forecasts to be made, with updates every six hours. In order to evaluate the model's performance, a period running from May 25, 2019, to August 24, 2019, was chosen. It is particularly interesting because it included different types of days, including public holidays, school vacations and a period when establishments were closed during the August vacations.

In addition, a power outage on August 17, 2019, during preventive maintenance of the medium voltage (MV) source substation was included to test the robustness of the algorithm. Figure 3.17 shows the overall forecasts and real consumption of the demo block over the 13-week test period. The overall nRMSE and nMAE obtained were 3.78% and 2.81%, respectively.

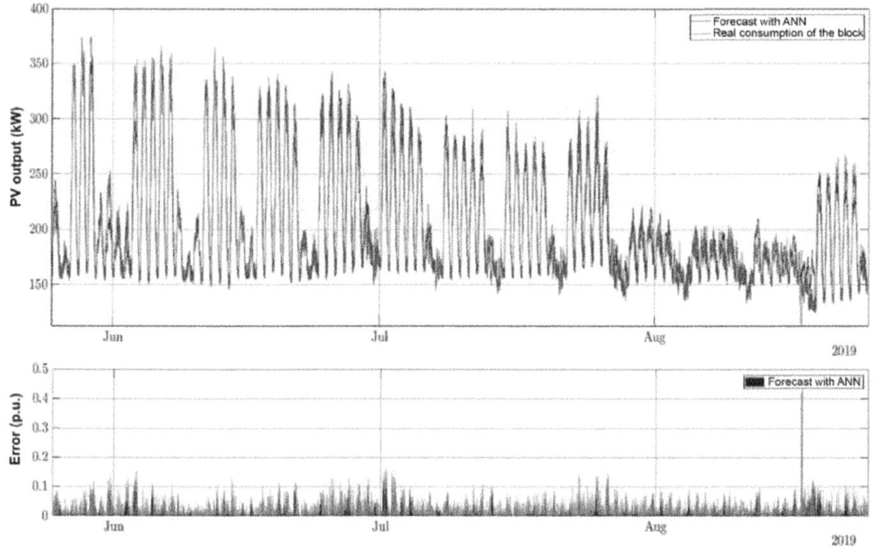

Figure 3.17. *Results of electricity consumption forecasts for the block from May 25th to August 24th. The error displayed is the absolute error. For a color version of this figure, see www.iste.co.uk/robyns/smartgrids.zip*

Next, three weeks of consumption behavior on the block are analyzed in detail:

– May 27th to June 2nd, which was particularly marked by Thursday May 30th, the national holiday for Ascension Day;

– June 17th to June 23rd, a period of normal summer activity;

– from July 29th to August 5th, which was marked by the closure of establishments for the holidays.

For ease of analysis, the three cases studied were full weeks, starting on Monday and ending on Sunday.

The first case characterizes normal block activity, with a high periodicity during the week, when electricity consumption values are higher than at weekends (Figure 3.17). The second case deals with consumption during the period when the facilities are closed, representing approximately 50% of peak activity (Figure 3.18(a)). In these first two cases, our prediction model adapted well to real consumption, with an absolute error of less than 0.12 p.u. in all cases.

The third case (Figure 3.18(b)) was the most complex, as Ascension Day had a double effect on consumption behavior. On the one hand, as expected, demand on this day was lower than on non-holiday days. On the other hand, it also had an influence on the following day, which constituted a bridge between the Thursday holiday and the weekend. The latter situation is more difficult to predict, as not all plants react in the same way. The various performance indicators obtained are shown in Table 3.5.

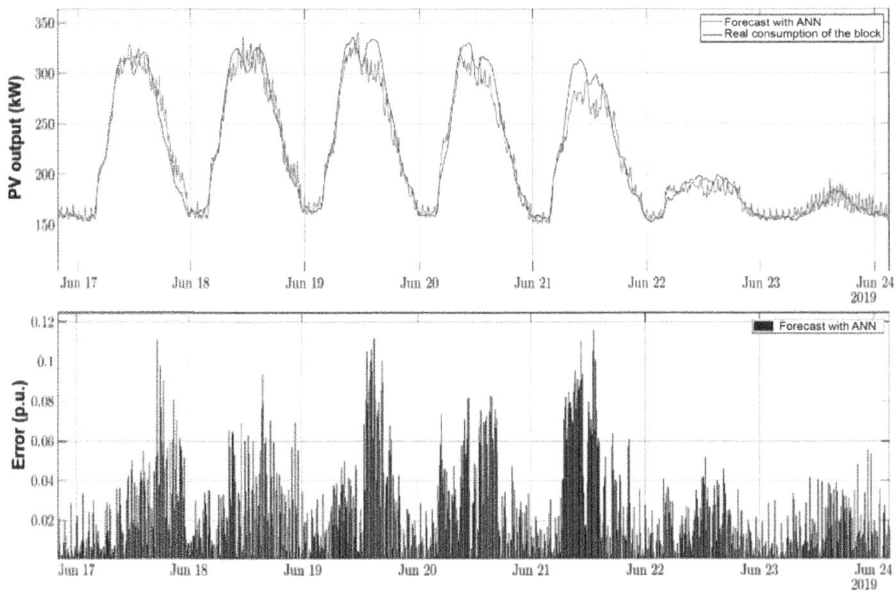

Figure 3.18. *Results of consumption forecasts for three different periods from June 17th to June 24th, normal activity. For a color version of this figure, see www.iste.co.uk/robyns/smartgrids.zip*

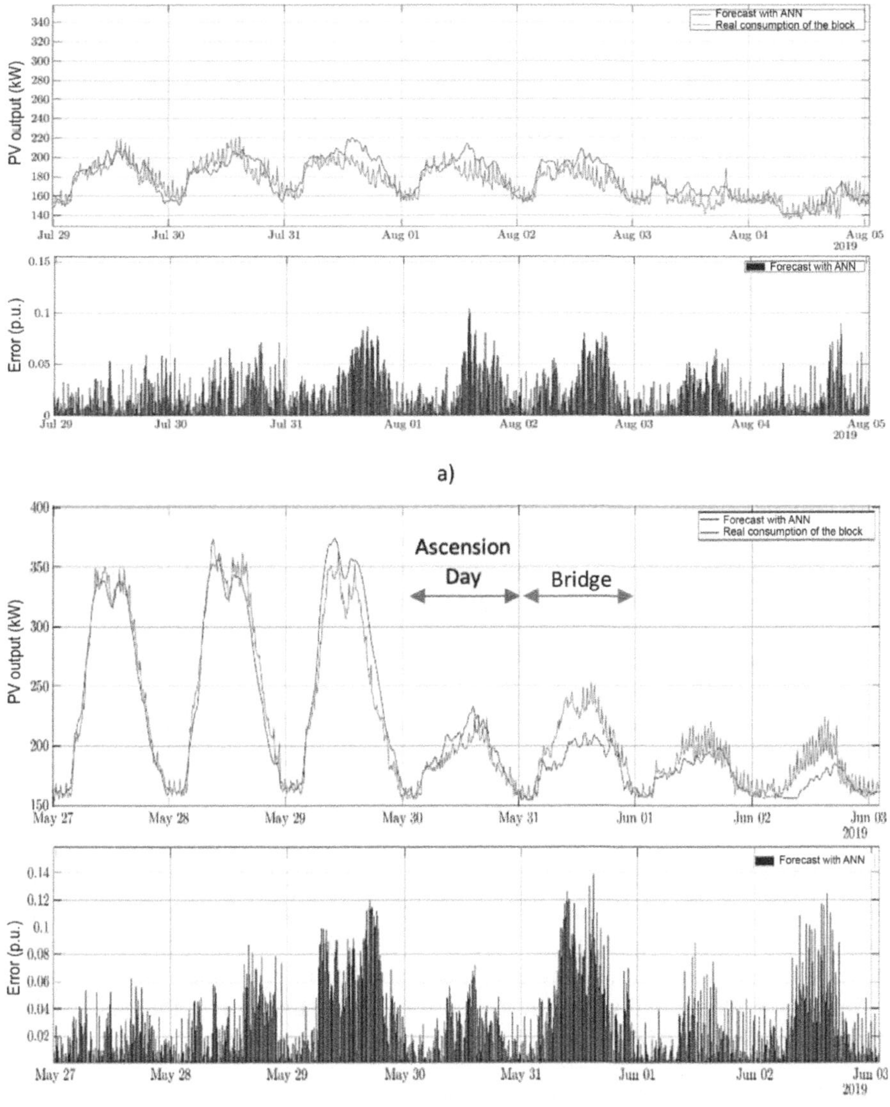

Figure 3.19. *a) From July 29th to August 5th, closed for August vacations; b) from May 27th to June 2nd, counting a public holiday. For a color version of this figure, see www.iste.co.uk/robyns/smartgrids.zip*

Case study	Activity mode	Errors	
		nRMSE	**nMAE**
1	Normal summer	3.51%	2.61%
2	School closures	3.24%	2.52%
3	Public holiday	3.98%	2.90%

Table 3.5. *Performance comparison for different study cases*

3.5. Valorization of forecasts and feedback

3.5.1. *Using forecasts to manage energy for the Université Catholique de Lille smart grid demonstrator*

The photovoltaic production and consumption prediction algorithms run continuously and autonomously on Matlab, on the Université Catholique de Lille smart grid demonstrator control platform. They connect to databases to retrieve historical PV consumption and production data. Communication with these databases is via Open Database Connectivity (ODBC), and the retrieval frequency is one hour for consumption and one day for production.

Meteorological data is automatically retrieved from two sources accessible via the Internet: CAMS Radiation Service and GFS. In the first case, data are accessed via a Wget-type application developed by the SODA PRO platform. This source provides historical data on Global Horizontal Irradiation (GHI) and cloud cover. Access to these parameters is free for up to 75 requests per day. In our case, the forecasting algorithm refreshes its historical databases once a day.

The algorithms developed communicate with PC-Vue via Open Platform Communications (OPC) so that forecasts can be displayed on the platform. Two types of display are proposed: one for D-1 forecasts, which is generated automatically at midnight (Figure 3.19), and another in real time, which compares the forecasts with the real values measured for the consumption of the block and the PV production of the RIZOMM building (Figures 3.20 and 3.21).

Inspired by SteadySat's commercial solution, the results of the two prediction models are presented in three curves: an average curve, a maximum curve and a minimum curve. In the case of building block electricity consumption, the mean curve is the forecast derived from the ANN algorithm. The other curves are constructed from the mean forecast, plus or minus a constant deviation. The latter is estimated as a function of the nRMSE of the previous day's forecast versus real consumption (test step in the ANN process). In this case, we assume that this error will remain constant for the day of the forecast and is generally between 3% and 5%.

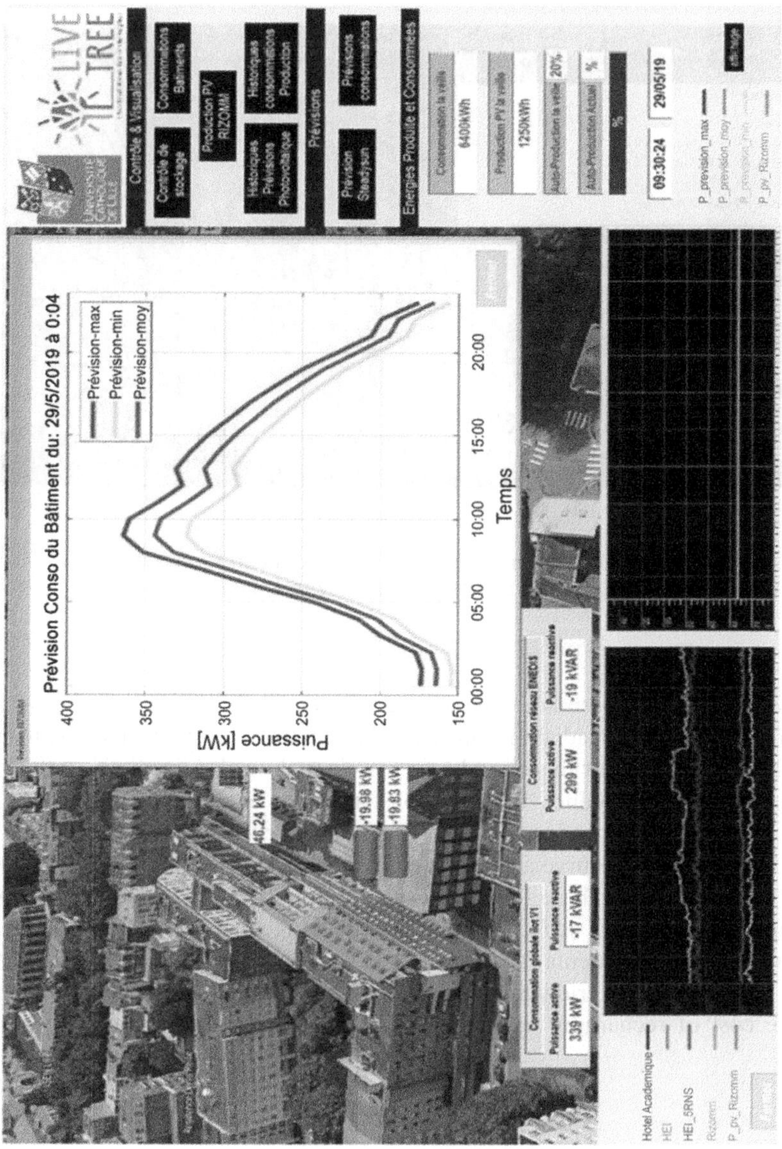

Figure 3.20. *Display of the D-1 forecast of overall consumption for the block on the control platform*

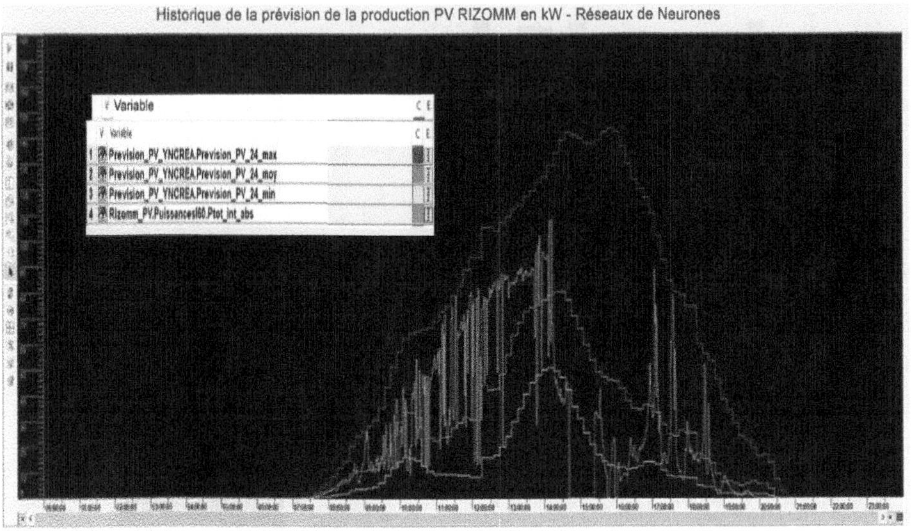

Figure 3.21. *Real-time display of PV production forecast for the RIZOMM building on the control platform*

For PV production forecasts, the average curve is the result of the ANN calculation with the solar irradiation forecast. The maximum curve is also the result of the ANN calculation, but this time based on clear-sky solar irradiation on D-1. We assume that this irradiance remains constant from one day to the next and represents maximum PV production. The minimum curve, on the other hand, is constructed from the mean curve and assigned a coefficient based on the cloud cover forecast at each time step. These forecasts are used to improve the use of renewable energy sources and optimize self-consumption and storage management to minimize the economic and environmental cost of electricity. These are also used to develop recharging services for electric vehicles using non-carbon energy, enabling users to save on the cost of recharging by targeting the hours when electricity prices are lowest.

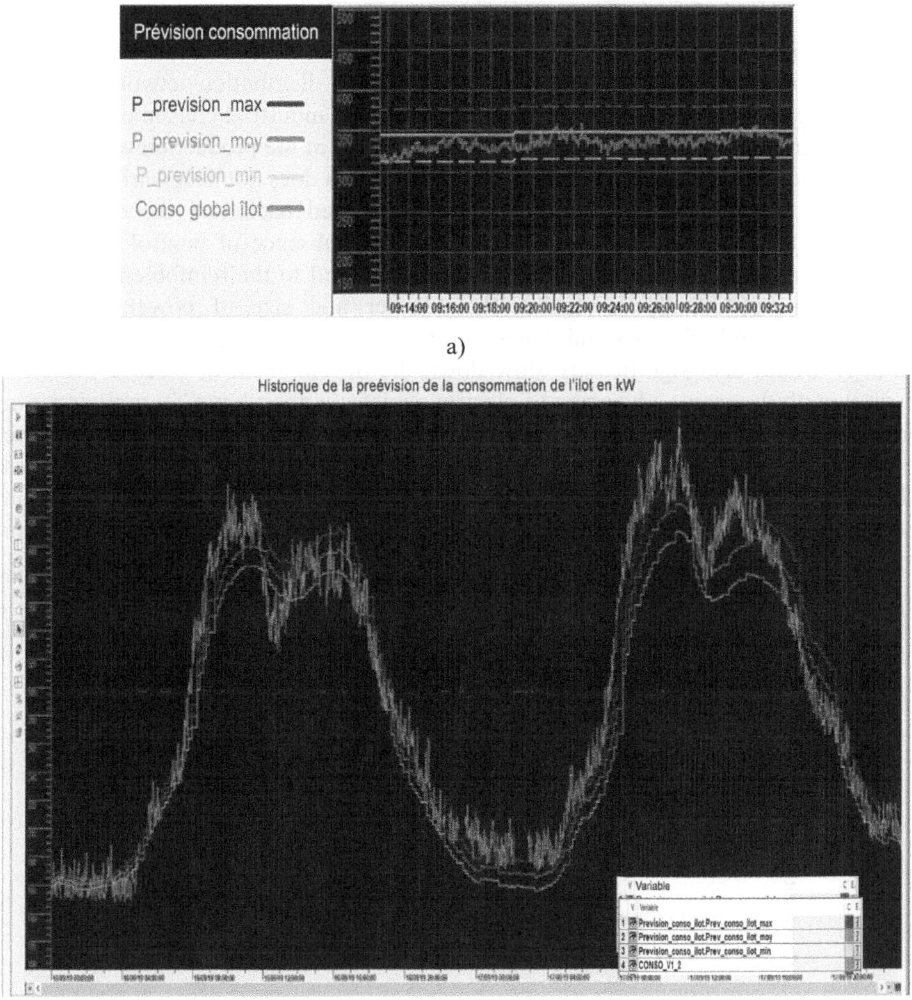

a)

b)

Figure 3.22. *Real-time display of forecast and real consumption on the control platform. Panel a) shows recent forecasts and panel b) shows a historical record*

3.5.2. *Load forecasting in a distribution network at a high-voltage/ medium-voltage (HV/MV) source substation*

For some years now, electrical power distribution networks have been at the heart of the energy and environmental changes affecting our society. Indeed, many applications exploit the electrical vector to increase their energy efficiency. This is particularly true of electric vehicles and heat pumps for heating. As the energy consumption of households and industry declines, the electrical share of this consumption often increases. As a result, the electrical distribution network is seeing an increase in its use. At the same time, domestic and industrial consumers supplied by the distribution network are increasingly involved in the production of electrical energy, often from renewable sources. This production does not necessarily coincide with electricity consumption, and energy is then fed back into the distribution network, or even the transmission network. In the absence of control over these developments, the constraints on structures would lead to the reinforcement of the distribution network by increasing the number and size of power lines and substations, with high costs and damaging consequences for the environment. This strategy would also significantly slow down the development of electric vehicles, even though the major advantage of electric vehicles is their low CO_2 emissions. In fact, the CO_2 emissions of this type of vehicle depend on the CO_2 content of the electrical energy used to power them, and ideally the charging of these vehicles should be based on renewable energy. For this reason, it is crucial to plan consumption on different levels of the electrical distribution chain, that is, an HV/MV source substation.

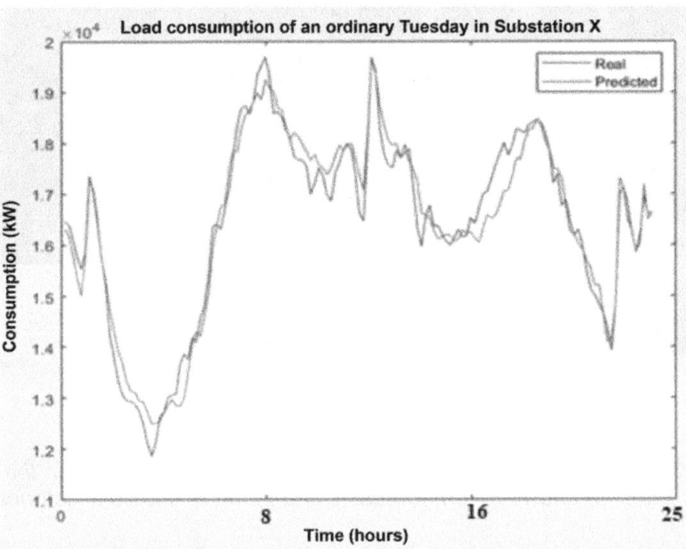

Figure 3.23. *Real (blue) and forecast (orange) consumption curves on a typical Tuesday for a group of municipalities in the Deux-Sèvres department. For a color version of this figure, see www.iste.co.uk/robyns/smartgrids.zip*

Therefore, using the backpropagation ANN (BP-ANN) prediction methodology presented in section 3.3.3, we were able to efficiently forecast the load at an HV/MV source substation in a rural distribution network serving a group of municipalities in the Deux-Sèvres department in France, which has the particularity of having a significant integration of renewable energy as well as a mesh of recharging stations.

To validate the prediction model, the three existing types of days – working days, weekends and public holidays – were tested. Figure 3.22 shows the real and predicted consumption curves on D-1 for a working Tuesday. The prediction model we have developed follows the reference and predicts the day's consumption peaks with precision. According to Table 3.6, the mean absolute error did not exceed 1.2%, while the root-mean-square error remained below 1.5% for this ordinary day.

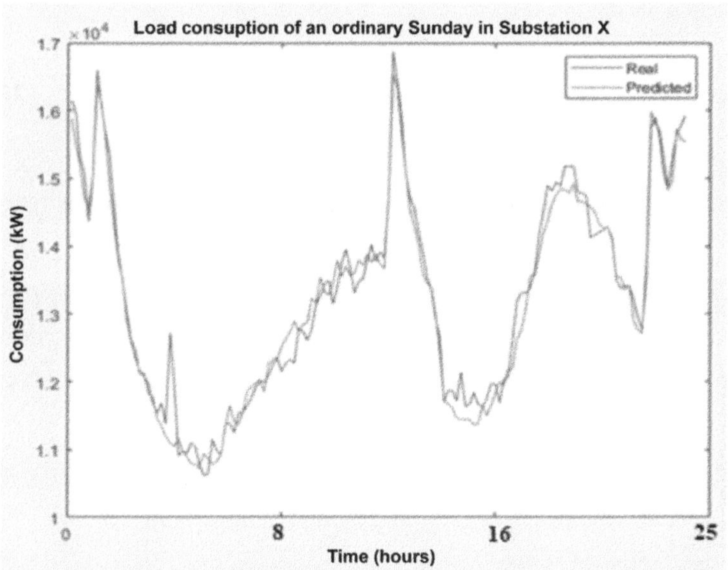

Figure 3.24. *Real curve (blue) and prediction curve (orange) of consumption on a typical Sunday. For a color version of this figure, see www.iste.co.uk/robyns/smartgrids.zip*

The model proved its effectiveness for a non-working Sunday, as shown in Figure 3.23. The prediction curve is very close to the real curve, with a mean absolute error of less than 1.8% and a root-mean-square error of less than 2.3% (Table 3.6). In addition, the model predicted the day's peaks with notable accuracy.

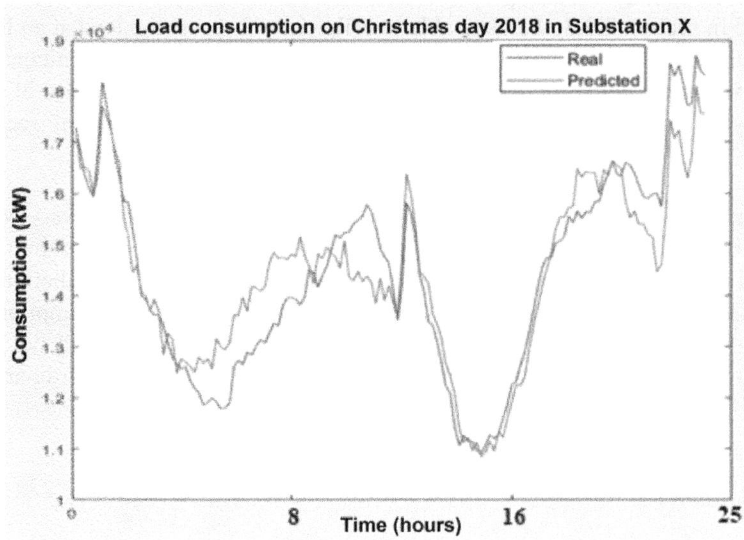

a)

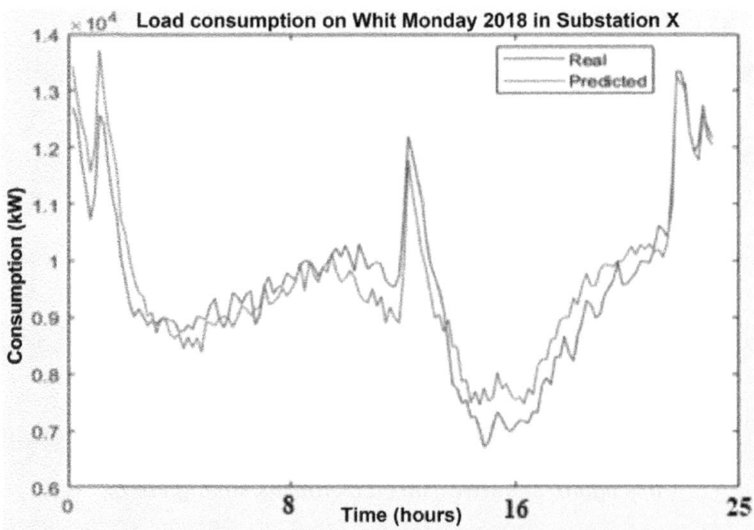

b)

Figure 3.25. *a) Real (blue) and forecast (orange) consumption curves for Christmas Day; b) real (blue) and forecast (orange) consumption curves for Whit Monday. For a color version of this figure, see www.iste.co.uk/robyns/smartgrids.zip*

For public holidays, the model was tested on Christmas Day 2018 (December 25, 2018) and Whit Monday 2018 (May 21, 2018). This approach enables the model to be tested in both winter and late spring. According to Figure 3.24, the model is not as accurate as in the above cases, due to the unpredictable behavior of the load curve on these days. However, its performance remains decent, with a root mean-square error of less than 3% for Christmas Day and 2.5% for Whit Monday.

Day	Date	Training test			
		RMSE (%)	**MAE (%)**	**RMSE (%)**	**MAE (%)**
Tuesday	December 4, 2018	2.65	1.97	1.48	1.13
Sunday	December 2, 2018	2.59	1.93	2.28	1.75
Christmas Tuesday	December 25, 2018	2.53	1.88	2.92	2.56
Whit Monday	May 21, 2018	2.64	1.96	2.38	1.87

Table 3.6. *Calculation of errors in the predicted consumption at an HV/MV source substation in the Deux-Sèvres department, France*

3.5.3. *The importance of meteorological forecasting*

As explained above, electricity production and consumption depend on several parameters, mainly related to climatic conditions and the days of the week and year. Access to reliable meteorological databases and the use of meteorological data prediction are therefore essential for predicting electricity consumption and production. Meteorology has a direct influence on consumer behavior, and therefore plays an important role in predicting electricity demand and supply. Forecasting weather data can therefore have a considerable impact on electricity consumption and production, helping to reduce costs and improve efficiency. In addition, weather forecasts are important for understanding seasonal variation and long-term trends in consumers and production systems. They can help companies make strategic and informed decisions, such as planning energy purchases and energy storage, which can help reduce operating costs. They can also help determine the best fuels and technologies for energy production, which can reduce costs and improve production efficiency. It is also worth noting that climate change is changing the weather faster and faster, which is likely to impact on current weather prediction models.

3.5.4. *The importance of uncertainty analysis*

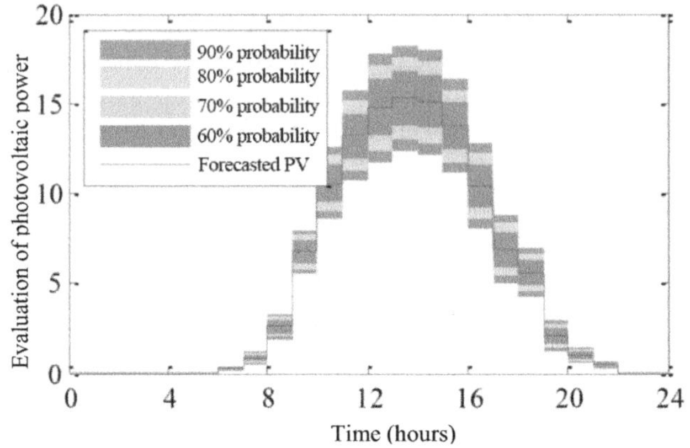

Figure 3.26. *Example of a PV forecast with uncertainty (on a random day) [XIN 17]. For a color version of this figure, see www.iste.co.uk/robyns/smartgrids.zip*

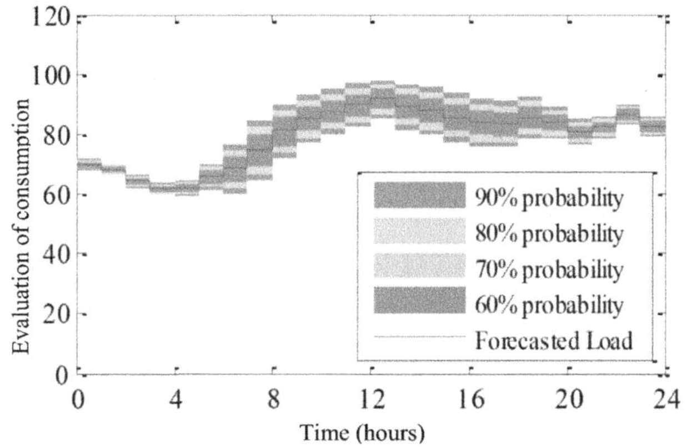

Figure 3.27. *Example of a consumption prediction with uncertainty (on a random day) [XIN 17]. For a color version of this figure, see www.iste.co.uk/robyns/smartgrids.zip*

Forecasting is not an exact science: it is important to associate an expected variability (a percentage of error) with it. What is more, uncertainty increases with the prediction horizon. Prediction uncertainty can be represented by upper and lower margins around the prediction values. A normal probability density function with mean and standard deviation can be applied to obtain a forecast error model. To calculate the margin limit, an inverse normal cumulative distribution function (CDF) with a desired probability index x can be used. The CDF then represents the probability of the random variable (in this case, forecast errors) being less than or equal to x. This method can be applied to PV power forecasting and to the evaluation of load prediction uncertainty. In this way, uncertainties can be determined according to different desired probabilities. Consequently, the bands of PV power forecast errors and load forecast errors can be obtained for a random day, considering different probability indexes (e.g. from 90% to 60% as shown in Figures 3.25 and 3.26).

3.5.5. *Importance of database size and quality*

The size of the database is very important when it comes to forecasting photovoltaic production and electricity consumption. The larger the database, the more accurate the prediction. Historical data contained in the database can provide valuable information on past levels of production or consumption and help to better predict future levels. The data can also help determine the best locations for photovoltaic panels and evaluate their performance. A comprehensive database can be used to compare and analyze the performance of similar photovoltaic installations to determine the best models and technologies to use to optimize future production. More accurate and reliable prediction models can also help companies to better manage energy consumption and reduce operating costs. On the other hand, forecast accuracy also depends on the quality of the various data used to train the neural network. This explains the discrepancies between the forecast and real curves if the database contains several measurement errors.

3.6. Conclusion

This chapter is dedicated to the problem of predicting photovoltaic production (using satellite images, cameras and ANNs) and consumption. The ANN method has been selected and applied to the forecasting of production and consumption in a block of buildings at the Université Catholique de Lille, France (smart grid demonstration) and to the predicted consumption in an HV/MV transformer substation in the Deux-Sèvres department, also in France. By using up-to-date databases and taking into account past trends and current conditions, ANNs can provide accurate forecasts (error of less than 10% for production and less than 5% for consumption), helping to optimize the energy production and consumption system to reduce costs and improve efficiency. ANNs are particularly useful for long-term forecasts, as they can integrate complex data and long-term trends to provide more accurate predictions.

In addition, neural networks can consider factors such as technological and social changes, which can have a significant impact on long-term electricity production and consumption. Finally, ANNs can adapt to changing conditions and adjust to new data to provide more accurate forecasts. This enables businesses and consumers to make informed decisions about electricity production and consumption. Nevertheless, it is essential to have a good quantity of data (at least one year) without many measurement errors.

3.7. Acknowledgments

The Live TREE demonstrations received funding from the Hauts-de-France region, in particular the photovoltaic production prediction camera, and from the European Regional Development Fund (ERDF) for the RIZOMM photovoltaic power plant. The photovoltaic forecasts received funding from ADEME via the So Mel So Connected project.

The authors would like to thank Khaled Almaksour and Grégory Vangreveninge of Junia for their contributions to the development of production and consumption predictions used by the Live TREE demonstrations.

3.8. References

[BAC 09] BACHER P., MADSEN H., NIELSEN H.A., "Online short-term solar power forecasting", *Solar Energy*, vol. 10, no. 83, pp. 1772–1783, 2009.

[ESP 10] ESPINAR B., AZNARTE J.L., GIRARD R. et al., "Photovoltaic forecasting: A state of the art", *5th European PV-Hybrid and Mini-Grid Conference,* Ostbayerisches Technologie Transfer Institut, p. 250, 2010.

[FEI 05] FEINBERG E.A., GENETHLIOU D., "Load prediction", in CHOW J., WU F., MOMOH J. (eds), *Applied Mathematics for Restructured Electric Electrical Systems: Optimization, Control, and Computational Intelligence*, Springer, New York, pp. 269–285, 2005.

[GLO 15] GLOBAL MODELING AND ASSIMILATION OFFICE, inst3_3d_asm_Cp: MERRA-2 3D IAU State, Meteorology Instantaneous 3-hourly (p-coord, 0.625x0.5L42), version 5.12.4, Goddard Space Flight Center Distributed Active Archive Center (GSFC DAAC), Greenbelt, MD, doi: 10.5067/VJAFPLI1CSIV, 2015.

[GRO 87] GROSS G., GALIANA F.D., "Short-term load prediction", *Proceedings of the IEEE*, vol. 12, no. 75, pp. 1558–1573, 1987.

[HUA 10] HUANG Y., LU J., LIU C. et al., "Comparative study of power prediction methods for PV stations", *2010 International Conference on Electrical System Technology*, pp. 1–6, 2010.

[HUA 16] HUANG N., LU G., XU D.A., "Permutation importance-based feature selection method for short-term electricity load prediction using random forest", *Energies*, vol. 10, no. 9, p. 767, 2016.

[KER 22] KERMIA M.H., BOSCHE J., ABBES D., "Comparison of photovoltaic production prediction methods", *International Journal of Renewable Energy Research*, vol. 2, no. 12, pp. 1041–1051, 2022.

[LAD 11] LADAN G., KALANTAR M., "Different methods of longterm electric load demand forecasting: A comprehensive review", *Iranian Journal of Electrical & Electronic Engineering*, vol. 4, no. 7, pp. 249–259, 2011.

[LIU 15] LIU J., FANG W., ZHANG X. et al., "An improved photovoltaic power prediction model with the assistance of aerosol index data", *IEEE Transactions on Sustainable Energy*, vol. 2, no. 6, pp. 434–442, 2015.

[PEL 13] PELLAND S., REMUND J., KLEISSL J. et al., "Photovoltaic and solar forecasting: State of the art", *IEA PVPS Task*, no. 14, pp. 1–36, 2013.

[ROB 23] ROBYNS B., LENGLET C., BARRY H. et al., *Smart Users for Energy and Societal Transition*, ISTE Ltd, London and John Wiley & Sons, New York, 2022.

[SHE 13] SHEELA K.G., DEEPA S.N., "Review on methods to fix number of hidden neurons in neural networks", *Mathematical Problems in Engineering*, 2013.

[VAZ 14] VAZ A.G.C.D.R., Photovoltaic forecasting with artificial neural networks, PhD thesis, University of Lisbon, available at: https://repositorio.ul.pt/bitstream/10451/11405/1/ulfc107351_tm_Andre_Vaz.pdf, 2014.

[WAN 12] WANG F., MI Z., SU S. et al., "Short-term solar irradiance prediction model based on artificial neural network using statistical feature parameters", *Energies*, vol. 5, no. 5, pp. 1355–1370, 2012.

[XIN 17] XIN Y., Energy management under uncertainty: Application to the day-ahead planning and power reserve allocation of an urban microgrid with active photovoltaic generators and storage systeme, PhD thesis, École Centrale de Lille, 2017.

[YAN 14] YANG H.T., HUANG C.M., HUANG Y.C. et al. "A weather-based hybrid method for 1-day ahead hourly forecasting of PV power output", *IEEE Transactions on Sustainable Energy*, vol. 3, no. 5, pp. 917–926, 2014.

4

Taking Actors into Account in Energy Management Strategies

4.1. Introduction

Smart grids will really deploy their potential for innovation by promoting interaction between the various actors in the power system (producer, consumer, storage operator, system operator), who themselves become "electrically" more intelligent actors. These actors have very different consumption and/or production profiles, and very different economic and societal objectives and/or constraints. New sets of actors may emerge, enabling us to target new business models, as well as to meet energy and climate challenges by promoting the development of REs. One of the challenges of this evolution is to ensure that all actors win, not forgetting those living in fuel poverty [ROB 19].

Legislative changes and technical explorations recognize the more active role that citizens can play individually or as a community, with or without the support of the local public or private sector. This redistribution of roles is leading to new thinking on multi-actor energy management, with full integration of network balancing tools (programmable generation plants, storage, demand management, interconnections). The question then arises as to how to take into account the behavior of actors in network management? Two visions are therefore possible for defining the intelligence displayed by future networks: either an intelligence based on the automation and optimization of tasks without human intervention or through the involvement of actors, which requires going beyond the technical aspects to put in place a framework that facilitates and enables this involvement.

Based on real-life experiments and a review of the literature, we have identified a need for a framework to address all the uncertainties raised by these developments. As the understanding and integration of energy behaviors in a network is one of the main challenges facing the energy issue, it will be approached in this chapter from an interdisciplinary angle. This exploratory work is therefore situated at the crossroads of electrical engineering, economics and sociology, with the aim of developing effective network management that can overcome these barriers, while respecting the sensitivities of each actor in these new smart grids.

The interweaving of different scientific disciplines, which are usually dealt with independently, represents a major challenge. To this end, the methodological aspect is very important, in order to define firstly the questions to be asked, and secondly the possible means and tools to answer them. The aim is to imagine a way of taking into account the different possible implications for actors in network management, by trying to understand the different profiles that may exist. Understanding does not just mean from a technical point of view, by studying a consumption profile, for example, but also endeavoring to understand a profile in terms of possible involvement, by seeking out its objectives and constraints, which may also be of a usage, social or other nature. In this way, we can find a framework for involvement that respects these rationalities, and in which everyone can find an advantage, while participating in the network's equilibrium. Acceptability can be seen as a framework within which technical solutions can be negotiated, depending on the nature of the transaction. For its part, involvement will be reflected in the concrete interactions between actors, and between the different involvement profiles that we will be trying to simulate. The ultimate goal is to find a network management system that can exploit these different implications.

In this chapter and in Chapter 5, answers are provided to the following three questions:

– *What are the involvement profiles of network actors?* Within a family of actors, there is a certain diversity of profiles, the result of different objectives and sensibilities. We want to understand them so that we can model them.

– *How to model these profiles?* Once understood, and with the aim of managing the network, these profiles need to be modeled in an appropriate way, so as to be able to imagine their interactions. Note that we are not trying to model the actor themself, but rather to model the components of their involvement that can be taken into account in energy management.

– *How can these profiles be used in energy management?* Since the aim is to develop an energy management strategy between all these actors, the final question is to allow each involvement profile to express itself by imagining their interactions within the network. The answer to this question is developed in Chapter 5.

To this end, this chapter presents a methodological proposal for defining actor involvement in the energy supervision of an electrical distribution network, before applying it to a simulated residential case in Chapter 5. This chapter is organized as follows:

– presentation of the various actors and roles in a power system;

– overall presentation of the three-stage methodology developed in this work and illustrated in these two chapters (stage 1 in Chapter 4 and stages 2 and 3 in Chapter 5);

– reflection on the modeling of residential consumer profiles for energy management. It should be noted throughout these chapters that the aim is not to model behavior, but rather the sensitivities of energy consumers in their act of consumption to the network;

– reflection on consumer profiles as seen through different approaches (engineering sciences and human and social sciences), and discussion of the interdisciplinarity required to accomplish this task;

– identification of the residential profiles to be taken into account, and presentation of these profiles.

The residential actor is chosen in this study, as it is the most interesting to observe due to the diversity of consumer profiles to be found there and the growing interest in it today as we think about integrating it into new so-called intelligent technologies, but for which *intelligence* must not be solely technical [ROB 23]. The main difficulty of this research is to identify a real case study and to retrieve the data (consumption, household characteristics, etc.). Since access to this sensitive information is very difficult, a simulated case is chosen, the possibilities of which will be discussed in this chapter, and the modeling of a relevant case study to illustrate the supervisor is provided in Chapter 5.

It is also worth highlighting the central concept underlying this work, that of *electrical flexibility*, which can be summed up, in simple terms, as the ability of an electrical system to adapt to variations in electricity demand and supply. This means responding to fluctuations in energy demand by adjusting electricity production or reducing electricity consumption. It is on this notion that the new approaches to taking into account network actors in network management are based, and which are developed in this chapter and the next. In so doing, this chapter positions itself in relation to the various SDGs mentioned in Chapter 1. The most obvious of these is goal 7, “Clean and affordable energy”, since energy management in networks is a key factor in the integration of RE into the energy mix (SDG 7.2). Such energy management also enables different actors to become genuine actors in the power system, rather than mere passive consumers or producers, by ensuring a fair distribution of gains (SDG 7.1). This links in with SDG 10 on reducing inequalities, since "taking actors into account" also implicitly includes the least favored, those with the most constraints (technical, economic, etc.).

What is more, enabling the most environmentally aware actors to express their views and contribute at their own level to energy management that is more closely linked to the production of RE is a direct response to the need to combat climate change (MDG 13). Lastly, from a technical point of view, improved energy management ensures the sustainability and resilience of today's power grids by acting directly on power flows, thus limiting the need for investment to manage critical situations (SDG 9 – "Industry, innovation and infrastructure").

Chapters 4 and 5 are based on the authors' own work, with all graphs and tables taken from their own work (unless otherwise stated). Interested readers wishing to go further may consult the editor's research articles or thesis for further details [DUR 19, DUR 20].

4.2. A system of actors in an electrical network

4.2.1. *The role of actors*

Power grids are constantly evolving. The roles that users and the grid operator can take on are changing with them: new uses and even new actors are emerging. To sum up, there are five main roles that influence day-to-day power flows on the electrical grid:

– manage the network (ensure balance);

– inject energy;

– extract energy;

– buy energy;

– sell energy.

These roles can be performed simultaneously by several actors. For example, an energy supplier buys and sells energy, a consumer withdraws and buys energy, a consumer with a photovoltaic installation can inject, withdraw, buy and sell energy, etc.

In order to identify these actors, we need to consider the time scale and the level of the network at which they are positioned. While the network is the place where energy is physically exchanged, there are many actors who have an influence on it: on the wholesale and retail markets, for example, with suppliers and traders, as well as in the longer term, such as the regulatory authority that defines the scope of energy exchange.

Looking only at daily power flows, the five actors considered in this study are shown in Figure 4.1: producers, consumers, producer–consumers, grid operators and aggregators. Other actors can be included later, following the same approach.

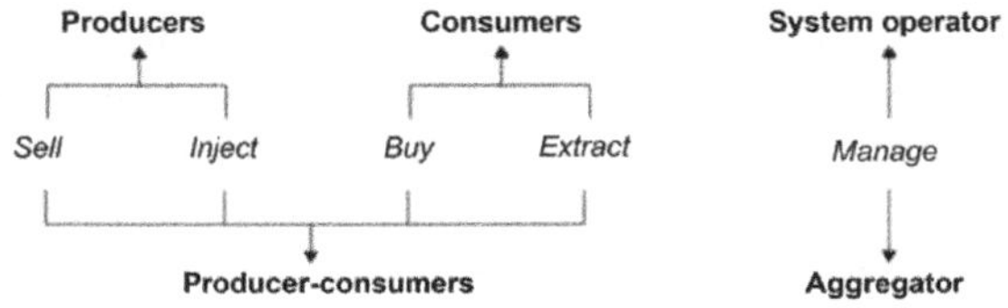

Figure 4.1. *Actors and roles considered in the methodology presented*

4.2.2. *System operator*

By ensuring the smooth operation of the network for which they are responsible, the system operator is the first to be interested in energy management, whether for transmission or distribution (described below). They are responsible for ensuring that frequency and voltage are maintained between contractually defined limits, mainly by acting on generation, since networks have up until now been largely demand-driven, as discussed in Chapter 3. Two mechanisms are available: forecasting energy flows (production and consumption) and adapting these flows. From a technical point of view, it is therefore of great interest to obtain greater flexibility for management; however, the difficulty of taking it into account lies in the contradictory interests that arise. Indeed, an energy management program would relieve the network at critical times, but at the same time reduce the possible economic returns if less energy is consumed and if the services provided must be remunerated. At various levels of the network, a distinction is made between the transmission system operator, an actor often with a national monopoly (as in France, but not in Germany, for example), and the distribution system operator, acting locally and whose role may be performed by multiple entities across the country.

4.2.3. *Aggregator*

Managing the network is complex, and ensuring its smooth operation is the role of the system operator, as described above. However, as more and more flexibility is possible, and in demand, a new actor is appearing on the grid: the aggregator. Although their role is not always well defined, they act as a central figure at the intersection of all the other actors and the energy market. The role can be summed up as a flexibility manager or operator. Their sole task is to manage network flexibilities, which can be contracted out. The difficulty of their task is to take into account everyone's constraints and objectives, which can be contradictory, while at the same time bringing more flexibility to the network.

The vagueness of the definition is due in particular to its concretization: it could be a person, a company or an algorithm, and on a global or local scale (an "aggregator" entity would then manage a piece of network and several entities could interact with each other). The role is to respond to the system operator's need to balance the network, by communicating with a portfolio of customers whose shaving or consumption capacities it knows, and which it then remunerates for the service rendered. The current example of this flexibility is the French technical and economic potential for load shedding in the steel industry, estimated at between 0.9 and 1.5 GW for a remuneration of EUR 30/kW [ECU 17]. The residential sector is also concerned by a historic tariff for load shedding on peak consumption days (EJP, a type of critical peak pricing). The legislative and contractual framework is evolving, and new flexibility mechanisms are appearing, such as the NEBEF[1] mechanism, open to residential and business consumers, introduced in 2015, or the capacity mechanism introduced in 2017, which enables operators to exchange capacity guarantees with suppliers on the capacity market.

4.2.4. *Producer*

The energy producer injects and sells energy to the grid, whether in small or large volumes. Their main objective is to ensure the profitability of their production in the face of market forces. To achieve this, they can rely on production forecasts, enabling them to sell energy on the wholesale market up to several years in advance, in the form of forward products. With recent developments in the electricity market, and environmental issues increasingly at the forefront of public debate, new constraints and uncertainties are complicating the company's strategy.

Indeed, rising production costs, the environmental impact of energy production coupled with growing environmental issues, and ongoing societal awareness are leading to a change in behavior with regard to energy use (whether economically or environmentally), and an increase in public policies pushing for more environmentally virtuous production.

1 In France, the Block Exchange Notification of Demand Response mechanism, known as NEBEF, allows third-party players (including aggregators) to offer demand response services in wholesale power markets. This mechanism was created in 2013 and defines the roles and obligations for the different market actors.

The resulting pressure on producers to limit the impact of production by improving processes or diversifying production from fossil fuels to renewable energy entails significant implementation costs. For renewable energy producers, over and above the economic aspect, production fluctuations and forecasting difficulties also entail major constraints on the optimal use of this production. The challenge is therefore to inject this production while guaranteeing grid balance and economic profitability for the producer.

4.2.5. *Consumer*

The consumer draws and buys energy from the network. Their main objective is to meet the need they are seeking to satisfy by maximizing their satisfaction, subject to economic and/or environmental constraints (if they attach importance to them). An important distinction to be made between the residential consumer and the company (industrial or otherwise) is the actor's behavior and its predictability.

For the former, forecasting is complex, since it depends on a number of parameters and is the result of behavior and social interactions within households, which are difficult to approach solely from a technical angle [CAY 11]. The economic impact, for its part, depends on the weight of energy expenditure in the household budget, and can therefore be a mechanism of involvement for the household. Similarly, the environmental impact of consumption can be a mechanism for changing behavior, provided that adequate information is available. While the aggregation of residential consumption can be predictable, the decentralization issue imposed by increasing RE production limits this possibility. What is more, the social benefits of energy management at household level should not be overlooked either, particularly in terms of encouraging people to get involved.

For companies in the primary and secondary sectors, the main consumption is the result of production and internal processes and as such is more predictable than for residential consumers. As the economic constraint is central for this type of consumer, it is easier to contractualize their energy behavior and the services they can render to the network. The environmental issue, on the other hand, is based on the company's image and the legal constraints to which it may be subjected to. The tertiary sector, however, raises a number of questions, because although internal organization and management enable consumption trends to be relatively predictable, they depend on a sum of individual behaviors, of consumers who do not pay the bill themselves, which is different from the behaviors encountered within a household setting.

Generally speaking, this actor is particularly interesting for energy management because the network was created for them, so their interest is in its smooth running, but their involvement is relatively little used and complex to exploit, especially for the residential case.

4.2.6. *Consumer–producer (prosumer)*

Although the technology to enable private individuals and businesses to produce their own energy has been available for many years, technical and, above all, legislative developments have only recently led to the emergence of new actors active on the grid, consuming and producing energy. As mentioned in Chapter 1 (see section 1.7.3), the issue of self-consumption is a topical one, with the most significant development being the emergence of the term "energy community". Whether acting individually or as a group, these actors are no longer simply consumers or producers, but both, simultaneously or at different times, and their interest then lies as much in achieving the desired consumption as in making optimum use of their production. This interest may be individual and/or collective, economic, or otherwise. This new complexity particularly calls for new thinking in network management, given the multiplication of two-way energy flows.

4.3. Methodology for managing energy flexibility involving actors

4.3.1. *Defining key concepts*

4.3.1.1. *Acceptance*

The term "acceptance" used in this study refers to the notion of adherence to an energy management program. It does not presuppose participation but is nevertheless necessary to consider before we can speak of involvement. For a residential consumer, for example, this implies a willingness to modify their behavior; for an energy-consuming company or producer, it means modifying its strategy or organization. It is important that the proposal made is desirable for the targeted actor, from their point of view and according to their own rationality, dependent on their values. It is therefore important to identify and understand different actor profiles.

From the point of view of the individual in society, we speak in particular of social acceptability: an interesting definition of this notion is that of Véronique Yelle[2], whose summary and linking of definitions in the literature is impressive:

> Social acceptability is the aggregation of individual judgments about the acceptance (or otherwise) of a practice or condition, whereby

---

2 Laval University, Canada.

> individuals compare it with its possible alternatives to determine its desirability. It is conveyed by politically significant groups within society who share the same judgment about this practice [ROB 19, authors' translation].

Within society, it is a collective process that requires a defined threshold of acceptance before it can spread to the target population, hence the notion of social acceptability.

4.3.1.2. *Involvement*

Once participation in network management has been accepted, different degrees of investment can be observed. These different degrees are defined as the actor's involvement. The higher the degree of involvement, the more the actor will be able to deviate from their unsolicited (or unsupervised, in the case of energy supervision) behavior in response to a network stimulus. In the remainder of this study, we talk about behavior in the energy sense of the term, which refers to the actor's consumption or production of electricity.

On the other hand, if acceptance is zero, involvement will also be zero. In this case, we speak of a passive actor who confines themself to their primary role of consumption or production and does not participate in network balancing. The signals received about the state of the network therefore have no influence on it.

4.3.1.3. *Flexibility*

Flexibility in the context of energy management is based entirely on the notion of shiftable energy. This is the volume of energy that the consumer or producer makes available to participate in the energy supervision program and achieve its objectives. It has two facets:

– Technical: the technology available to the actor determines the range of possibilities in terms of what they can or cannot shift.

– Organizational: the way in which the actor operates restricts or increases the possibilities of energy shifts.

In other words, having flexibility requires having energy to shift (in production or consumption), being technically able to shift it and being able to shift it at the right time.

The term "flexibility" used hereafter is linked to the idea of flexible consumption and production, but this is only from a network point of view. It is important to bear in mind that, from the point of view of the consumer or producer, making their flexibility potential available induces new constraints in the day-to-day performance of their tasks and, as a result, reduces the flexibility of the operations they wish to carry out. In the case of consumers, for example, a study of residential load shedding revealed that switching off appliances such as the dishwasher or washing machine was considered “manageable but seen as an extra effort and an extra thing to remember”, in an already busy schedule [FRI 16].

4.3.1.4. *Comfort and satisfaction*

The notion of comfort in this work refers to the idea of disposing of energy as desired, that is, injecting or withdrawing energy in the desired quantity and at the desired time for a producer or consumer, respectively. Discomfort therefore represents the inability of an actor to perform an a priori desired task. The term “comfort” should not be confused with “satisfaction”, which refers to the aggregation of the level of comfort obtained (the degree to which a desired task can be carried out at the desired time) and the moral or financial compensation obtained in parallel.

Nor should it be confused, for the residential actor in particular, with the idea of thermal comfort frequently evoked in the literature: in this work, we are not talking about physiological comfort, but rather psychological comfort, discomfort being the mental load possibly borne by an individual when changing behavior.

4.3.2. *Comprehensive methodology for energy supervision*

The central question of this chapter, “How can we take into account the behaviors of actors and include them in an energy supervision system?”, highlights two important notions: actors and energy supervision. The methodology reflects this division, with, on the one hand, the apprehension and understanding of actors, and on the other hand, the introduction of this reflection into the development of the energy supervisor, itself divided into a part involving prior optimization the day before for the next day and a real-time adjustment of flows. Each of these three stages must be seen as a component of the overall methodology, and not in isolation. This is one of the challenges of this study: the search for profiles is carried out with a view to supervision, and therefore guides the method, and the two stages of the supervisor follow on from this study of profiles. The overall vision and articulation of this methodology are shown in Figure 4.2.

The *first step* looks for the flexibility of the actors that it is possible to involve in the network, by answering the question: “How can we approach and define the profiles of actors present in a given network?” The aim is to model the components of this participation that can be taken into account in the energy supervisor, and to formulate them appropriately. Interdisciplinarity is essential at this stage, as the plurality and diversity of actors mean that approaches need to be combined according to their nature (individual, organization, industry, etc.), and the relevance of the supervisor subsequently developed for each actor depends on an appropriate definition of profiles.

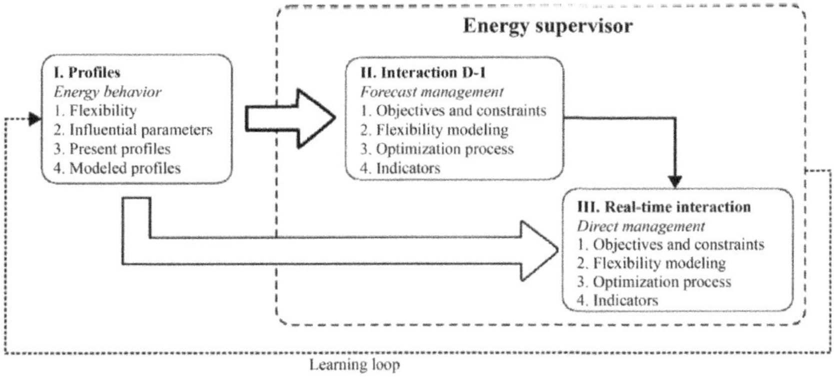

Figure 4.2. *Comprehensive energy supervision methodology*

The *second step* answers the question: "Knowing the actor profiles, what framework and optimized interaction can be used to manage energy flows, taking the day before the day under study?" The issue at stake at this stage is not only optimization, but also the approach chosen to model the profiles: thought is needed to ensure this integration, as their modeling will itself influence the supervisor architecture chosen and the trade-offs to be made.

The *third step* answers the question: "Knowing the actor profiles and the energy flow optimization performed the day before, how can we adjust these flows in response to hazards observed in real time?" The key points of this stage are, once again, interfacing with the first stage (in what form should the profiles selected in the first stage be introduced?), reflection on the supervisory structure and management arbitrations, but this time based on the output of the previous stage.

The second and third steps are developed in Chapter 5.

4.4. Modeling actor profiles

4.4.1. *An interdisciplinary approach*

The first stage of the methodology consists of studying and defining the actor profiles that will be considered in the energy supervisor developed later. The key points of this stage are summarized in Figure 4.3. The challenge is twofold: firstly, the existing profiles of the population of actors to be supervised need to be studied in order to understand the rationalities specific to each of them, in particular by determining how sensitivities and objectives are distributed; secondly, these profiles need to be formulated in a suitable way for input into the supervisor at the next stage.

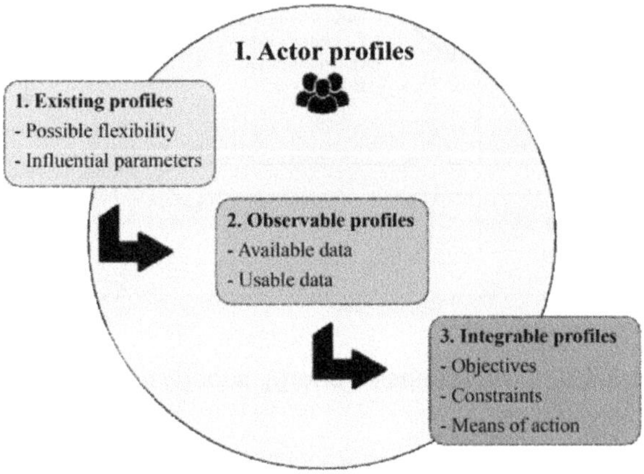

Figure 4.3. *Defining actor profiles*

First, the available technical and organizational flexibility is determined before the influencing factors are identified. From these profiles, the next step is to identify those that can be observed in a given population, based on available and usable data. The acquisition and use of data is a key stage in the approach, and careful thought must be given to it, as outlined in the previous section. Finally, these profiles must be sorted so as to retain only those that can be taken into account in the energy supervisor, by determining for each of them its objectives, constraints and means of action, arising from the study carried out.

The supervisor's objective is not to force the actors to participate, but to suggest a path for each of them to follow, enabling them to achieve their objectives while contributing to the balance of the network. The profiles must therefore be clearly defined, as their distribution may change over time. The most important thing is to consider all aspects of actor involvement profiles, because beyond energy management as such, its construction and implementation must go through a phase of consultation with the actors present to ensure its success with the greatest possible efficiency.

To meet these challenges, interdisciplinarity is essential. While network management is naturally driven by electrical engineering, this discipline is already opening up to other technical disciplines, drawing on AI and the Internet of Things, for example. This openness to interdisciplinarity is necessary, as network management increasingly involves a range of different actors as part of an effort to offer new services, and network architecture is set to evolve in the future. Beyond the smart grid, new concepts are gradually emerging, such as multi-energy networks that can be managed jointly and the concept of the smart building as a possible smart grid node (see Chapter 1, section 1.9). This concept brings together habits, usages, networks (IT, electricity, heat) and flows (people, air, water, heat), which need to be understood in order to manage the networks. Building management and control is therefore one of the keys to the success of the smart grid, which cannot be achieved without contributions from the other sciences.

A wide range of approaches can be called upon to understand the actors, their objectives and their constraints, and to provide a framework for their interaction: from the engineering sciences such as electrical, thermal, civil and computer engineering, through management sciences (production, organization, logistics, management, etc.), to the human and social sciences such as economics, law, sociology and political science. Depending on the context and the actors involved in the network under study, an upstream analysis must be carried out to determine which disciplines are best suited to the case under consideration. To this end, the authors of [PEL 17] show that understanding and including the involvement profiles of actors in an energy network is one of the main themes to be addressed by the interdisciplinary approach.

4.4.2. *Existing actor profiles*

4.4.2.1. *Defining actor flexibility*

The first step is to determine the technical and organizational flexibility of each actor. The aim here is to *understand* flexibility and then *determine* the extent to which these actors can participate in the network, whether by shifting consumption or production, by defining the quantity of energy that can be made available and the constraints attached to it.

If an actor cannot be modeled in all their complexity, the question arises as to their involvement in an electrical network. In the case of a residential consumer, for example, we need to consider the ins and outs of energy consumption.

From a methodological point of view, we need to understand the activity system that characterizes each actor [ROB 19], as this determines their use of the energy resource and enables us to highlight the possible flexibilities that they may have at a later stage. For this purpose, an activity system is defined according to four components:

– vocation: What is the actor's activity or activities?

– personnel: number, type, activity, etc.;

– work situation: indoor/outdoor, static/mobile, time of day, etc.;

– regularity of activities: time slot, frequency, regularity, etc.

For each of these components, it is then possible to determine its impact on energy consumption or production, resulting from the operation of the actor, its needs and its organization.

The flexibility of each actor can then be determined using a multi-disciplinary approach, looking for potential sources of mismatch. As mentioned above, this involves studying both aspects of flexibility.

On the one hand, we need to study the technical system in place at the actor's premises to determine the technical possibilities available, drawing on the engineering sciences. This capacity depends on the energy devices owned, their conditions of use and the environment in which they operate. Modulating them requires knowledge of their nature (production of cold, heat or work). The question is: "Is it possible to modulate processes in terms of power and time, while still providing the required service?" This can be done by relying on improved efficiency, inertia (thermal or mechanical), or a dedicated storage medium. We also need to define the actor's control over their process. To give a simple example: a photovoltaic producer cannot technically inject energy into the grid if there is no sun and there is no means of storage, unlike a producer operating a gas-fired power plant, who can inject and modulate power.

Secondly, we need to study the organization of the actor, this time drawing on the human and social sciences. Beyond the technical aspect, the aim is to determine to what extent the actor can modulate their energy (injected or extracted) by modifying their organization. This simply boils down to the ability to shift a task in time, from an organizational point of view, to perform it at another time. This ability is based on the actor's strategy and the constraints they face. For example, a consumer company may have the technical capacity to shift a process but be unable (or unwilling) to do so due to organization, staff availability or another process depending on it or to retain a strategic advantage, for example.

4.4.2.2. *Influencing factors*

Once the volume of energy that can be shifted has been defined, we need to think about the following issue, the influencing factors: "What influences (encourages or constrains) the use of this flexibility?" The aim is to define the actor's sensitivities to external factors, to be able to manage this flexibility in the best possible way. The ultimate aim is to highlight actor profiles, which will be defined in terms of flexibilities, influencing factors and associated sensitivities.

At this stage, individual and differentiated preferences are less important than their overall distribution within the population of actors. The intelligence underlying this work can be both technological and collective, as flexibility can be enabled by control from the manager or a central entity, as well as by individual behavior in response to a given piece of information.

These factors may be meteorological (external luminosity, irradiation, temperature, wind speed, etc.), economic (energy price, remuneration for flexibility), technical (information received, energy mix) or social (image, group pressure, event, culture), and may or may not be interrelated, depending on the strategy and rationality of the actor.

At this level, it is important to separate the factors of acceptance, which enable profiles to be defined, from those of involvement, which define the framework to be respected for the supervisor's development. The objective here is day-to-day energy management, but we must also consider all the factors that influence acceptance, in particular regulations and public policies that may constrain or encourage certain behaviors and evolve over time [LOS 13, YIN 13], or even help to contractualize certain services provided. The supervisor developed must be able to provide a framework conducive to fostering this acceptance but must also be part of a wider societal project, developing the legal context in parallel [BEN 12, LIN 13].

To understand influencing factors, it is important to first ask this question alongside the disciplines concerned, as discussed above.

In addition to the engineering sciences, we can mention, for example, for residential consumers, microeconomics, which can be used to understand economic behavior in energy consumption trade-offs, and social sciences (including energy sociology, psycho-sociology, anthropology, etc.), which can be used to understand the social interactions that take place within the home setting and in the actor's social environment.

Logically, economics plays an important role, as relationships to energy are mainly monetary, as illustrated by the contracts offered historically, which have various implications. In addition to understanding profiles, the methodologies used in economics can also be used to break down profiles in a given population, and to consider the model to be put in place to manage flexibility. The challenge of a new economic model is then to encourage participation, but without placing a heavier burden on those who participate less [SAL 13, HAI 16].

4.4.3. *Observable profiles*

Knowing the existing profiles, their use for energy management requires the ability to observe them using available data. Data concerning the various aspects of flexibility and the factors influencing it may not exist or may not be accessible, which may be the case for large populations of actors or for actors who do not wish to divulge their data (personal or strategic data).

We then need to think about recovering observable actor profiles from the set of known profiles, for which three situations are possible:

– the data required to define existing profiles exist and are accessible;

– actors are involved in the construction of the supervisor and provide the information needed to understand their flexibility, or actors are given the technical option of declaring their profile and are then called upon once the system has been set up;

– the data do not exist or are inaccessible.

In the latter case, if a sufficiently accurate model cannot be used, the corresponding flexibilities and sensitivities cannot be used in the supervisor. In the other two cases, the question arises of their recovery and use, as mentioned in the key point of the method.

4.4.4. *Integrable profiles*

This final stage enables us to characterize the existing profiles of the actors, defining their objectives, constraints and means of action. This final step aims to recover the volume of flexible energy and the factors influencing it that are observable and that it is physically possible to take into account when managing the network. These are either directly measurable signals (price, temperature, current production, schedules, etc.), or individual or actor-specific preferences, which, if they cannot be measured, must be anticipated or declared. Conversely, factors that cannot be translated into precise information should be discarded.

The degree of precision in defining flexibility then depends on the actor and the data obtained in the previous steps. The more precise the definition of possible flexibility, the easier it will be to rely on it, and the more easily network management will be able to take it into account. Otherwise, an uncertainty study will have to be carried out in parallel, considering the possibility of failure or increasing reliability requirements.

To define this ideal case, we can draw on the criteria defined by the rules for participation in electricity market mechanisms [ECU 17]. These criteria are defined for consumption load shedding but can also be used to precisely define flexible generation capacities for producer actors. The five main ones are explained in Figure 4.4:

1) power loss P_{flex};

2) mobilization time (DMO);

3) minimum duration of use (DU_{min});

4) maximum operating time (DU_{max});

5) activation and deactivation time (DA, DD).

These criteria can then be used to discriminate which flexibilities can be taken into account for management purposes. To do this, it is necessary to establish the constraints to which these flexibilities will have to conform, as is the case, for example, for the contractual valuation defined by the Energy Code.

At the end of this stage, the profiles obtained for each actor are summarized in two steps:

– The flexibility chosen is described as a means of action, alongside the *constraints* associated with it.

– The actor's *objectives* and *sensitivities* that will define the use of this flexibility are gathered so that they can be included in the supervisor.

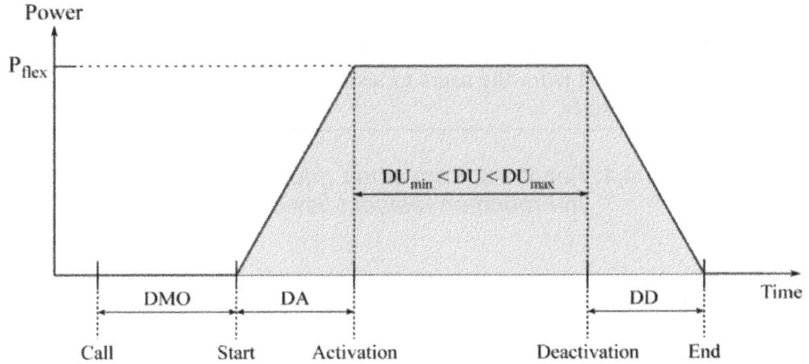

Figure 4.4. *Characteristic criteria for flexibility [ECU 17]*

4.5. Residential actor profiles

4.5.1. *Return feedback from experimentation/scale-one projects*

The search for consumer profiles in future smart grids is carried out through experiments carried out on demonstration projects at a scale of one. In France, to begin with, it is interesting to study the progress made in this field thanks to the feedback from a few demonstrations, listed in Table 4.1. The arrival of smart meters (called Linky in France) in households is highlighted as a key factor in the democratization of these experiments and the study of consumer involvement in these new networks. This feedback not only allows us to highlight participation profiles, but also to discuss potential flexibility and the key points that make it possible.

Project	Key Results
GreenLys	Real interest in flexibility on the part of residential consumers: 63% decided to keep the control equipment at the end of the experiment Reflection on the evolution of flexibility through a real option approach developed in [ACC 16]
REFLEXE	Residential flexibility prediction algorithms [DIN 10, BOR 11] Algorithms for optimizing the choice of flexibilities to maximize revenue, uptime or peak erasable power
Nice Grid	Four profiles observed: technophiles (willing to adopt new technologies in advance), energophiles (attentive to their energy use), economists (aiming to limit or even reduce their bills) and ecophiles (sensitive to the environmental impact of their consumption) Participation of 13% for households and 50% in industry and the public sector (out of eligible potential)
Condisplay	Influence of consumption display on the development of an energy culture, involving the active involvement of the user in these issues
Issy Grid	While the technical conditions seem to be in place to get the actors involved, all that is needed is for the users to take ownership of the project to achieve convincing results.

Table 4.1. *Feedback from smart grid demonstrations in France on resident flexibility*

On the other hand, a summary by ADEME of these various demonstrations [BER 16] shows that informing residential consumers about their consumption can save between 1% and 10% of energy, but that it is important to provide a suitable model and infrastructure (easy to use for the user). Another interesting fact to note is that, once the principle of automated deletion has been accepted, user involvement is at its highest: in other words, they hardly ever deviate from the order given. The important thing, then, is to set the framework for maximum acceptance, and thus optimize involvement.

4.5.2. *Consumer profile research*

Real-world demonstrations make it possible to observe behavior in real-life situations, but only on profiles that have already volunteered and within a predefined framework. For the more general case, and to imagine a framework designed to encourage this involvement, it is interesting to discuss the possible participation of actors, as well as the factors that influence it. Many examples show the technical aspects of this consideration, but do not discuss the concrete study of these profiles and their integration. However, understanding consumer profiles cannot be done from a purely technical point of view, and integrating this issue into the purely electrical engineering approach requires an openness to interdisciplinarity.

Consumer types can be observed in a population based on the observation of energy consumption, but this does not allow the objectives to be determined. The main studies aimed at segmenting residential consumption on the basis of socio-demographic variables, electrical appliances or the adoption of new energy technologies (solar panels, batteries, heat pumps, etc.) show that the main socio-economic explanatory variables are household size, income and employment status.

Using surveys conducted around the world, [FRA 13] shows the heterogeneity of consumer commitment by establishing six involvement profiles, based on price and environmental sensitivities, as well as on interest in adopting new technologies, sensitivity to external control or sensibility to social pressure. [FRA 13] proposes five, in order of flexibility, according to the criteria of sensitivity to price, the environment and interest in new technologies.

Using a data partitioning method, six profiles are also selected by [BER 11]: beyond the main demographic characteristics of each group, their flexibility is discussed in terms of accepted economic gains or losses of comfort.

The most representative profiles from the last three studies, and their correspondence, are presented in Table 4.2, in descending order of flexibility.

[FRA 13]	[FRA 13]	[BER 11]	Commentaries
Ecoconscious	Green advocate energy savers	Idealistic energy savers	The most sensitive to the environment and the most flexible in their consumption, with a strong belief in their ability to influence the state of the world
Economical	Traditionalist cost-focused energy savers	Materialistic energy consumers	Behaviors highly sensitive to the cost of their consumption, but little interested in new technologies
Proactives	Home-focused selective energy savers	–	Interested in energy savings, sensitive to new technologies and the proportion of their bills
Pragmatists	Non-green selective energy savers	Thrifty energy savers	Not very concerned by environmental issues, but adept at set-and-forget behavior, as long as it does not negatively impact their bill
–		Selfless inconsequent energy savers	Very willing to accept restrictive regulations, yet aware of their impact, they make little effort
Indifferent	Disengaged energy wasters	Convenience-oriented indifferent energy consumers	They have little interest in cost, the environment or new technologies, and believe that their behavior has little impact on the state of the world
Skeptics		Problem-aware well-being-oriented energy consumers	Ditto, but with a particular reluctance toward programs requiring participation that could potentially cause them to lose comfort or freedom in their energy management

Table 4.2. *Comparison of energy consumer profiles in the literature*

4.5.3. *Sociological approaches for accepting participation in network management*

The social conditions of acceptability seem to be based on the simplicity of the devices, the control left to the actor (all the more important for an industrial actor) and finally the sharing of costs/benefits, insofar as actors who were once simple consumers become co-producers [ZÉL 15b]. Moreover, understanding a human acceptance decision is not always easy, and we can draw on behavioral theories to understand the basics. These concepts seek to determine a consumer's commitment through their own perceptions. With regard to the smart grid, in [PAR 14], from the theory of technology acceptance, which postulates that acceptance is based on the perception of usefulness and the perception of ease of use, a development was added with the perception of risk. The exogenous criteria influencing these perceptions in the case of a smart grid were then evaluated, these criteria being: compatibility, understanding, reliability, savings, ecology, data protection, electromagnetic radiation and performance. Similarly, [BRO 14], combining two approaches, proposes the Responsible Technology Acceptance Model 1, based on the premise that smart grid technologies will not only be accepted for the personal benefits they would generate, but also because they are perceived, to some extent, as a duty to society and the environment. The results show that, as the short-term benefits for a household are sometimes low, the environmental and societal aspects of smart grids also need to be emphasized.

Taking an interest in consumer profiles and new energy networks also means studying the factors driving changes in these behaviors, in order to take them into account or create a context conducive to the adoption of new models (technological, economic, social). The approach is therefore based on the social sciences: sociology, psycho-sociology, behavioral science, etc. The encouragement of pro-environmental behavior, for example, is discussed by ADEME [ADE 16]. What emerges from these multidisciplinary analyses is that we must not forget that behavior is a complex process resulting from the interaction between the individual, their immediate technical and social environment, and finally the society in which they evolve. Although we need to find relevant physical indicators to imagine energy management, the rationality of each individual and the choices they make cannot be summed up in simple variables.

Last, but not least, in a paradigm where all participation is automated, the question of a retroactive indication of the actor's consumption arises. This automated vision of energy management is not insignificant as a societal choice and has been studied in numerous research studies. If manual or semi-automatic consumer participation is envisaged, another important issue, which is one of the hypotheses used in the development of the present supervisor, is the ability of consumers to modify their consumption, and therefore to receive the appropriate information to carry out these actions [FRA 13]. The interest is twofold: in the short term, it informs and enlightens the user on the best consumption choice given their interests; in the longer term, it creates the energy culture needed to generate commitment (even beyond the energy domain) by making electricity consumption, which is difficult for consumers to estimate, tangible/palpable [ZÉL 15b]. The form of this information is therefore a subject in its own right, underlying every demand response program, upon which the effectiveness of consumer engagement in the short and long terms partly depends [MA 18]. For greater effectiveness, this communication should not be of a single form for all households but adapted according to preferences.

4.5.4. *Economic approaches for consumer involvement*

An overview of the literature on demand management programs shows that most are price-driven. Beyond the aspect of optimizing the demand of each consumer, the construction of a price scheme is then proposed. On the one hand, this presupposes that price is the main motivation for change, and on the other hand, it entails the risk of negative externalities for small consumers on the network, who then see their costs rise as a result of larger consumers. Particular attention must therefore be paid to the lowest consumption levels (which may include consumers in fuel poverty), as well as to the largest consumers, who may be responsible for the biggest imbalances on the network. Pricing schemes must therefore be carefully designed to ensure acceptance and fair access to energy. It is therefore natural to turn to economic approaches to consumer involvement in demand-side management programs. Following on from these considerations, the first question that needs to be addressed in the search for profiles and economic models is: Do we want to target all consumers, or just those most likely to contribute to network balance?

Since price is an observed barrier to the adoption of pro-environmental behavior, another approach to this economic-environmental duality of profiles is to ask this question: How much are consumers willing to pay to consume RE as a priority? Numerous studies have sought to determine this propensity to accept an additional cost, not only in terms of this cost, but also in terms of the parameters influencing it. For example, [MA 15] concatenates 29 studies using this approach. What emerges from all these studies are the socio-economic factors indicating the greatest propensity to accept: income, level of education, age, as well as energy-related knowledge within the household, showing that there is a need to financially encourage virtuous behavior, as well as to increase the energy culture as mentioned above.

Beyond the price–environment link, a more global approach, used in microeconomics, is that of research into demand systems (link between consumption, income, prices and socio-demographic variables). The contribution of this discipline to the subject is threefold:

– Firstly, the aim is to understand household consumption by searching for explanatory variables in available socio-economic databases.

– Secondly, the aim is to find a demand system that allows sufficient flexibility for regression. Its use in the search for demand systems can be found in [REA 18], for example, based on German data.

– Thirdly, in a broader perspective, the aim is to find the price/income[3] elasticities of energy consumption in a given population, to determine the interest in participating in a consumption control program: this issue is addressed in [AJM 13] for all G7 countries.

4.5.5. *The need for interdisciplinarity*

This overview of the literature clearly shows the scope of the problem: the integration of profiles into an energy management program therefore requires a precise methodology to consider all aspects, while shifting the paradigm from the techno-centric approach, observed in engineering sciences, to an interdisciplinary approach necessary to apprehend the various facets of the problem [PEL 17]. As the residential case study considered in this work is fictitious, it is impossible to study the profiles of a precise population. However, it is possible to discuss the crucial points to be taken into account and the tools available to define the profiles to be considered. The necessary inputs are as follows:

– assess what each science can contribute to the profile research approach, and how these can be articulated with each other;

– assess what each science can contribute to the definition of profiles, their breakdown in a given population, and what can be taken into account in an energy supervisor;

– discuss actor objectives, acceptance and involvement factors;

– in the absence of a specific residential area, study the case of a region or country;

– determine the profiles that will be considered depending on the desired application for the development of the supervisor.

While it is interesting to discuss the acceptance and involvement of residential consumers, it is all the more important to not characterize each household, but to define the individuals in a population likely to participate, so as to foster sufficient flexibility.

Finally, although we are seeking to introduce these actor profiles, it should be made clear that we are not modeling human behavior: what we are interested in here is discussing and defining the objectives of each individual in relation to electrical energy consumption, and with which it is possible to build to modulate this consumption to bring more flexibility to the network than if we did nothing. Indeed,

3 Elasticity is the quantity that answers the question: "How does consumption of a good evolve when price (or income) increases or when price (or income) decreases?"

it is impossible to be exhaustive in considering and understanding behaviors and participation within a network, especially as the changing context is leading and will lead to new uses. The aim of this chapter is therefore to provide as comprehensive a reflection on the subject as possible since a new technical system only makes sense in a considered context.

4.5.6. *Characterizing flexibility*

4.5.6.1. *A multidisciplinary approach*

As a result, today's consumers are more solicited than in the past, and are tending to become active network users in their own right. The much more complex environment in which they operate is summarized schematically in Figure 4.5. It includes centralized and decentralized energy production, new energy technologies (storage, electric vehicles, etc.), as well as changing legislation (possibility of self-consumption of our own production or of forming a community to do so). The challenge is therefore to anticipate the profiles of the actors in this changing environment. The state of the art clearly shows this complexity, as well as the multidisciplinary aspect of the problem. Applying the methodology to the case of the residential actor raises several questions, which can be answered by the three questions shown in Figure 4.6:

– What are the profiles of the residential actors?

- to answer this question, we need to begin by characterizing flexibility, both technically and socially, drawing on the engineering sciences, on the one hand, and the human and social sciences, on the other hand;

- we also need to define the parameters that influence this flexibility and that need to be taken into account in energy management. The contributions of sociology and economics are important for studying the extent to which changes in consumption can be observed according to the signals received by users;

– What profiles can we observe?

- the idea is to imagine how to find them in a given population (based on available data) by studying the contribution of microeconomics, which considers that a household's behavior is the result of the reconciliation between two fundamental elements: the marginal benefit and the marginal cost associated with a change in its behavior. Behind this benefit–cost ratio lies the notion of propensity to accept a behavioral change, the magnitude of which needs to be measured at household level;

- the sociological approach and smart grid feedback can then be used to determine the non-economic sensitivities of the profiles observed.

– Which profiles can be taken into account? Of all the profiles that exist and can be observed, not all can be introduced into energy management (D-1 and real time), as measurable characteristics are needed over the timeframes of management (24 h at most). For example, while temperature can be measured, the corresponding sensitivity of a household in real time is difficult to integrate, except for the most technophile among them.

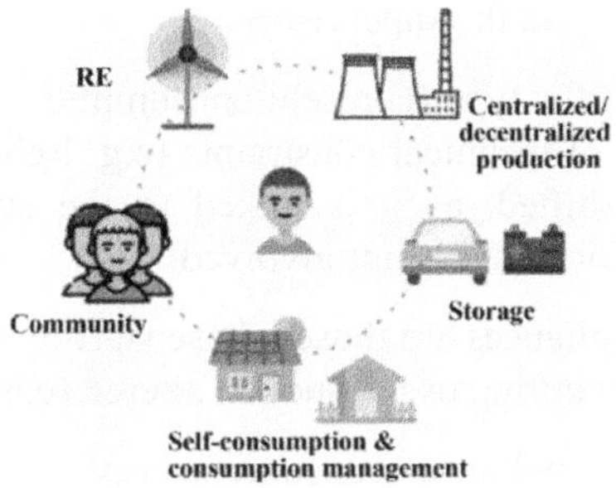

Figure 4.5. *Residential actor's environment*

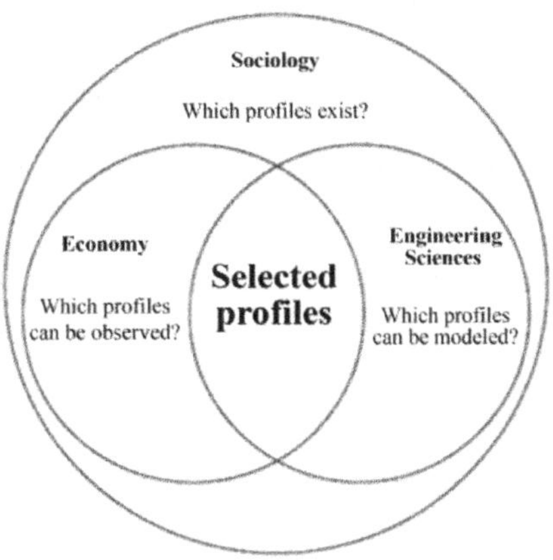

Figure 4.6. *Interdisciplinary coordination for profile research*

4.5.6.2. *Technical flexibility*

One of the major trends in energy use is the move towards electrification of many activities: transport, with electric vehicles, and increasing dematerialization, as evidenced by the growing number of electronic devices in the home. This growth leads to two parallel effects: an increase in electricity consumption, and greater possibilities for flexibility through appliances that allow their use to be programmed.

The number of electrical consumption points within households is therefore increasing, but not all appliances can technically take part in consumption shifts with the same amplitude. To represent these differences, appliances are classified into four main categories according to their contribution to the actor's flexibility, and then modeled separately in the supervisor:

– *Fixed*: appliances of this type represent consumption that does not take part in flexibility, either because of technical constraints (e.g. lighting or television, whose consumption cannot be shifted, as it is linked to the simultaneous use of their benefits), or because the consumer is not involved.

– *Cycle*: cycle-type appliances are those whose start-up time can be deferred, but whose power consumption during use cannot be altered (e.g. washing machines).

– *All-or-nothing*: these appliances consume energy only at their rated power but can be shifted during the day as long as they consume the energy required to perform their task (e.g. hot water tank).

– *Flexible*: the most flexible devices are those that are fully adaptable. Their power can be modified during use, and they can be shifted throughout the day, as long as they fulfill their task. What is more, the most flexible devices can also be used to feed energy back into the grid (e.g. batteries, electric vehicles).

From this initial segmentation, the devices involved in flexibility can be further differentiated into two sets:

– those whose time lag is visible to the user: that is, if their consumption is delayed, this will be physically perceived by the consumer. This category includes all cycle-type appliances (dishwashers, washing machines and tumble dryers), whose flexibility requires a particular effort on the part of the user;

– those whose time lag is invisible to the user: that is, those whose service to the consumer is not simultaneous with the extraction of power from the network. As long as this service is provided, the date (in the sense of "time of day") has no effect on user comfort. Shifting this date can therefore bring flexibility to the network. Typically, the two main devices are hot water tanks and EVs.

If devices can be categorized by flexibility, the actor's total technical flexibility may change with the environment in which it evolves, as well as its penchant for new technologies. As discussed in sections 4.5.6.3 and 4.5.7, a demand-side management program requires a more comprehensive approach than simply building a supervisor, by not only encouraging participation but also increasing the energy culture within the population concerned. Increased understanding of the issues at stake, and of the actions that can be taken to address them, will have an impact on flexibility at different timescales:

– In the short term: more active participation in controlling energy consumption, and therefore increased daily flexibility.

– In the medium term: the purchase of less energy-intensive or more controllable appliances, reducing consumption and increasing the possibility of offsetting it.

– In the long term: home improvements or the search for less energy-intensive, more energy-efficient housing, leading to lower consumption and greater flexibility for the network.

4.5.6.3. *Social flexibility*

While there are technical possibilities for shifting the consumption of certain uses, this shift must be accepted and implemented by the household's inhabitants. Although there may be an obvious financial or environmental advantage to doing so, an actual change in planned consumption is not always observed. Collective rhythms within a household have a major influence on the actual flexibility of a household's consumption. Certain practices, such as showering or cooking, are directly linked to professional rhythms (work, school), and can therefore only be taken into account with difficulty when considering the flexibility of a home. The gain from the occasional shift in these activities is small compared to the social cost (in the sense of discomfort) paid [MAT 18]: modifying these rhythms is therefore a matter of lifestyle and goes beyond this work. What is more, calling on flexibility relies in part on anticipating activities, so planning consumption entails a mental burden that acts as a brake on the actor's effective involvement.

The gap between the devices involved in the actor's flexibility, while it may be large in theory, will therefore be limited in practice:

– on the one hand, by the time slots that the user agrees to leave for shifting their appliances: either calculated by consumption analysis, declared ("between such and such a time") or specified when the appliance is launched ("over the next x hours" or "must be done before such and such a time");

– secondly, by the user's acceptance of their participation, particularly with regard to the treatment of their invisible charges. Indeed, while some consumers are not involved in network participation, they may accept that their invisible load is controlled by a third-party entity, since this has no impact on their comfort. Conversely, others may refuse to have their loads controlled at all. These different types of acceptance can have a major influence on the flexibility available to the network, and we will need to distinguish between them later on.

To understand the variety of social practices and their roots at different scales (micro-individual, micro-social or macro-social), a number of human and social sciences (psychology, psycho-sociology, sociology, economics) are called upon by ADEME [ADE 14]. The results show that each individual has their own rationality, and that two very important factors are *perceived utility* and the *behavior of those around them* (neighbors, neighborhood, etc.). Indeed, controlling consumption within a household is a complex matter, since energy expenditure depends on the user's relationship with their appliances, taking into account their perception of comfort, household configuration, standard of living and social context. Interaction options can therefore play on the *spirit of competition* or *cooperation*, or on the consumer's *sense of control* and *ownership of management*.

Although this flexibility of use within households can only be modified with difficulty within the framework of the development of an energy supervisor alone, as this requires measures that go beyond the simple network, it should be kept in mind for future development around these issues. It can, however, be steered by certain parameters, which we discuss in section 4.5.7.

4.5.7. *Parameters influencing flexibility*

4.5.7.1. *Discussion of factors influencing flexibility*

In addition to flexibility itself, there are many factors that can influence and direct it, and these need to be taken into account within the framework of an energy supervisor. Consumer segmentation studies have highlighted two main measurable parameters of involvement: cost and environment. These will be discussed later, but other parameters should also be discussed, because even though they cannot be taken into account directly by the supervisor, they must nevertheless be considered in its implementation to avoid the rejection of a technical solution that then seems imposed. Two initial points precisely illustrate the importance of taking consumers into account in the construction of a supervisor and emerge from existing demonstrations: the notions of control left to the actor and the sharing of costs/benefits, insofar as actors who were once simple consumers provide a service to the network and can become co-producers. Not taking them into account, or adding uncertainty in their regard, means running the risk of disengaging consumers from the network in question.

To understand the factors influencing consumer involvement, it is necessary to draw on the contribution of the human and social sciences, as the rationality leading to energy consumption depends not only on the individual, but also on the context in which they find themselves. We need to bear in mind that household energy behavior is the result of interaction between needs, on the one hand, and technical, legal and financial possibilities/constraints, on the other hand, as well as *beliefs* (knowledge, values, morals), *habits* and *culture*. So, the challenge for smart grids is not to focus on each individual actor, but to imagine and foresee the complex interactions that will result from this growing involvement. Energy consumption, at the household level, for example, is more a collective and social process than at the individual level [HAR 10].

The influence of feedback (within a household, between households or between households and suppliers, whose form is also to be studied) on daily social practices has also been studied. Consumption reductions of the same order as for a price increase between 11 and 20% were observed. It has been shown that the rebound effect can even be limited by appropriate non-economic measures. Similarly, [EK 10] shows that adapting the level of detail in information is an important factor in supporting behavioral change.

To study behavior, social psychology theories can be used, with a particular focus on social representations in communication, contractualization and the implementation of new network management models, to facilitate actor involvement. Regarding their application to smart grids, [BRO 14], for example, highlights the criteria of understanding, performance, reliability and ease of use as favoring involvement.

The implementation of an energy supervisor must therefore consider each of these aspects throughout its development, and once in place, rely on the two factors that can influence a consumer's use of flexibility (price and environmental impact), which we now discuss.

4.5.7.2. *Price*

The most obvious and widely used parameter with which to play with the various actor flexibilities is price: it is direct, meaningful for all actors, and easy to set up and modify (subject to good communication). In fact, in Germany, price is used by law as a factor influencing demand. To this end, there are four main pricing models (although many variations exist [MOS 15]):

– time of use pricing (TOU): principle of the French HP/HC tariff;

– critical peak pricing (CPP): principle behind the French EJP tariff (peak-shaving tariff), now known as the Tempo tariff;

– real-time pricing (RTP): energy price varying in real time according to the market (need for devices capable of managing consumption in real time);

– peak time rebates (PTRs) (also Peak Load Reduction Credits): based on the same principle as CPP, a baseline consumption is measured, and consumption downgrades (relative to the baseline) on peak days are rewarded.

A summary of various studies analyzing the impact of these models on peak reduction shows that the CPP model is the most effective and can achieve up to 30% peak reduction [NEW 10]. It is shown that the best results are obtained when loads are controlled automatically, but that the effectiveness of these solutions relies on the involvement of the actor in their definition, and on their freedom to refuse load shedding at any time. The point to bring consumers back into the discussion is: What would they choose if they were given the choice? Research seems to show that the answer leans towards less dynamic models. However, consumers are generally in favor of more automatic control systems because they are reluctant to change their habits themselves, and the trend is still towards "high comfort at low cost".

This question of gains is complex, because the form of the incentive is debatable: Should we reward good use or punish overruns and peaks in consumption? Rewarding could be misleading (shifting consumption is good for the network, but not consuming is even better for society), while punishing would be more effective [ADE 16]. The ability of households to manage their consumption must also be taken into account. This differs from one household to the next (depending on the technical, pricing, financial and cognitive context), which makes it impossible to set up a single pricing model to manage this flexibility in the best possible way [DLA 12, VEN 18], as this would represent an injustice for the most precarious consumers. Thus, the more the choice is towards dynamic (several price levels) and changing tariffs (price levels at different times depending on the state of the network), the more consumers need to be able to manage their consumption (automatically or remotely) [NIC 18].

4.5.7.3. *Environmental impact*

While the price paid is easily measurable, the impact of energy consumption on the environment and the perception of those involved in the residential sector make this a more complex issue. When energy is consumed on a daily basis, without external indicators or incentives, reducing environmental impact means reducing energy consumption. Here, we are interested in the possible shifts in appliances that contribute to the flexibility of actors, in a fixed-energy consumption pattern.

Environmental impact is an important factor that can help explain behaviors that cannot be understood by economic considerations alone. The complexity of this parameter in its influence on consumption lies in consumers' perception of environmental impact. [GUO 14] shows, for example, that participation in energy management in a smart grid is, in part, perceived as a duty in view of the environmental impact it represents.

Two aspects are therefore important for using flexibility to influence environmental impact:

– The form of feedback, because the form of feedback will lead to a new (non)understanding, and new uses that are difficult to define a priori. We need to think of feedback as personalized rather than generic [MIL 17], because efficiency also comes from the user's need and interest in the impact of their own consumption.

– The data sent: real environmental impact is a notion that is difficult to summarize in a single variable. However, too much complexity can lead to disinterest on the part of less involved consumers [MIL 17]. These data must show the added value possible for the user, while being apprehensible and accessible to them, to make a change of behavior in line with their desired principles.

Finally, it is difficult to use environmental impact as a basis for the corresponding flexibility, by defining its influence a priori. In addition to the information provided, users' perceptions and beliefs about low-impact solutions play an important role in the acceptance of a change in behavior [NAV 19].

4.6. Identification of residential actor profiles

4.6.1. *Introduction*

To understand existing profiles and flexibilities, we need to be able to find them in the population of the network under consideration. As the case study applied in this second part is prospective, no local data from a specific population can be used; nevertheless it is possible to discuss approaches for finding these profiles based on national observations.

4.6.2. *A microeconomic approach to price sensitivity*

4.6.2.1. *Introduction to microeconomics*

To find out how to determine price sensitivity and cut out profiles accordingly in a given population, we need to turn to economic approaches. In the specific case of the residential actor, microeconomics is highly appropriate, since it studies the behavior of economic agents, as opposed to macroeconomics, which is concerned with the major economic aggregates. For this work, the economic approach applied to the databases available for the French case is described in Figure 4.7, explained throughout this section. To understand this approach, a brief history of the foundations of microeconomics is in order.

The basis of economics is that each individual acts rationally. Starting from this premise, Adam Smith in the 18th century[4] presupposed that individual interest is the foundation on which all economic interaction rests. Breaking away from Smith's purely material concept of wealth, Jeremy Bentham initiated the theory of utilitarianism in 1780:

> By the "principle of utility" is meant the principle that approves or disapproves of every action according to the tendency it appears to have to increase or lessen – i.e. to promote or oppose –the happiness of the person or group whose interest is in question [BEN 80, Chapter 1].

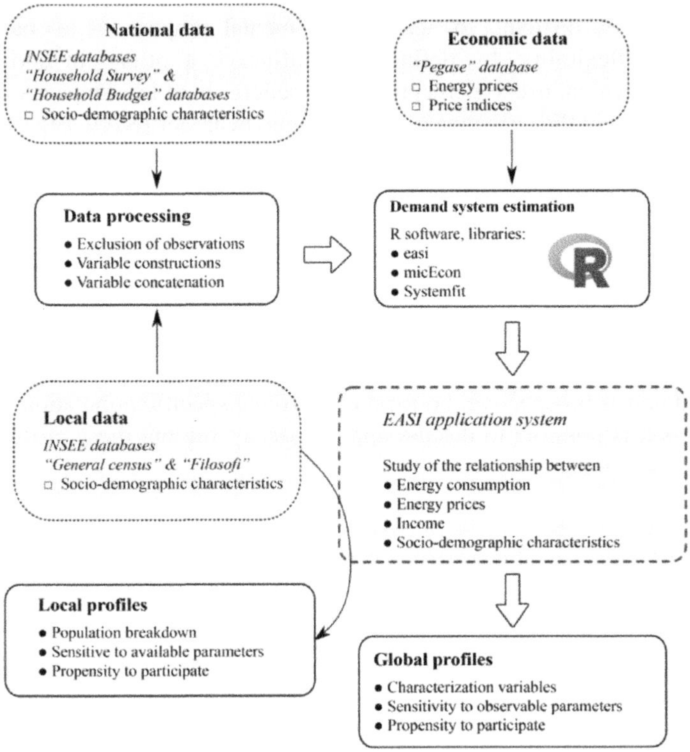

Figure 4.7. *Summary of the microeconomic approach to researching from French databases*[5]

4 See: *An Inquiry into the Nature and Causes of the Wealth of Nations*, 1776.

5 INSEE: *Institut national de la statistique et des études économiques*, the national statistics bureau of France; the French Institute Pégase: monthly energy database; EASI: Exact Affine Stone Index; Filosofi: *fichier localisé social et fiscal*, localized social and tax file.

In other words, utility is the propensity of a thing (object, service, etc.) to produce benefit/profit/pleasure or avoid harm/suffering for those whose interest is at stake. In developing this notion, the economist Stanley Jevons redefined wealth in 1871[6] by no longer considering that the individual consumes a good for its own sake, but for the satisfaction that its consumption generates. Utility can thus be seen as a scale for ranking possible consumption choices: it is not so much its absolute value that is important, but the relative value of the different goods considered. For the development of economic theory, and to account for the phenomenon of satiability, utility is subject to the first law of demand, or the law of the "satiability of needs", embodied by the concept of diminishing marginal utility: the ratio of the variation in utility to the variation in the number of goods is an always positive (non-saturation of needs) and diminishing quantity (satiability).

The microeconomic approach, both in its philosophy and in its applied mathematics, is therefore based on this notion of utility, and the hypothesis of rationality translates into the search for maximum utility under budget constraints. By focusing on consumers, this science studies the behavior of individuals, in particular their consumption and what affects it.

4.6.2.2. *Demand model and EASI*

For the economic approach, individuals' choices to consume a quantity q_x of a good X are affected by their income R, the price of good p_x, the price of other goods p_y, and, secondly, socio-demographic characteristics. Studying their behavior therefore requires the development of tools to describe the relationships between these factors: demand functions. There are three types of demand functions:

– simple: $q_x = f(p_x)$, linking the consumption of a good to its price;

– Engel: $q_x = f(R)$, linking the consumption of a good to the individual's income;

– Marshalian: $q_x = f(p_x, p_y, R)$, linking the consumption of a good to the set of prices, and to the individual's income.

The successful application of these functions provides a wealth of information useful for economic analysis, such as the definition of elasticities, types of good, pricing and the study of market equilibria. For a basket of goods, the set of demand functions for each good is called a demand system.

The most complete and interesting of these are the Marshalian demand functions, from which the others can be deduced. The aim of economic analysis is therefore to find these functions by regression, which can then take any form and cannot be presupposed. However, these functions and demand systems must satisfy a number

---

6 See: *Theory of Political Economy*, 1871.

of constraints, not only to remain consistent with reality and the hypothesis of consumer rationality, but also to enable them to be calculated and exploited [PEN 09, ROB 99]:

– Additivity: given two goods x and y, with q_x (p_x , p_y , R) and q_y (p_y , p_x , R), the additivity constraint imposes:

$$p_x \cdot q_x (p_x , p_y , R) + p_y \cdot q_y (p_y , p_x , R) = R \quad [4.1]$$

– Zero-degree homogeneity: this constraint states that if prices and income increase by the same proportion α, then consumption does not change:

$$q_x (\alpha.p_x , \alpha.p_y , \alpha.R) = \alpha^0 . q_x (p_x , p_y , R) = q_x (p_x , p_y , R) \quad [4.2]$$

This constraint reflects the assumption of the absence of monetary illusion, which assumes that an individual reasons in real values (inflation-adjusted values) and not in nominal values[7].

– Slutzky condition: letting b be the minimum budget required to achieve a level of utility u in the consumption of J goods, the function satisfying [4.3] is called the C cost function:

$$b = C(p_1,..., p_J , u) \quad [4.3]$$

– The Slutzky condition imposes the concavity of this function with respect to prices, valid for most goods and based on the assumption of consumer rationality mentioned above.

– Symmetry property: the relative variation between q_x and p_y is the same as that between q_y and p_x.

– The mathematical functions used must be as unrestricted as possible in their form, in order to translate observations as flexibly as possible.

– These functions must be easy to estimate using iterative linear estimation methods.

The idea is that there is a statistical link between the propensity to participate in an energy-saving program and the impact of energy consumption on household expenditure. An effective measure of this link is provided by the relationship between energy expenditure (or its budget coefficient) and re-income, known as the Engel curve. Measuring income elasticities thus enables us to define categories according to their propensity to be accepted.

7 An example of monetary illusion is the belief that wages increase when their nominal value rises, while prices as a whole rise simultaneously.

However, the condition for using Engel curves to measure this inclination to accept a change in behavior is the possibility of defining curves in the relationship between budget coefficient and income. Classical methods of estimation, such as the almost ideal demand system (AIDS) [DEA 80], are characterized by the definition of linear Engel curves, and therefore do not allow the definition of contrasts in situations.

Among existing demand systems meeting all these conditions, the approach chosen is that of the Exact Affine Stone Index (EASI) introduced by [LEW 08, PEN 09]. The expression of the budget share w_j for a good j among J−1 other goods is given by the relation in equation [4.4]. The interest of this system in particular is that it enables nonlinear Engel curves to be defined, based on the calculation of budget shares allocated to each type of energy (five for this study: electricity, fuel oil, wood, gas and liquid petroleum gas (LPG)):

$$w_j = \sum_{r=1}^{R} b_{jr} y^r + \sum_{t=1}^{T} g_{jt} z_t + \sum_{k=1}^{J} a_{jkt} \ln(p_k) + \sum_{k=1}^{J} b_{jk} \ln(p_k) + \epsilon_j \quad [4.4]$$

where y is the total real expenditure (income), p^k is the price of each good, z_t is one of the T demographic characteristics, ε_j is the error term interpreted as the heterogeneity of unobservable preferences, and R is the maximum degree in $\underline{y}$ (determined empirically).

The terms in equation [4.4], respectively, represent:

– income elasticity of demand: ratio of the growth rate of quantities consumed to the growth rate of income;

– dependence on demographic variables;

– direct and cross-price elasticities: ratio of the growth rate of quantities consumed of a good to the growth rate of the price of the good (direct cases), or of other goods (cross cases);

– the interaction between prices and income;

– unobservable preferences.

4.6.2.3. *Description of case study data*

In this study, the analysis of residential consumption is based on data from the National Housing Survey[8] conducted by The French National Institute for Statistics and Economic Studies (INSEE) [INS 13]. This institute, which is responsible for producing and publishing official statistics in France and conducted the major surveys between June 2013 and June 2014, provides highly detailed information on

8 Translation of: Enquête nationale logément (ENL).

household composition, household characteristics, spending, income and other demographic variables. A total of 33,670 households were observed with regard to 1,500 variables for this survey. In particular, this survey provides information on energy-related expenditure, such as electricity, fuel oil, wood, gas and LPG.

Category	Ave.	Med.	Std. dev.	Variable
Year of construction (binary):	-	-	-	z_1
Before 1990	0.801	-	0.400	-
1990 and after	0.199	-	0.400	-
Number of children per household:	0.60	0.00	1.02	-
≤ 6 yrs	0.21	0.00	0.54	z_2
7–18 yrs	0.39	0.00	0.79	z_3
Age of reference person	53.98	53.00	16.82	z_4
Binaries:	-	-	-	-
< 25 yrs	0.018	-	0.134	-
25–29 yrs	0.051	-	0.220	-
30–34 yrs	0.073	-	0.260	-
35–39 yrs	0.081	-	0.273	-
40–44 yrs	0.100	-	0.300	-
45–49 yrs	0.099	-	0.299	-
50–54 yrs	0.101	-	0.302	-
55–59 yrs	0.100	-	0.300	-
60–64 yrs	0.095	-	0.294	-
65–69 yrs	0.083	-	0.276	-
70–74 yrs	0.114	-	0.318	-
75–79 yrs	0.056	-	0.229	-
≥ 80 yrs	0.083	-	0.275	-
Number of people per household	2.42	2.00	1.37	z_5
Socio-professional category:	-	-	-	-
Farmer operators	0.009	-	0.095	z_6
Craftsmen, shopkeepers, company directors, self-employed professionals	0.048	-	0.213	z_7
Managers and intellectual professions, academics, except practitioners	0.118	-	0.323	z_8

Category	Avg.	Med.	Std. dev.	Variable
Mid-level professions	0.146	-	0.353	z_9
Employees	0.143	-	0.350	z_{10}
Workers	0.164	-	0.370	z_{11}
Retirees	0.327	-	0.469	z_{12}
Unemployed	0.039	-	0.193	z_{13}
Not specified	0.007	-	0.082	z_{14}

Table 4.3. *Selected household characteristics (26,225 households surveyed between June 2013 and June 2014)*

Category	Avg.	Med.	Std. dev.	Variable
Actual household income	40,299	32,180	47,474	-
Proportion of budget for electricity	0.028	0.019	0.030	p_{elec}
Wood as a portion of budget	0.002	0.000	0.007	p_{wood}
Heating oil as a portion of budget	0.006	0.000	0.023	p_{fuel}
Gas as a portion of budget	0.011	0.000	0.023	p_{gas}
LPG as a portion of budget	0.002	0.000	0.10	p_{lpg}
Share of budget for other uses	0.952	0.964	0.044	p_{other}

Table 4.4. *Household economic variables (26,225 households observed between June 2013 and June 2014)*

From this raw database, 26,225 observations[9] are exploitable. The database gives access to a set of variables which require preliminary processing to retain the most relevant and eliminate those causing multi-collinearities. The second step is to eliminate variables that do not respect the additivity constraint required by economic analysis. As the aim of the study is first to be able to apply the approach locally, on accessible data for a population in a given perimeter, the reference for variables usable for the model is taken from the 2013 census data, and those of the French 2013 Filosofi (localized social and tax file).

The final variables and their statistics finally used for the analysis are presented in Table 4.3, with the economic variables concerning household income and expenditure summarized in Table 4.4.

9 An observation corresponds to a household's response to all the variables.

4.6.2.4. *Preliminary calculations*

From the nominal total expenditure x, the stone index P^* introduced by [STO 53] is used to control the endogeneity of the system and calculate the real total expenditure y as defined in the EASI demand system, and as described by the following equation:

$$y = \ln(x) - \ln(P^*) \quad [4.5]$$

Stone's index is calculated using Equation [4.6]: for a consumer of J goods facing prices p_k for each good k, their consumption choices are then described by the budget shares w_k:

$$\ln(P^*) = \sum_k^J w_k . \ln(p_k) \quad [4.6]$$

In order to correct for the effects of inflation for the observation period (June 2013 to June 2014), the Consumer Price Index (CPI) is obtained from INSEE statistics [INS 15]. Price indices for each type of energy (relative to the CPI) based on LHV[10] values are then calculated using the French Pégase database (oil, electricity, gas and other energy statistics) made available by the Ministry of Ecological Transition and Solidarity [MIN 17]. Figure 4.8 shows the evolution of the monthly values of the five energy price indices alongside the Consumer Price Index.

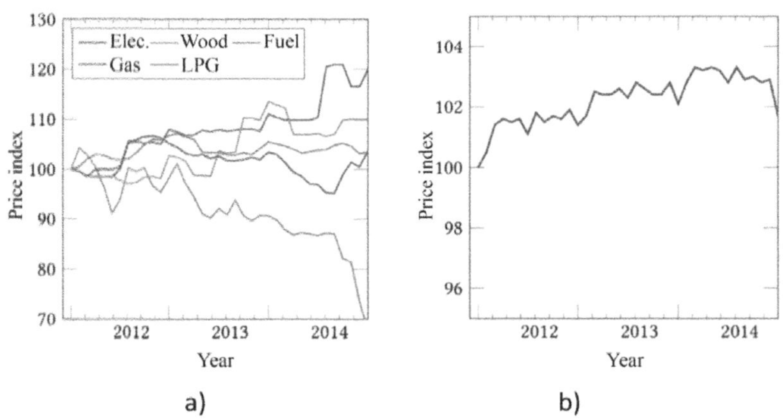

Figure 4.8. *a) Deflated price index for electricity, wood, heating oil, gas and LPG; b) Consumer Price Index for 2012–2014, normalized to 100 in January 2012. For a color version of this figure, see www.iste.co.uk/robyns/smartgrids.zip*

10 Lower heating value.

4.6.2.5. *Results*

This demand system is then estimated with the R programming language, using the triple least squares method. The resulting Engel curve of the EASI model for electricity expenditure is shown in Figure 4.8 (with the median values of the budget coefficients and the smoothed Engel curve), and the coefficients of the corresponding demand system are collected in Table 4.5. This curve provides valuable information on the segmentation of the population and its economic sensitivity, and therefore on its propensity to take part in a demand control program. Three profiles can be seen in Figure 4.9:

– *Precarious*: a strong decreasing income elasticity is observed on the lowest incomes, up to 23%. This sign of fuel poverty shows the extent to which energy expenditure by these households is a point of tension, which should not be exacerbated by costly or discriminatory measures in the definition of new economic models. The elasticity observed here is more of a constraint that should not be relied upon in network management.

– *Strong interest*: the second part of the curve shows an initially increasing positive income elasticity, then decreasing after the inflection point to 70%. There is therefore a strong economic interest in participating in an energy management program, in order to control the rising weight of energy expenditure within the household.

– *Disinterested*: above 70%, income elasticity decreases sharply, and the weight of energy expenditure collapses. The economic interest here is therefore very weak, and only the environmental aspect can be a determining factor acting on flexibility. Investment in RE is easier for the better-off, as financial considerations no longer act as a brake, as shown, for example, in [SAR 13].

Apart from profiles for power grid management, it is interesting to note the negative sign of the cross-price elasticities in Table 4.5 of the coefficients, showing that the different types of energies are complementary and non-substitutable goods. From a political and economic point of view, the most effective actions are those that take energy into account as a whole: a good economic policy is therefore one that gives priority to insulating and increasing the energy efficiency of buildings, in order to reduce the overall energy consumption, rather than seeking to substitute one type of energy for another.

This initial study enables us to determine the profiles that exist in the French population, along with their key variables. To find these profiles locally, and to apply this work to a precise perimeter, it is necessary to be able to observe the selected characteristics. To this end, local databases with the desired data are available on request, at the scale of an urban unit using the general population census (RGP) [INS 18a] and the localized social and tax file (Filosofi) [INS 18b], or at a smaller scale by making a specific request for the desired segmentation.

	Elec.	Wood	Fuel	Gas	LPG
ϵ	0.0587 ***	0.0024 ***	0.0079 ***	0.0135 ***	0.0038 ***
ln(y)	−0.0424 ***	−0.0009 ***	−0.0016 **	−0.0124 ***	−0.0027 ***
$\ln(y)^2$	0.0117 ***	−0.0001	−0.0017 ***	0.0005	0.0010 ***
$\ln(y)^3$	−0.0010 ***	0.0001	0.0005 **	0.0009 ***	−0.0003 ***
z_1	−0.0020 ***	−0.0005 ***	0.0003	−0.0006	−0.0002
z_2	−0.0018 ***	−0.0002 **	0.0002	−0.0001	−0.0002
z_3	0.0049 ***	0.0002	−0.0024 ***	−0.0026 ***	0.0000
z_4	−0.0007 ***	0.0001 ***	0.0011 ***	0.0009 ***	0.0002 ***
z_5	0.0032 ***	0.0005 ***	0.0003	0.0020 ***	0.0004 ***
z_6	0.0016	0.0025 ***	0.0105 ***	−0.0087 ***	0.0045 ***
z_7	0.0033 ***	0.0004	0.0026 ***	0.0013	0.0006 *
z_8	−0.0023 **	−0.0001	0.0003	0.0018 **	0.0004
z_9	−0.0016 *	0.0000	0.0003	0.0010	0.0002
z_9	−0.0029 ***	−0.0005 **	−0.0015 *	0.0004	−0.0002
z_{10}	−0.0017 **	−0.0001	−0.0010	−0.0003	0.0002
z_{11}	−0.0006	−0.0006 *	−0.0020 *	0.0035 ***	−0.0001
z_{12}	0.0003	0.0008	−0.0008	−0.0010	−0.0008
p_{elec}	−0.0742 *	−0.0183 ***	−0.0259 *	−0.0111	−0.0414 **
p_{wood}	−0.0183 ***	−0.0071 ***	−0.0049	0.0007	−0.0059 *
p_{fuel}	−0.0259 *	−0.0049	0.0249 **	0.0033	−0.0082
p_{gas}	−0.0111	0.0007	0.0033	0.0618	0.0121
p_{lpg}	−0.0414 **	−0.0059 *	−0.0082	0.0121	−0.0053

Statistical significance: no sign = $p < 1$; * = $p < 0.05$; ** = $p < 0.01$; *** = $p < 0.001$; standard deviation not shown for clarity.

Table 4.5. *Coefficients of the energy demand system*

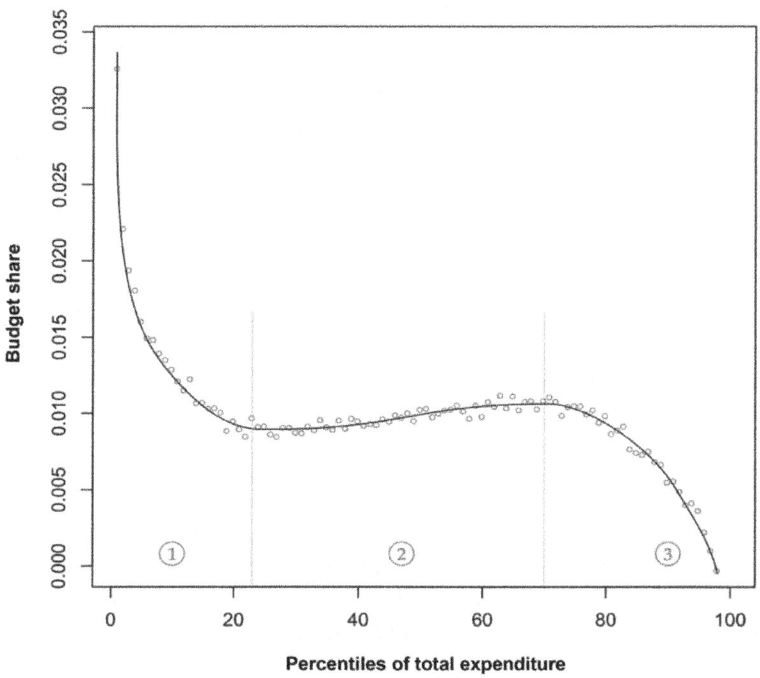

Figure 4.9. *Engel curve for electricity consumption in France: share of electricity expenditure according to percentiles of total expenditure (median values and smoothed curve)*

4.6.3. *A sociological approach to environmental awareness*

In a context where renewable energy consumption entails higher energy costs, the populations with the highest propensity to get involved in consumption shifts will be those for whom the economic impact has the least effect. These populations can initially be addressed using the economic approach presented above.

Knowing the existence of profiles and their corresponding factors, finding them in a given population or territory can be executed in several ways:

– by surveying the target population:

- either through a questionnaire directly linked to energy use and possible consumer involvement;

- or by questioning values, which can then be used for other projects with environmental aims, to gain a broader understanding of the population in question;

– through individual household declarations, for example, by recording preferences directly in the local individual consumption management system;

– through statistical analysis of consumption and socio-demographic variables, as for the economic aspect, but this requires the availability and accessibility of relevant data on pro-environmental behavior, which often means carrying out a data survey.

In order to do this, it is important to think in terms of the group and not the individual, as the behavior of the social group in which a household is located has a strong impact on it [SAN 17, ZHA 18]. The study of the group clearly shows the interest of studying the local population and not of defining a global energy management, hence the interest of a methodology rather than turnkey energy management.

As application to a real-life case study is not possible here, defining the environmental sensitivities of a specific population with the aim of introducing it into an energy management system cannot be studied or achieved. As discussed in section 4.5.7, however, it is possible to define the context in which to encourage involvement by playing on this sensitivity, which can be summarized as follows: increasing and fostering the energy culture and reflecting on and personalizing the feedback sent back to households.

In view of the price and environmental sensitivities that encourage or limit flexibility on the part of consumers, and which can be used by the supervisor, it is now possible to define the involvement profiles that need to be taken into account in network supervision. It should be borne in mind, however, that the distribution and definition of these profiles can be done locally, and subsequently adjusted.

4.7. Profiles of selected residential actors

The five residential actor profiles selected for this study are described below.

4.7.1. *Economical*

For these profiles, the main influencing factor is price, and the consumer's objective is to reduce their electricity bill. To achieve this objective and take advantage of price incentives, their flexibility is variable but rather high, given that low flexibility is associated with indifferent or disengaged profiles.

4.7.2. *Eco-sensitive*

Their main objective is to reduce their environmental impact. For the purposes of this study, the ratio of renewable energy consumed is considered to be the main factor in energy management. Their flexibility is also variable, but with a strong tendency, given that low flexibility is associated with indifferent or disengaged profiles.

4.7.3. *Technophiles*

The objective can be as much to reduce the environmental impact of these profiles as to cut costs. The interest shown in new technologies and the possibilities they offer place them in the highly flexible categories, thanks to the high penetration rate of programmable devices and dedicated applications. There is therefore a high level of involvement for this type of profile.

4.7.4. *Indifferent – moderate opportunists*

These profiles have no particular objective in mind. If an environmental or economic preference can be observed, the commitment to achieve the corresponding objective is weak. On the other hand, control is accepted for devices whose consumption is invisible to the user (e.g. hot water tank). Their flexibility therefore depends on the devices they own, which may or may not allow such flexibility. This type of actor is sensitive, for example, to set-and-forget actions, where an action (accepting a commitment or setting an application) has to be performed once, but then requires no further commitment.

4.7.5. *Disengaged*

These profiles are not involved in energy management, and do not allow for any shifting or control of their consumption. They are therefore insensitive to external signals, either through indifference, ignorance or simply rejection. Whatever the values of their price sensitivity and environmental impact, their involvement is nil, and their flexibility is very low. Their overall consumption can be considered fixed.

4.7.6. *Discussions*

To conclude, the segmentation of these profiles is presented in Table 4.6. This arbitrary, evenly distributed breakdown of the population makes it possible, above all, to present the other stages of the methodology and the interest in each profile group. But the underlying assumption is that any real profile can be a mixture of these. It will therefore be interesting in the next chapters to discuss the integration of these specific profiles into the supervisor. In addition to this, other population segmentations will be tested in the supervisor simulations in Chapter 5.

Profiles	Population	Cost	Envir.	Flex.
Economical	28	High	Low	High
Eco-sensitive	28	Low	High	High
Technophile	28	Non-comital	Non-comital	High
Indifferent	28	Non-comital	Non-comital	Low except for invisible load
Disengaged	28	Non-comital	Non-comital	Low

Table 4.6. *Selected profile for a population of 140 modeled households*

While the three axes of sensitivity selected cover a wide range of households to be taken into account, the representation of a particular household by one of these profiles also depends on the context. For example, a household in fuel poverty will not be represented by the same profile (as defined above), depending on the network's business model. In the case of a highly price-constrained network with no special tariffs for the most vulnerable, it could be considered thrifty, reacting strongly to possible cost variations. On the other hand, it could be seen as a disengaged consumer in an economic model where basic energy is provided at a fixed price for each actor, and only overruns are subject to variable costs.

Finally, as already mentioned, demand response programs do not lead to a reduction in energy, but to a better distribution of power over time [HOL 15]. The integration of these profiles will be carried out at constant energy, an assumption which will be recalled and discussed in the following chapters.

Objectives	Satisfy energy demand Increase your satisfaction with your sensitivities (lower your bill and/or reduce your environmental impact)
Constraints	Technical and social constraints Fluctuating energy prices Availability of renewable energies (actual production)
Means of action	Offsets (in power and time) of devices made available for flexibility

Table 4.7. *Description of objectives, constraints and means of action of a residential actor profile*

Each profile can be summarized in three dimensions: objectives, means of actions and associated constraints, as illustrated in Table 4.7.

4.8. Conclusion

The first stage of the methodology (see section 4.3.2) was applied to the search for residential user profiles. The preliminary literature review showed the complexity of the problem, as well as the multiple possible approaches that were subsequently developed. Firstly, it was necessary to characterize the flexibility of residential consumers, both technically and socially, since behavior at the household level is a complex process that cannot be approached from a single point of view. The factors behind this flexibility were then observed, enabling involvement profiles to be defined. Finally, possible approaches to finding these profiles in a given population were discussed, based in particular on the microeconomic approach.

The results of this analysis included the five involvement profiles that were retained through three criteria, at the borderline between existing, observable and technically incorporable objectives in an energy supervisor. These three criteria were sensitivity to costs, sensitivity to environmental impact and sensitivity to consumption shifts (or flexibility). The assumption of this approach is that any other observable profile that can be technically incorporated into the supervisor is a combination of these. One of the aims of this work is to show how they can be taken into account.

In the event of application to a real case, however, a feedback loop (learning loop) must be considered to perfect this segmentation. Indeed, other factors influence the flexibility of these actors; however, it is not possible to take them all into account in the supervisor, particularly those relating to social representations or interactions within the household or community.

What is more, while it is possible to find the profiles with the greatest propensity to commit to an energy management system, the approach cannot be reduced to this, because as we have seen throughout this chapter, a feeling of injustice or isolation can lead to outright disengagement on the part of the actor. The approach, while technical (with supervision focusing on the physical aspects of consumption), needs to be more global than simply looking for willing profiles to support in energy management. A real company project, or at least a community project, must be part of the network environment in question.

Before moving on to the next chapter, it is important to bear in mind that there are two possible paths to this new flexibility available for managing network balance, when we talk about "taking these behaviors into account". Indeed, defining whether behaviors are *driven* or *informed* is very important because, as discussed above, the control left to the actor is a factor playing on their involvement. In the end, these two visions are the same as those found in the scenarios studied by ADEME in Chapter 1: a bias towards information, leaving the actors to decide for themselves and in full knowledge of the facts, is in line with scenario S2 of "Territorial Cooperation". Everything can be done in consultation, in shared governance, which can lead to more efficient management if each actor accepts being linked to the others in their energy consumption/production. On the other side of the spectrum, the *fully controlled* scenario is closer to ADEME's S3 "Green Technologies" scenario, in which all energy management is carried out automatically by smart technologies, leaving it up to technology to do everything for the actors, taking into account their needs and those of the network, and seeking to increase energy efficiency and the integration of renewable sources. Of course, there is an in-between situation, and the approach may differ according to the actor or the situation under consideration (technological, economic, etc.), and be contractualized where necessary, as is the case with current load shedding schemes.

Following on from this reflection on consumer profiles, Chapter 5 develops the energy supervisor itself, with, firstly, the optimization of consumption the day before for the next day, known as D-1, and, secondly, the real-time management of energy, necessary for adapting to any contingencies encountered. Chapter 5 introduces the search for balance in overnight interactions between the objectives and constraints of the profiles described in this chapter and the needs of the network.

4.9. Acknowledgments

The authors would like to thank the Hauts-de-France region for its contribution to the funding of this research, and the economist Jean-Pierre Marissal for his invaluable help with the economic modeling.

4.10. References

[ACC 10] ACCENTURE, Understanding consumer preferences in energy efficiency, Report, 2010.

[ACC 16] ACCOUCHE O., "Évolution du gisement de flexibilité : une approche options réelles", *Symposium de Genie Électrique*, Grenoble, 2016.

[ADE 14] ADEME, *Changer les comportements, faire évoluer les pratiques sociales vers plus de durabilité*, ADEME Éditions, Paris, 2014.

[ADE 16] ADEME, *Changer les comportements, faire évoluer les pratiques sociales vers plus de durabilité*, ADEME Éditions, Paris, 2016.

[AJM 13] AJMI A.N., EL MONTASSER G., NGUYEN D.K., "Testing the relationships between energy consumption and income in G7 countries with nonlinear causality tests", *Economic Modelling*, vol. 35, pp. 126–133, 2013.

[BAL 18] BALDWIN E., ROUNTREE V., JOCK J., "Distributed resources and distributed governance: Actor participation in demand side management governance", *Energy Research & Social Science*, vol. 39, pp. 37–45, doi: 10.1016/j.erss.2017.10.013, 2018.

[BEN 80] BENTHAM J., *An Introduction to the Principles of Morals and Legislation*, Ecolib, Paris, 1780.

[BEN 12] BENEDIKT R., REICHHART P., KRANZ J. et al., "The role of smart metering and decentralized electricity storage for smart grids: The importance of positive externalities", *Energy Policy*, vol. 50, pp. 486–495, doi: 10.1016/j.enpol.2012.07.047, 2012.

[BER 11] BERNADETTE S., BRUNNER T.A., SIEGRIST M., "Who puts the most energy into energy conservation? A segmentation of energy consumers based on energy-related behavioral characteristics", *Energy Policy*, vol. 39, pp. 8137–8152, doi: 10.1016/j.enpol.2011.10.008, 2011.

[BER 16] BERTHOLON M., KEROUEDAN A.F., REGNER M., Systèmes électriques intelligents : premiers résultats des démonstrateurs, Report, Open Energies, 2016.

[BOR 11] BORGES C.E., PENVA Y.K., FERNANDEZ I., "Optimal combined short-term building load forecasting", *Proceedings Innovative Smart Grid Technologies Asia*, 2011.

[BRO 14] BROMAN TOFT M., SCHUITEMA G., THØGERSEN J., "Responsible technology acceptance: Model development and application to consumer acceptance of smart grid technology", *Applied Energy*, vol. 134, pp. 392–400, doi: 10.1016/j.apenergy.2014.08.048, 2014.

[CAY 11] CAYRE E., ALLIBE B., LAURENT M.-H. et al., "There are people in the house! How the results of purely technical analysis of residential energy consumption are misleading for energy policies", *European Council for an Energy Efficient Economy (ECEEE) 2011 Summer Study*, pp. 1675–1683, 2011.

[DEA 80] DEATON A., MUELLBAUER J., "An almost ideal demand system", *The American Economic Review*, vol. 70, pp. 312–326, 1980.

[DIN 10] DING L., LV J., LI X. et al., "Support vector regression and ant colony optimization for HVAC cooling load prediction", *2010 International Symposium on Computer Communication Control and Automation (3CA),* Tainan, pp. 537–541, 2010.

[DLA 12] DLAMINI N.G., CROMIERES F., "Implementing peak load reduction algorithms for household electrical appliances", *Energy Policy*, pp. 280–290, 2012.

[DUR 19] DURILLON B., ModAICSS : Modélisation de l'acceptation et de l'implication de producteurs et consommateurs d'énergie électrique en vue d'une stratégie de supervision énergétique multi-acteurs, PhD thesis, University of Lille 1, 2019.

[DUR 20] DURILLON B., SALOMEZ F., DAVIGNY A. et al., "Consumers' sensitivities and preferences modelling and integration in a decentralised two levels energy supervisor", *Mathematics and Computers in Simulation*, doi: 10.1016/j.matcom.2020.02.003, 2020.

[DUR 21] DURILLON B., DAVIGNY A., KAZMIERCZAK S. et al., "Decentralized neighbourhood energy management considering residential profiles and welfare for grid load smoothing", *Sustainable Cities and Society*, vol. 63, p. 102464, doi: 10.1016/j.scs.2020.102464, 2021.

[DÜT 13] DÜTSCHKE E., PAETZ A.G., "Dynamic electricity pricing – Which programs do consumers prefer?", *Energy Policy*, vol. 59, pp. 226–234, 2013.

[ECU 17] E-CUBE STRATEGY CONSULTANTS, CEREN, ADEME, L'effacement de consommation électrique en France, Report, 2017.

[EK 10] EK K., SÖDERHOLM PATRIK P., "The devil is in the details: Household electricity saving behavior and the role of information", *Energy Policy*, vol. 38, no. 3, doi: 10.1016/j.enpol.2009.11.041, 2010.

[ENE 16] ENEDIS, Nice Grid, un quartier solaire intelligent, GRID4EU demonstrator, available at: http://www.nicegrid.fr, 2016.

[FRA 13] FRANKEL D., HECK S., TAI H., *Using a Consumer-Segmentation Approach to Make Energy-Efficiency Gains in the Residential Market*, McKinsey & Company, London, 2013.

[FRI 16] FRIIS F., CHRISTENSEN T.H., "The challenge of time shifting energy demand practices: Insights from Denmark", *Energy Research and Social Science*, vol. 19, pp. 124–133, doi: 10.1016/j.scs.2016.08.016, 2016.

[GUO 14] GUO X., LIU H., MAO X. et al., "Willingness to pay for renewable electricity: A contingent valuation study in Beijing , China", *Energy Policy*, vol. 68, pp. 340–347, doi: 10.1016/j.enpol.2013.11.032, 2014.

[HAI 16] HAIDER H.T., SEE O.H., ELMENREICH W., "A review of residential demand response of smart grid", *Renewable and Sustainable Energy Reviews*, vol. 59, pp. 166–178, 2016.

[HAR 10] HARGREAVES T., NYE M., BURGESS J., "Making energy visible: A qualitative field study of how householders interact with feedback from smart energy monitors", *Energy Policy*, vol. 38, no. 10, doi: 10.1016/j.enpol.2010.05.068, 2010.

[HOL 15] HOLLADAY J.S., PRICE M.K., WANAMAKER M., "The perverse impact of calling for energy conservation", *Journal of Economic Behavior & Organization*, vol. 110, pp. 1–18, 2015.

[INS 13] INSEE, ADISP, Enquête logement 2013, Report, 2013.

[INS 15] INSEE, Indice des prix à la consommation – base 2015 – ensemble des ménages – france métropolitaine, Report, 2015.

[INS 18a] INSEE, Recensement de la population – fichier détail, Report, 2018.

[INS 18b] INSEE, Dispositif fichier localisé social et fiscal (FiLoSoFi), Report, 2018.

[LEW 08] LEWBEL A., PENDAKUR K., "Tricks with Hicks: The EASI demand system", *American Economic Review*, vol. 99, pp. 827–863, 2008.

[LIN 13] LIN C.C., YANG C.H., SHYUA J.Z., "A comparison of innovation policy in the smart grid industry across the Pacific: China and the USA", *Energy Policy*, vol. 57, pp. 119–132, 2013.

[LOS 13] LO SCHIAVO L., DELFANTI M., FUMAGALLI E. et al., "Changing the regulation for regulating the change: Innovation-driven regulatory developments for smart grids, smart metering and e-mobility in Italy", *Energy Policy*, vol. 57, pp. 506–517, 2013.

[MA 15] MA C., ROGERS A., KRAGT M. et al., "Consumers' willingness to pay for renewable energy : A meta-regression analysis", *Resource and Energy Economics*, vol. 42, pp. 93–109, doi: 10.1016/j.reseneeco.2015.07.003, 2015.

[MA 18] MA G., LIN J., LI N., "Longitudinal assessment of the behavior-changing effect of app-based eco-feedback in residential buildings", *Energy and Buildings*, vol. 159, pp. 486–494, doi: 10.1016/j.enbuild.2017.11.019, 2018.

[MAT 18] MATTLET B., Potential benefits of load flexibility: A focus on the (future) Belgian distribution system, PhD thesis, Université Libre de Bruxelles, Brussels, 2018.

[MIL 17] MILLER W., SENADEERA M., "Social transition from energy consumers to prosumers: Rethinking the purpose and functionality of eco-feedback technologies", *Sustainable Cities and Society*, vol. 35, pp. 615–625, doi: 10.1016/j.scs.2017.09.009, 2017.

[MIN 17] MINISTERE DE LA TRANSITION ECOLOGIQUE ET SOLIDAIRE, Base pégase, available at: https://www.statistiques.developpement-durable.gouv.fr/donnees-mensuelles-de-lenergie, 2017.

[MOS 15] MOSER S., "Optimisation of small customer network tariffs for smart grids", *8th International Conference on Energy Efficiency in Domestic Appliances and Lighting*, 2015.

[NAV 19] NAVRATIL J., PICHA K., BUCHECKER M. et al., "Visitors' preferences of renewable energy options in "green" hotels", *Renewable Energy*, vol. 138, pp. 1065–1077, doi: 10.1016/j.renene.2019.02.043, 2019.

[NEW 10] NEWSHAM G.R., BOWKER B.G., "The effect of utility time-varying pricing and load control strategies on residential summer peak electricity use: A review", *Energy Policy*, vol. 38, pp. 3289–3296, 2010.

[NIC 18] NICOLSON M.L., FELL M.J., HUEBNER G.M., "Consumer demand for time of use electricity tariffs: A systematized review of the empirical evidence", *Renewable and Sustainable Energy Reviews*, vol. 97, pp. 276–289, doi: 10.1016/j.rser.2018.08.040, 2018.

[PAR 14] PARK C.-K., KIM H.-J., KIM Y.-S., "A study of factors enhancing smart grid consumer engagement", *Energy Policy*, vol. 72, pp. 211–218, doi: 10.1016/j.enpol.2014.03.017, 2014.

[PEL 17] PELLEGRINO M., MUSY M., "Seven questions around interdisciplinarity in energy research", *Energy Research & Social Science*, vol. 32, pp. 1–12, doi: 10.1016/j.erss.2017.07.007, 2017.

[PEN 09] PENDAKUR K., "EASI made easier", in SLOTTJE D. (ed.), *Quantifying Consumer Preferences*, Emerald Group Publishing, London, pp. 179–206, 2009.

[REA 18] REAÑOS M.A.T., WÖLFING N.M., "Household energy prices and inequality: Evidence from German microdata based on the EASI demand system", *Energy Economics*, vol. 70, pp. 84–97, 2018.

[RHO 14] RHODES J.D., COLE W.J., UPSHAW C.R. et al., "Clustering analysis of residential electricity demand profiles", *Applied Energy*, vol. 135, pp. 461–471, 2014.

[ROB 99] ROBIN J.-M., "Économétrie des systèmes de demande", *Économie et statistiques*, vol. 324, pp. 135–142, 1999.

[ROB 19] ROBYNS B., DAVIGNY A., BARRY H., *Electrical Energy Storage for Buildings in Smart Grids*, ISTE Ltd, London and John Wiley & Sons, New York, 2019.

[ROB 23] ROBYNS B., LENGLET C., BARRY H., *Smart Users for Energy and Societal Transition*, ISTE Ltd, London and John Wiley & Sons, New York, 2023.

[SAL 13] SALIES E., "Real-time pricing when some consumers resist in saving electricity", *Energy Policy*, vol. 59, pp. 843–849, doi: 10.1016/j.enpol.2013.04.050, 2013.

[SAN 17] SANGROYA D., KUMAR J., "Factors influencing buying behaviour of green energy consumer", *Journal of Cleaner Production*, vol. 151, pp. 393–405, doi: 10.1016/j.jclepro.2017.03.010, 2017.

[SAR 13] SARDIANOU E., GENOUDI P., "Which factors affect the willingness of consumers to adopt renewable energies?", *Renewable Energy*, vol. 57, pp. 1–4, doi: 10.1016/j.renene.2013.01.031, 2013.

[STO 53] STONE R., *The Measurement of Consumers' Expenditure and Behavior in the United Kingdom, 1920–1938*, Cambridge University Press, 1953.

[VEN 18] VENIZELOU V., PHILIPPOU N., HADJIPANAYI M. et al., "Development of a novel time-of-use tariff algorithm for residential prosumer price-based demand side management", *Energy*, vol. 142, pp. 633–646, doi: 10.1016/j.energy.2017.10.068, 2018.

[YIN 12] YIN MAH D., VAN DER VLEUTEN J., HILLS P. et al., "Consumer perceptions of smart grid development: Results of a Hong Kong survey and policy implications", *Energy Policy*, vol. 49, pp. 204–216, 2012.

[YIN 13] YIN MAH D., WU Y., MAN IP J. et al., "The role of the state in sustainable energy transitions: A case study of large smart grid demonstration projects in Japan", *Energy Policy*, vol. 63, pp. 726–737, 2013.

[ZÉL 15a] ZÉLEM M.-C., BESLAY C., *Sociologie de l'énergie. Gouvernance et pratiques sociales*, CNRS Éditions, Paris, 2015.

[ZÉL 15b] ZÉLEM M.-C., "Les smart meters peuvent-ils contribuer à infléchir les usages domestiques vers plus de sobriété?", *La Revue de l'Énergie*, 2015.

[ZHA 18] ZHANG C., YU B., WANG J. et al., "Impact factors of household energy-saving behavior: An empirical study of Shandong province in China", *Journal of Cleaner Production*, vol. 185, pp. 285–298, doi:10.1016/j.jclepro.2018.02.303, 2018.

5

Energy Supervision of a Local Residential Network with Actor Involvement

5.1. Introduction

As the methodology for taking actors into account and searching for profiles, in particular that of the residential actor, was developed in Chapter 4, this chapter presents a simulated case study for applying, testing and evaluating this method. To this end, the chapter is structured as follows:

– discussion of the points of the overall methodology (see Chapter 4) specific to energy supervision applied in this chapter;

– presentation of the residential case study model used to illustrate the application of the method;

– development and application of D-1 supervision for the case study (supervision the day before for the following day). Once the specifications have been presented, game theory and dynamic programming are used to optimize and coordinate consumer flexibilities;

– development and application of real-time supervision for the modeled case study. Once again, specifications are drawn up, then a fuzzy logic supervisor (a method developed by the L2EP[1] network team) is used to manage flexibility in response to contingencies (forecast errors on energy prices, consumption and renewable energy production);

– description of a two-week simulation of the supervisor to test against any developments within the case study;

1 Laboratoire d'électrotechnique et d'électronique de puissance de Lille: the Lille Laboratory of Electrical Engineering and Power Electronics.

– a discussion of the results and perspectives for the methodology and management of power grids taking into account the involvement of actors;

As this chapter is a continuation of Chapter 4, the SDGs to which this work responds are the same as for Chapter 7 (clean and affordable energy), Chapter 9 (industry, innovation and infrastructure), Chapter 10 (reducing inequalities) and Chapter 13 (measures to combat climate change). The key points of this approach enable us to take greater account of the technical, economic and social dimensions of network actors, to better integrate renewable energies, and to improve the resilience of existing network infrastructures.

This chapter, like the previous one, is based on the author's own work, and *all graphics and tables are taken from this work (unless otherwise stated)*. Readers wishing to go further should consult the research articles or the author's thesis for further detail [DUR 19, DUR 20].

5.2. Energy supervision methodology

The three-step energy supervision methodology was presented in Chapter 4, section 4.3.2. It is shown in Figure 5.1.

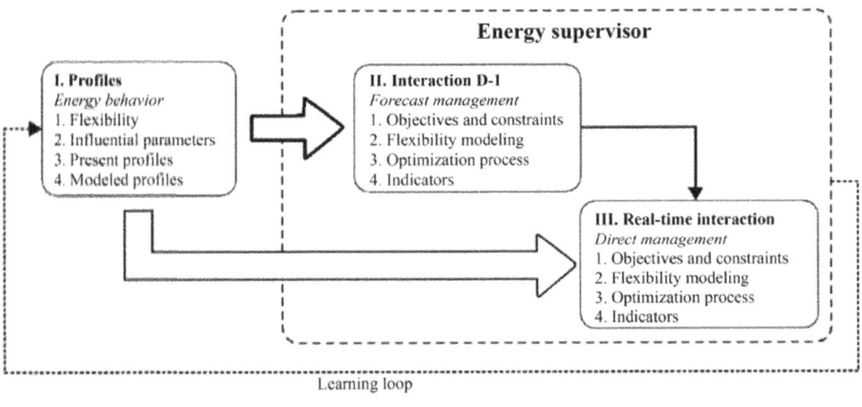

Figure 5.1. *Energy supervision methodology in three steps*

The *first step* is to determine the flexibility of the actors who can be involved in the network, by answering the question: "How can we approach and define the actor profiles present in a given network?" This step, shown in black in Figure 5.1, was developed in Chapter 4.

The next two steps, shown in gray in Figure 5.1, are developed next in this chapter.

The *second step* answers the question: "Knowing the actor profiles, what framework and optimized interaction can be used to manage energy flows, taking the day before the day under study?" The issue at stake at this stage is not only optimization, but also the approach chosen to model the profiles: thought is needed to ensure this integration, as their modeling will itself influence the supervisor architecture chosen and the trade-offs to be made.

The *third step* answers the question: "Knowing the actor profiles and the optimization of energy flows carried out the day before, how can we adjust these flows in response to the hazards observed in real time?" The key points of this stage are, once again, interfacing with the first stage (in what form should the profiles selected in the first stage be introduced?), reflection on the supervision structure and management arbitrations, but this time based on the output of the previous step.

5.3. Modeling a residential case study

5.3.1. *Electricity network under consideration*

To study and demonstrate the relevance of this approach, let us begin with a specific case study that will serve as an application example. We will consider the diversity of the objectives, while integrating them into a network seeking balance, represented here by the objective of peak reduction and load smoothing. The actors involved will therefore be as follows:

– The *distribution network operator*: its objective is to reduce fluctuations in the total load of part of the network, and to reduce power peaks.

– The *aggregator*, represented as a central entity in charge of dispatching information on the state of the network (more precisely, price, status of RE generation and status of total load on the network).

– *Residential consumers*: 140 households were modeled to enable us to study the impact of taking them into account on the state of the network, as well as whether or not they achieve their own targets. This section is dedicated to modeling their consumption.

The objectives of each, their constraints, the information available to them and the flow of information between them are explained as we go along, during the implementation of the D-1 and real-time interaction models.

5.3.2. *Modeling consumption*

Studying the flexibility of network actors requires precise knowledge of the appliances used, so as to be able to characterize them and define what can and cannot be taken into account. Various consumption modeling methods exist in the literature but will not be detailed here. As this study seeks to draw on the flexibilities of actors to model possible interactions, details of appliances and their consumption are necessary. In addition, a realistic distribution is required to simulate a scenario that not only highlights the problems observed on the network (congestion, peaks, etc.), but also simulates the complexity of domestic appliances. The aim is to *observe* a posteriori the practical implications of the methods developed. Indeed, an optimization resulting, for example, in all appliances being shifted over a reduced period of the day would make little sense. We therefore opted for a bottom-up model, detailed at the level of household appliances. The CREST[2] model [MCK 16], developed by a team at Loughborough University in the UK, was chosen. It aims to produce residential load curves in order to assess demand for distribution networks. The advantage of this model, which has been used in other demand-side management studies, is that it is available open source [MCK 20].

Based on demographic data, household appliance penetration rates, activity probabilities, climatic data and residential building types, the CREST tool simulates a daily load curve for a user-defined number of households. It is a *timetable*-type model, the principle of which is illustrated in Figure 5.1. The method, developed in [RIC 10], is based on associating the use and start-up of an appliance with the probability of carrying out an activity during the day. In addition, appliance start-ups are calibrated with their average annual energy demand, in order to obtain a consistent result over simulations covering several days.

At the heart of this model, dwellings are thermally modeled (several building types are considered), and hot water and heating requirements are taken into account. A good approximation of the associated energy demand is thus obtained by modeling air and water masses with thermal capacities, whose corresponding system of equations can be solved simply by iterative methods.

This model has been adapted to the French situation, based on existing data from INSEE and ADEME, highlighting in particular the specificity of electric heating. A fleet of electric vehicles has also been added to these consumption figures to reflect the possibilities offered by these new technologies. The technical details of this model can be found in [DUR 19].

---

2 Center for Renewable Energy Systems Technology.

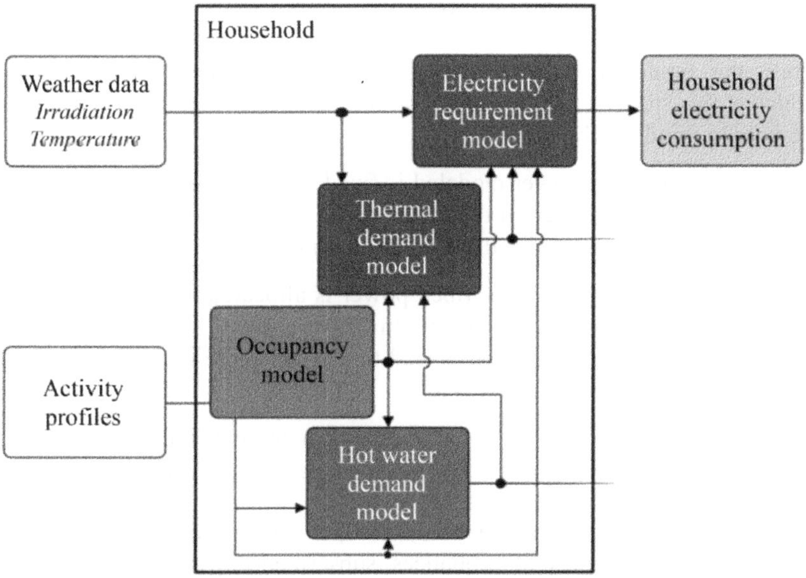

Figure 5.2. *Schematic diagram of CREST model, version 2.2*

Based on the model selected and adapted, the consumption of 140 households is simulated over a month, with a time interval of 10 min, ensuring in particular the continuity of the interior temperature of the dwelling and the temperature of the hot water tank. This aggregate load is shown in Figure 5.3.

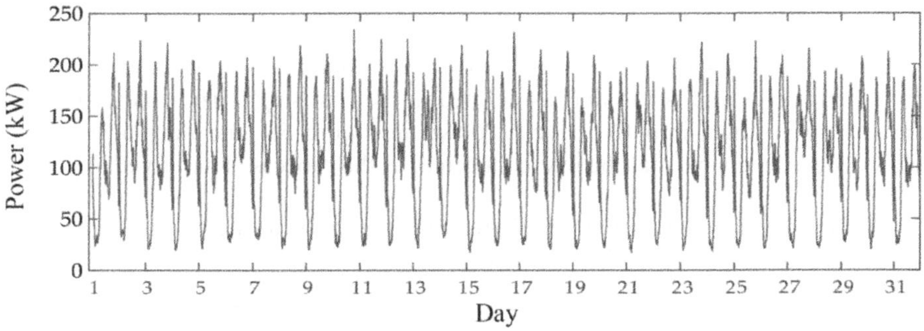

Figure 5.3. *Aggregate consumption of 140 households over one month*

5.3.3. *Discussion of model limitations*

While the model selected and adapted provides interesting details for the remainder of this study, a number of limitations can nevertheless be raised; the main ones to bear in mind are the following:

– The electric heating power modeled is fixed (here at 2 kW) and therefore does not depend on the surface area of the dwelling. In addition, room temperature is not considered, and heating is modeled as "all or nothing".

– For each household, the subscribed power is allocated taking into account the French retail market (in 3 kVA steps). In this way, the heating model and the simultaneous use of certain appliances lead to an over-representation of households with 9 kVA subscribed power.

– Modeling over the month is not sliding, which prevents the use of an appliance straddling two days, although this could be envisaged under normal circumstances.

Although this model has a few limitations, it does reflect the issues involved (peak consumption, diversity of appliances, subscribed power constraints, technical constraints and, subsequently, usage constraints) and can therefore serve as a benchmark for testing the proposed approaches. We start with the next step in the methodology: supervision the day before for the next day.

5.4. Day ahead supervision (before D-1)

5.4.1. *Discussion of predictive supervision*

5.4.1.1. *Steps in the predictive supervision process*

The question this step answers is: "How can we optimize energy flows given the actors' objectives and constraints as defined in the first step?" To achieve this, this step is broken down into five levels, shown in Figure 5.4 and developed further below. These levels must all be interconnected, as the process is iterative.

The first level is the definition of specifications prior to the construction of the supervisor, the aim of which is to bring together:

– assumptions that will influence decisions and trade-offs. Assumptions may have to be modified once the supervisor is up and running, which is why it is so important to get them right;

– the objectives and actor profiles that will be taken into account and that need to be integrated into the supervisor (→ outputs from the previous step);

– the constraints of all the actors defined in the previous step, which will delimit the optimization;

– the defined indicators, which will be fed back during the optimization phase and communicated to the actors. They must be relevant to assess the contribution of supervision and ensure a gain for all actors.

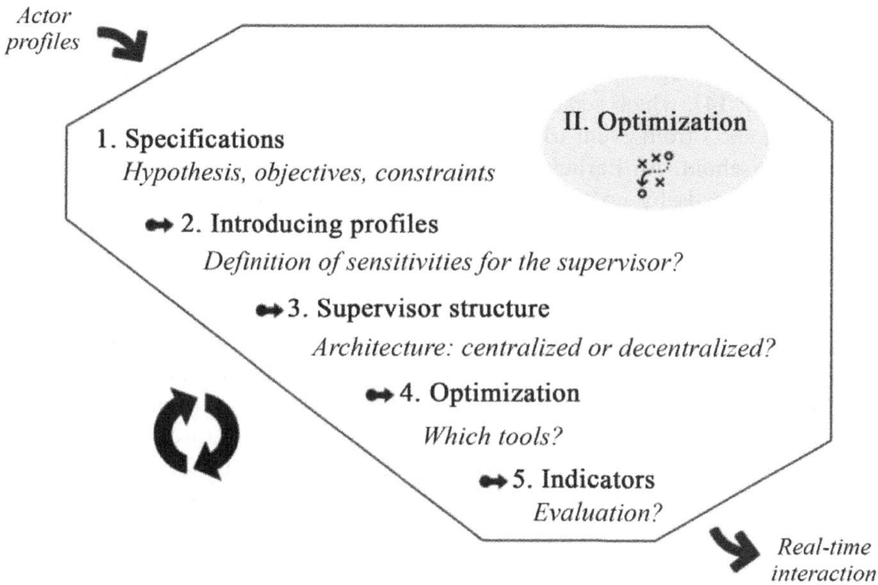

Figure 5.4. *Second stage of the methodology, optimization of upstream energy flows*

The optimization of energy flows carried out at this stage takes place ahead of the day in question and is therefore referred to as D-1 supervision, carried out by network operators. The timeframe of this supervision can be discussed at this level, considering the information available for forecasting, the computing and communication capacities influencing optimization time, and the ability of actors to anticipate and plan their energy behavior. This chosen timeframe will influence the flexibility that can be used in the supervisor, given the ability of the actors to integrate the requested shifts, as well as according to the degree of forecast accuracy, which decreases the further away from the day in question we get.

Whether or not to automate flexibility must also be decided at this stage, as the impact will not be the same for the actor. By means of storage, for example, flexibility management can be carried out transparently according to [LUT 16, REI 16].

The economic viability of this type of management is shown by the authors of [ZHE 14, ZHE 15] for different residential storage technologies, for example. As far as time-shift control is concerned, we can separate two categories which can also be used simultaneously: *automatic control*, where consumption or production can be remotely controlled by a control unit, and *manual control*, where the actor decides whether or not to participate and operate their own flexibility. While the reasoning remains the same for production, examples of both approaches can be seen in the literature for the residential consumer actor:

– In the first category, an example of a real-time scheduling algorithm can be found in [CAP 14], which separates time-triggered loads (heating, refrigeration, air conditioning, etc.) from event-triggered loads (oven, dishwasher, washing machine, etc.) in a household. An Earliest Deadline First scheduling algorithm is then used, reducing peak loads by up to 46% by pooling load shedding. Genetic algorithms can be used to manage load shedding, combined with real-time pricing [LOG 12, SHA 16];

– for manual load shedding, consider the results of the Écowatt project [RÉS 16]: by disseminating best practices and providing alerts in the event of peaks, peak consumption was reduced by 3%. For more daily and systematic action, we can use daily forecasts to indicate in advance the best times of day to consume (in periods of low consumption or renewable energy production), and thus let users program their appliances [GEE 13]. Taking this a step further, we could imagine a system that suggests the best possible path to reach an objective (e.g. charging your car at the lowest cost over the course of a day);

– between these two categories, and without the use of storage, the study [DLA 12] proposed three management algorithms for time-able loads (defined in advance by the consumer), exploring the relationship between the possible automation of loads within a household (or group of households) and occupant behavior. In order of preference for households, the proposals included automatic management of all appliances by specifying operating times in advance (the algorithm then allocates them), coordination of frequent intermittent loads (refrigerator, water heater, etc.) and prohibition of additional loads if the total load exceeds 70% of the maximum load, which is then put on hold (and managed in First-In-First-Out mode).

These two points – time and load management – must therefore be addressed in the specifications, as the possibilities offered for network management depend on them, as does the willingness of actors to participate.

5.4.1.2. *Introducing actor profiles*

The second level questions the introduction of profiles: “In what form will the profiles be modeled in the D-1 supervision?”

While the volume of shifts accepted is precisely described in the previous step, along with the associated constraints, introducing the sensitivities of each actor profile requires special thought. The sensitivities selected and the corresponding factors must be translated and physically integrated into the supervisor. The following questions arise from this and need to be addressed:

– How to introduce influencing signals: the form of influencing signals (price, temperature, etc.) helps define the range of possibilities. Indeed, if several sensitivities are taken into account, the trade-off that will take place during the optimization phase requires that a comparison be possible between them, and that they retain an apprehensible meaning for the actors involved.

– How to introduce sensitivities: as each actor does not react in the same way to influence signals, the sensitivities selected in the first stage need to be translated into the objective function used for optimization.

– How sensitive is the time lag to external factors: if the time lag is linked to external factors, how should this relationship be translated?

These considerations should enable profiles to be translated in a coherent way, both mathematically and in terms of the interaction of the actors whose strategies we are seeking to optimize [MAR 10].

5.4.1.3. *Supervisor structure*

The supervisor structure to be put in place is the result of a combination of two aspects: the infrastructure available for management (either existing or that is possible to put in place), and the conclusions drawn from the analysis of the data. Indeed, the technical possibilities will define the supervisor's ability to calculate optimal solutions to the problem at hand, as well as to circulate this information. In the same way, the results of the data analysis will guide the possible configuration of the supervisor, particularly for the most sensitive data, which will limit the possible circulation of data.

In the first step, we defined the involvement profiles that are technically possible to model, and the information required to take them into account. Defining the supervisor structure then requires determining which actor possesses what information, and what exchanges will take place, depending on the physical optimization location. There are two opposing structures:

– *Centralized approach*: this enables the central entity (aggregator) to have all the parameters at its disposal, the main advantage being to achieve the best possible optimization. The three key points of this approach are: the considerable computing power required, the capacity and reliability of the system for exchanging the large amount of information required, and the sensitivity of the data, which leads to a strong need for data protection.

– *Decentralized approach*: optimization is carried out by each actor, with the aggregator simply collecting the results of successive optimizations. The key points of this approach are the need to ensure the computing capacity of each actor, the technical cost of equipping all actors and the global convergence of the optimization, despite local calculations.

The trade-off is between data protection, required or desired performance, speed, reliability and the willingness of actors to participate and share their data.

5.4.1.4. *Optimizing energy flows*

Techniques for energy management and flow optimization are well known and depend on the variables of the problem, the way it is posed (whether or not a mathematical model is known), the mathematical form of the functions used on the solution space and the populations of known solutions that can be used to solve it. Of the two main families of approaches, the most widely used in the literature include:

– continuous optimization: linear, nonlinear or dynamic programming;

– combinatorial optimization: using approximate, heuristic or meta-heuristic methods, including genetic algorithms, particle swarms and all the combinations that follow.

However, the problem in this work is not so much the technique, since numerous studies listing possible optimization techniques and approaches already exist [MEY 17, FAH 18], as the spin-offs for the actors: the choice of optimization technique must also be based on the possibilities of taking actor profiles into account, particularly their sensitivities, by studying how each approach will call on the flexibility potential of each of them, as well as the spin-offs by means of indicators (see section 5.4.1.5). The challenge is also to define how network and actor objectives are intertwined, the relationships between the different optimization terms and the parameters representing the profiles.

The structure of the chosen supervisor will then influence the optimization technique:

– Centralized: in this case, the quantity of data and computing power will guide the optimization choice towards more or less exact methods, in order to limit the computing time required.

– Decentralized: with this approach, each actor can calculate the exact solution to its optimization problem itself, taking into account the signals received from the network, but a global framework must be put in place to ensure the overall convergence of all the optimizations carried out separately.

This reflection on the possible approach may require going back to the previous step of introducing the actors, as the mathematical framework that must be respected influences the form of the parameters and functions that can be used.

While the centralized case can be related to a classical optimization framework, the decentralized approach requires each actor to optimize its own strategy (see the optimization approaches mentioned above) and an aggregator to oversee the global optimization for the entire network. An interesting approach to managing energy flows in a decentralized way is that of multi-agent systems, whose most promising tool is game theory [LON 17, FAH 18].

Game theory provides a formal analysis of the problems posed by the strategic interaction of a group of rational agents pursuing their own goals. Game theory is appropriate here, as many agents (consumers, producers, network managers, aggregators, etc.) are involved in the network "game", exchanging energy. It provides a framework for the interaction of actors, to ensure, under certain conditions, the existence and uniqueness of equilibrium points in the sharing of the resource. This framework is mathematical in nature and dictates the form of the functions used to model the different involvement profiles, as well as the space of constraints, the image of technical limits and actor preferences.

The challenge in using this tool lies in introducing sensitivities while respecting the imposed global framework, since the local calculation to be carried out by each actor can then be reduced to a classic optimization problem.

5.4.1.5. *Indicators*

The final phase of this stage is the evaluation of the supervision carried out. From this perspective, two aspects need to be considered with regard to the indicators to be calculated:

– During the design phase: indicators must assess the supervisor's ability to integrate profiles, verifying that it has indeed enabled each actor to improve their satisfaction in relation to their objectives and sensitivities. If not, it will be necessary to return to the first stage for the definition of hypotheses, or even to the second stage of introducing profiles to modify the way they are integrated into the supervisor.

– During the operating phase: communicating with each actor on their gains and developing the energy culture and skills for each of them, by ensuring and verifying the proper utilization of their flexibility. We therefore need to think about the indicators we need to be able to send up to determine who participates and how, particularly if a reward (or malus) system is set up.

The study of *feedback* is an important field of study because, beyond their calculation method and the displayed value, the technical realization of feedback is equally significant to delve into (frequency, normalization, form, etc.), and must be implemented according to the flexibility characteristics of each actor [HAR 10].

5.4.2. *Implementing the D-1 supervisor*

Numerous demand management formulations are under study, and there exist in the literature many algorithms to solve them. The question of how to take actor profiles into account is not dealt with as such, and some points can be highlighted:

– The only control variable on the network is price, on the assumption that all users involved will seek to reduce their bills.

– When it is taken into account, the notion of comfort (see section 4.3.1.4) is approached in several ways. If comfort is considered solely as whether or not a previously defined strategy is followed[3], the literature proposes two solutions for its optimization:

- comfort is an additional objective on par with price and/or peak reduction, which implies no guarantee of being limited;

- comfort is considered as a deviation from an initially planned strategy, in which case it is considered in a dichotomous way: *respected/not respected*, which makes it impossible to assess the actor's real satisfaction.

These points pose a problem, as the notions of control and gain regarding users are essential: not taking them into account, or adding uncertainty to them, means running the risk of their disengagement [MAR 10].

Based on the profiles presented above, and presenting different scenarios for their integration into a D-1 supervision system, the contributions of the proposed supervisor are thus manifold:

– offer decentralized, multi-actor, multi-appliance energy management, with no restrictions on the type, number or distribution in the population;

3 In other words, in terms of energy consumption and timeframe, without considering the notion of thermal comfort, for example.

– represent profiles observed in the field, with the objectives presented in section 4.7 of Chapter 4. In addition, sensitivity to appliance mismatch, often associated with comfort, is here introduced independently of the other two sensitivities: economic and environmental. Indeed, a low level of accepted flexibility in no way presupposes sensitivity to costs, for example. It should be noted that we are not modeling behavior, but rather trying to take into account the main objectives of the profiles observed;

– sensitivity parameters are introduced in such a way as to retain an apprehensible meaning, keeping their value between 0 and 1 (from non-sensitive to highly sensitive);

– the impact of the proposed energy management strategy is assessed not only from the point of view of the network, but also from that of the users, by means of six indicators.

5.4.3. *Scope statement*

5.4.3.1. *Introduction*

The technical scope is defined on the basis of the network under consideration but must evolve if other actors are taken into account. This is shown in Figure 5.5, with the aggregator on one side, capable at least of communicating with consumers, to calculate the total load on the network. On the other hand, each consumer is assumed to know their load one day in advance, either by declaration or by automatic forecasting [ALQ 18]. Indeed, recent developments in non-intrusive load management (or non-intrusive load monitoring (NILM)) [LIU 18] enable the identification, estimation and forecasting of equipment consumption, as well as the evaluation of their energy shaving potential. This information is therefore assumed to be available locally (i.e. at the household level) as inputs to the D-1 energy supervisor.

The underlying assumption of the approach used is also the ability of households to manage their loads manually or through a centralized automatic appliance (smart home appliances) [ALQ 18], both technically and implicationally. Using dedicated applications, and without the user having to know the technical details, users can, for example, declare their activities, which will then be translated into terms of appliances and consumption, as in [CHR 19]. Without complicating the existing installation, the transmission of relevant information to voluntary users manually controlling their loads is already leading to new services for the network.

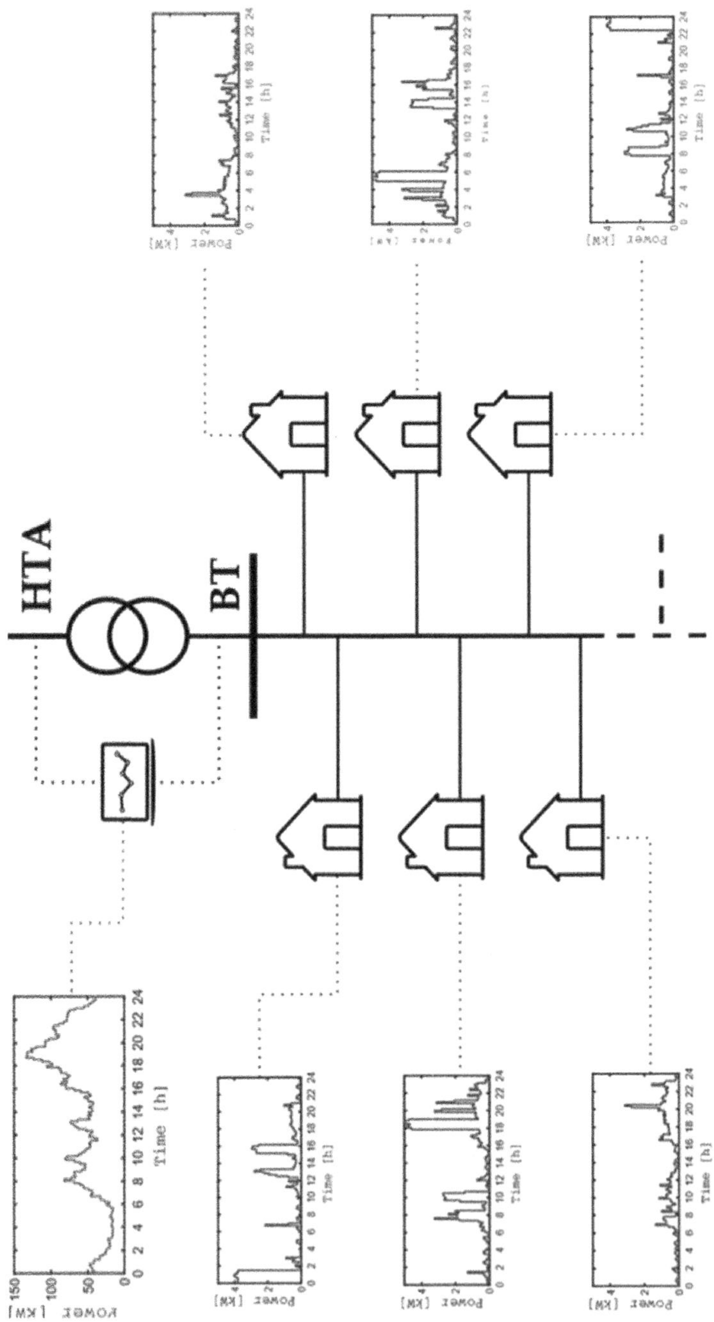

Figure 5.5. *Diagram of the network under study.*

The last assumption is that of constant energy. Each consumer carries out a certain number of activities, which may or may not involve the appliances they own: the task of the supervisor developed is then to calculate the best consumption strategy (i.e. the power to be consumed for each time interval) to carry out their activities, given the constraints imposed. Demand response programs do not in fact reduce the energy consumed on the network, but only modify power over the course of the day.

5.4.3.2. *Objectives*

As discussed in Chapter 4, section 4.7, a number of objectives need to be taken into account, not only for the various consumer groups, but also for the network. While price, smoothing and time lag objectives are technically and mathematically easy to implement, the same cannot be said for environmental sensitivity. There are many possibilities for this objective, and for the sake of representativeness and simplicity, renewable energy consumption has been chosen as the environmental impact indicator. However, this can be easily modified with any other relevant variable, for example, by replacing the corresponding model input with energy-equivalent CO_2 emissions, primary energy consumed or any other indicator derived from life-cycle analysis of electricity production that is comprehensible to the consumer.

Five objectives are therefore introduced into the supervisor. On the one hand, there are those specific to the network itself:

– reduction of power peaks;

– load smoothing.

On the other, there are those specific to residential consumers, weighted for each profile:

– lower bills;

– increased consumption of renewable energy;

– compliance with accepted appliance time lag (which impacts comfort).

5.4.3.3. *Constraints*

The constraints aim to take into account both technical and usage constraints. The challenge for the supervisor developed is therefore to respect what is technically feasible, as well as what is acceptable to consumers. The constraints must represent this complexity:

– On a technical level, the following must be entered: subscribed power, power consumption intervals, appliance rated power, standby power, any characteristic cycles (duration, power profile) and charging times for storage appliances (thermal or electrochemical) to ensure the effective performance of residential activities.

– In terms of usage, the following must be taken into account: the user's preferred times for shifting appliances, the possible charging times for the electric vehicle, depending on the arrival and departure times of the household, and the battery's state of charge, as well as the appliances that do or do not contribute to flexibility.

The definition of user preferences, as well as their constraints and the optimization of their consumption, must be carried out locally and not by a centralized entity, in order to respect consumer privacy and limit data exchange with the aggregator.

5.4.3.4. *Means of action*

Consumers can take action by shifting the energy consumed by their appliances:

– invisible consumption: the consumption of appliances whose time lag does not create any discomfort for the user, typically the hot water tank load;

– visible consumption: that which requires an effort on the part of the user, such as a mental load or a change in activity, typically cycle appliances.

Table 5.1 summarizes the technical scope.

Objectives	Smooth out total network load
	Increase consumer satisfaction by considering: – the costs – the consumption of RE
	Respecting accepted consumption timeframes
Constraints	Techniques: – subscribed power – appliance power rating – duration of appliance use – any cycles characteristic of the appliance
	Usage: – authorized time ranges – possible charging times – possible characteristic cycles – participating appliances
	Decentralized approach
	Constant daily energy
Means of action	Offsetting cycle-type appliances
	Power modulation for on/off and flexible appliances

Table 5.1. *D-1 supervisor's technical scope*

5.4.3.5. *Indicators*

Beyond these various aspects, a crucial point of the proposed methodology is the calculation and study of relevant indicators for each objective, enabling us to assess the impact of the proposed supervision strategy and the achievement or non-achievement of the objectives taken into account, for each actor:

– From the network point of view, whose objective we are considering is load smoothing, two corresponding indicators are introduced:

- measurement of power peaks;

- measurement of power fluctuations around the mean value.

– From the point of view of consumers, whose interests are many and varied, three indicators will be calculated for each of the sensitivities defining the profiles, as well as a more global indicator. The four indicators are as follows:

- evaluating overall satisfaction;

- changes in electricity bills;

- evolution of renewable energy consumption;

- measuring appliance lag, and therefore potential discomfort.

5.4.4. *Modeling actor profiles*

5.4.4.1. *Introducing sensitivities*

Consumer profiles will be introduced in this chapter by means of three α parameters ranging from 0 to 100%, one per sensitivity: α_{price} for the economic aspect, α_{re} for the environmental aspect and α_{flex} for the accepted flexibility aspect.

The problem with the simple weighted summation of the three actor objectives lies in the fact that the attainment of an objective is often unpredictable and dependent on the corresponding coefficient (which can vary arbitrarily with no upper limit). What is more, comfort is an objective that may not be met. However, as introduced in the discussion on actors, one guarantee of acceptance and involvement on the part of actors, particularly residential ones, is the control they can retain over the consumption process. Even if optimization remains a guide to be followed, and the user can choose whether to respect it or not, optimizing overall network consumption and having only a low participation rate in relation to the planning obtained would make little sense.

To achieve this, the actor's comfort is taken into account as a reflection of their flexibility. The associated sensitivity is therefore introduced as a constraint for the supervisor, and thus always respected. The hypothesis is that an actor who accepts a modification of their comfort to achieve their objectives will allow a greater range of maneuver to shift their appliances than a user who is reluctant to get involved. A posteriori, all will have their appliances consuming within the authorized time range, which is either declared or defined by their sensitivity. The notion of comfort is thus considered and always respected, since if it is not convenient, this range can be restricted at will. We consider here that users adequately determine their flexibility in the optimization phase of their strategy. This model is presented in section 5.4.4.2.

With regard to the other two actor objectives (economic and environmental), one of the problems highlighted by the state of the art is the representativeness of the coefficients with regard to sensitivities. In this study, environmental sensitivity is modeled as sensitivity to the rate of renewable energy in the electricity available on the grid, as discussed above, and this rate is normalized over the day. The price is also normalized over the day. These normalizations make it possible to introduce targets on an equal footing, so that they can be weighted by the sensitivities adopted for each actor. These introduced sensitivities are also bounded between 0 and 100% (from insensitive to fully sensitive), and their sum is equal to 1. This approach allows us to have the same amplitude of sensitivities and objectives introduced to account for the relative importance of each in relation to the others, and thus to represent user preferences in a coherent way, as highlighted in [MA 10].

5.4.4.2. *Accepted consumption shift*

As the first profile-defining parameter discussed above, the flexibility accepted, α_{flex}, by each actor can either be declared, or defined mathematically from the corresponding parameter (more or less flexible). Therefore, profiles can be seen in two paradigms: profiles observed (subject to adequate information) on a part of the network and whose parameters would be calibrated automatically; or profiles declared by the users themselves, either fixed or dynamic, for the day or for the year. It should be remembered that these parameters can vary according to periods or situations, and the most important thing is not to have them fixed, but to know them the day before for the next day. They can be fixed for the day (or a longer period) or calculated according to the information available, both from the user and from the network. The flexibility parameter can thus depend on what the user stands to gain by making their flexibility available. If the price is fixed for the day, a user who is only sensitive to the price has no interest in shifting their consumption, so they may decide not to be flexible on that day. All this can be calculated or declared (or both) by posing $\alpha_{\text{flex}} = f(\text{price}, \text{RE})$. The same principle can be adopted for sensitivity to RE consumption, which varies much more widely.

The time interval allowed for shifting an appliance during D-1 consumption optimization is defined as a percentage of the maximum possible time (between 0 and 24 hours), using the coefficient α_{flex} determined in the previous step of the methodology and modeling the actor's flexibility (from nil, "0%", to total, "100%"). We aim to show how to integrate this notion of comfort, but it is entirely possible to imagine this interval simply being declared by the user, which would not change anything in the optimization process. The use of this coefficient to determine the range of possible flexibility is shown graphically in Figure 5.6, and therefore introduced as a constraint in the problem formulation, for each cycle-type appliance. For example (Figure 5.6), for an appliance initially scheduled between 8 a.m. and 12 p.m., a user with a flexibility coefficient of "0.4" accepts an offset of "16%" of the total time available (here between midnight and 8 a.m. and between 12 p.m. and midnight), that is, possible scheduling between 6:40 a.m. and 2 p.m.

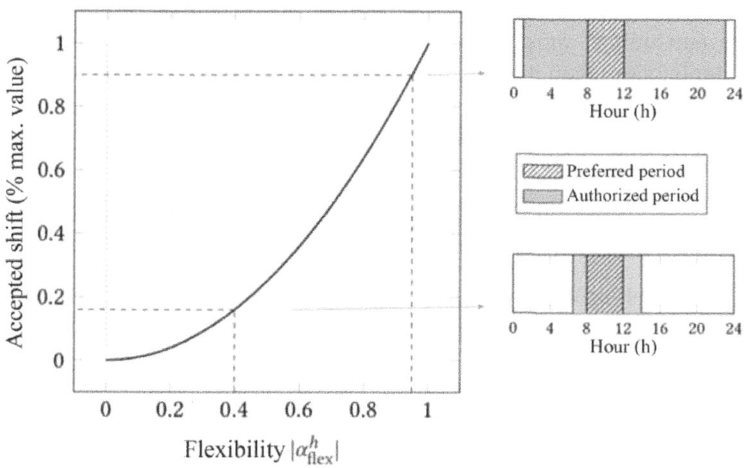

Figure 5.6. *Flexibility modeling for a household* ***h***

For other appliances, such as EVs or hot water tanks, their consumption is invisible to users. The only things that matter are whether they can use their car at the right time (i.e. have a minimum state of charge at a given time), or whether they have hot water throughout the day.

However, using your appliances to serve a third-party entity, or having them controlled by that same entity, is not accepted by everyone. Similarly, *indifferent* consumers who do not wish to actively participate in an energy management program may still be willing to make their invisible loads available (as is the case with hot water tanks today, for example). To differentiate between these various implications in the model, the sign of the flexibility coefficient reflects this acceptance of these appliances, enabling all involvement profiles to be placed on three distinct axes (price, RE consumption and flexibility). A negative sign for α_{flex}^n means that the user does not make this type of load available, and that the flexibility they use to improve their satisfaction is due solely to the cycle-type appliances they themselves decide whether to activate or not. Conversely, a positive sign for this parameter indicates acceptance of the use of all the user's appliances in the quest for greater satisfaction. For the population considered in this work, this graph is presented in Figure 5.7.

5.4.4.3. *Appliance constraints*

The flexibility of each actor, discussed above, represents the possible (or accepted) consumption shifts that they can make. However, these shifts also depend on the appliances owned and their capabilities. The different types of loads that may or may not contribute to the actor's flexibility, with their corresponding technical and usage constraints, must therefore be modeled. Table 5.2 summarizes all the domestic appliances taken into account for the actors, sorted by order of flexibility.

In order of increasing flexibility, the constraints associated with each type of appliance are as follows:

– *Fixed*: fixed consumption (e.g. lighting, TV) is that which does not take part in the optimization process (but which must nevertheless be taken into account).

– *Cycle*: cycle-type appliances (e.g. washing machines) have a fixed consumption profile over their operating time T_a. Optimization therefore only modifies the cycle start time.

– *All or nothing*: the instantaneous power of all-or-nothing appliances (e.g. BECs) with a rated power P_a is constrained to be either 0 or P_a.

– *Flexible*: the most flexible appliances (e.g. electric vehicles or storage) are constrained by their power step and rated power.

– *Subscribed power*: the constraint of each actor is due to the subscribed power of the household, and each individual power must remain below the subscribed power threshold.

Type	Appliance
Fixed	Lighting, refrigerator, heating, hotplates, etc.
Cycle	Dishwasher, tumble dryer, washing machine
All or nothing	Hot water tank
Flexible	Electric vehicle

Table 5.2. *Type of appliances modeled*

5.4.4.4. *Representation of selected profiles*

The profile groups presented in section 4.7 of Chapter 4 are therefore finally introduced into the energy supervisor through the three image coefficients of their sensitivities. The profile groups are summarized in Table 5.3 and shown in Figure 5.7, along the three axes considered. To introduce variability within the population, a parameter randomization of "+/-20%" around the values indicated for price and RE consumption sensitivities is introduced. Flexibility variability is shown directly in Table 5.3.

Profiles	Population	Cost α_{price}	RE α_{RE}	Flexibility α_{flex}
Economical	28	80%	20%	Between 75% and 100%
Eco-sensitive	28	20%	80%	Between 75% and 100%
Technophiles	28	50%	50%	Between 80% and 100%
Indifferent	28	50%	50%	Between 0% and 20%
Disengaged	28	50%	50%	Between - 20% and 0%

Table 5.3. *Profile distribution in the model population*

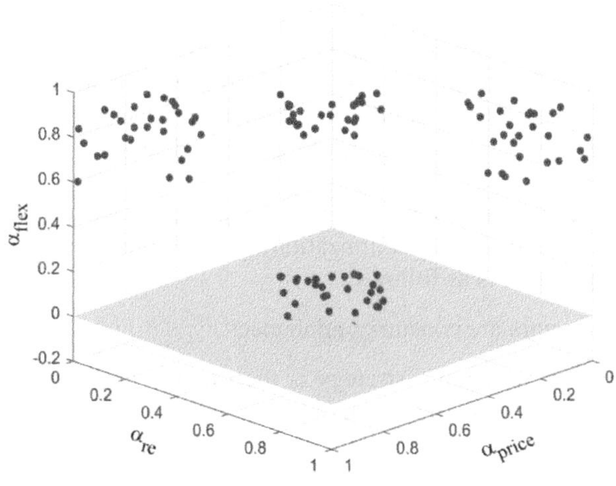

Figure 5.7. *Breakdown of involvement profiles along the three sensitivity axes*

5.4.5. *Supervisor structure*

5.4.5.1. *Methodology*

5.4.5.1.1. Game theory and decentralization

The study of interactions within a group falls within the scope of multi-agent models. Game theory offers a promising framework for decentralized reasoning.

Game theory is a branch of mathematics that studies decision-making in situations where several people (or "actors") are involved. It seeks to model these situations and identify the optimal strategies for each actor: in a game, each actor has a set of possible choices, called "actions". Each action has a consequence, which may be favorable or unfavorable for the actor. Actors may also have preferences for the outcome of the game, which can be expressed as a "payoff function". Game theory uses mathematical models to describe the interactions between actors, assuming that each actor will seek to maximize their gain or minimize their loss. It can be applied to many fields, such as economics, politics, biology, sociology, etc.

Game theory is therefore defined in simple terms as the *study of mathematical models of conflict and cooperation between rational agents*. Widely used in economics, it has also proved its worth in the context of smart grids [SAA 12, CHA 14, LON 17].

The literature offers examples of the use of this tool, but the various studies do not take into account the diversity of consumer profiles. The idea is therefore to use game theory to ensure network equilibrium by letting users independently optimize their own consumption in line with their objectives and sensitivities.

The concept used is that of a non-cooperative game known as an N-actor convex game (concave when seeking maximization). The application of game-theoretic formalism to our problem is as follows:

– Set of actors: actors are consumer appliances.

– Set of strategies: an actor's strategy is their power profile with respect to the network (power sent or withdrawn), and is defined using the corresponding constraints (see section 5.4.4).

– The objective function (called the utility function in economics) is used to prioritize the different possible strategies, with each actor seeking to find its *extremum*.

Each actor will individually seek to maximize their gain by optimizing their consumption to achieve their best strategy. For this type of interaction, game theory studies the various equilibria that can exist during the global optimization process.

The equilibrium of interest here is the Nash equilibrium: at this point, no actor has any interest in deviating from their strategy. In a decentralized way, if actors independently calculate their best strategy in turn, then the game will reach a Nash equilibrium. The existence of this equilibrium point is guaranteed under certain conditions (finite set of actors and conditions on the strategy sets and the form of the utility functions used).

5.4.5.1.2. Local optimization

The global framework ensures that if each actor proposes its best strategy, the equilibrium point on the network exists and will be reached. An exact solution to the local optimization problem faced by each appliance can be found, and as the number of time and power steps are limited, the use of meta-heuristic methods[4] classically used for energy optimization is not justified here.

What is more, since the time interval is fixed and the power consumed at each step is bounded, a dynamic programming algorithm[5] is used to calculate the best strategy. Indeed, the problem is then posed as a search for the best path among a set of intermediate energy levels, to reach a certain final energy level. The search for the exact solution to the problem posed is based on Bellman's principle of optimality: "An optimal policy has the property that whatever the initial state and initial decision are, the remaining decisions must constitute an optimal policy with regard to the state resulting from the first decision".

5.4.5.2. *Data*

Figure 5.8 summarizes all communications between the actors. At the aggregator level, the only information known is the total load on the network for the day (which is updated as soon as a household modifies its consumption), the price and the RE production for the day in question, which will then be transmitted to each household.

At the household level, each household is the only one to know its own sensitivity parameters, appliances and constraints. Based on the information received by the aggregator, the user then calculates their optimal consumption and sends it back to the aggregator.

---

4 A meta-heuristic method is an algorithmic approach to solving optimization problems that does not guarantee an optimal solution but is capable of finding acceptable solutions in a reasonable amount of time.

5 Dynamic programming is an algorithmic method for solving optimization problems using the technique of iterative programming. It is used to solve large optimization problems that can be decomposed into smaller, simpler sub-problems.

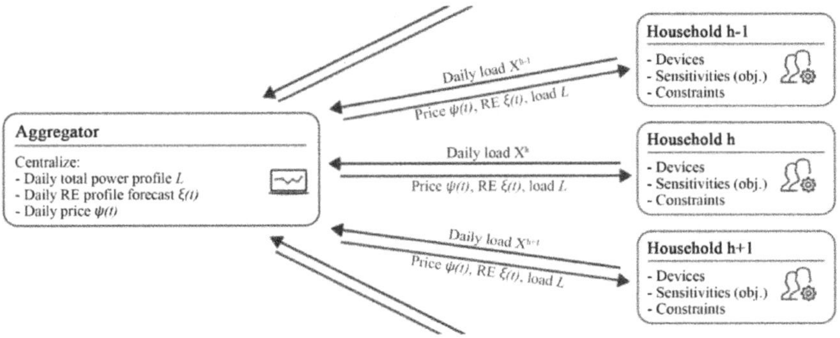

Figure 5.8. *Information flows for supervision*

5.4.6. *Global optimization and game theory*

5.4.6.1. *Mathematical formulation*

Once the overall framework and the various profiles and constraints to be taken into account have been defined, the mathematical formulation of the problem can be tackled. This formulation aims to introduce the network objective and those of the consumers in parallel. The network objective of reducing fluctuations must be integrated by all, since its state is a guarantee of the service expected by consumers. This objective will be introduced as the minimization of the square of the total load (equivalent to the minimization of the distance of the load from the mean) weighted by user preferences. The consumer's interest therefore lies in minimizing the square of the load, by shifting the flexible part of their load towards slots that suit their preferences and respect their constraints.

Each household h will therefore seek to minimize its utility function U^h by playing on its consumption X over the day, and according to its preferences ρ^h. The consumption of a household h for time interval t is denoted x_t^h. The mathematical problem including the reduction of fluctuations on the network and household preferences is therefore denoted, for a household h:

$$\min U^h(X) = \sum_{t=1}^{T} \left(1 - \rho^h(t)\right)\left(x_t^h + \sum_{j=1, j\neq h}^{N} x_t^j\right)^2 \qquad [5.1]$$

The function $\rho\rho^h$ represents the satisfaction of household h with respect to price, the environmental impact of its consumption and its sensitivities α defined in Table 5.3. It is expressed by equation [5.2]:

$$\rho^h(t) = \alpha_{price}^h \cdot \phi_{price}(t) + \alpha_{enr}^h \cdot \phi_{RE}(t) \qquad [5.2]$$

To define satisfaction with the external factors influencing the actor's objectives, the ρ^h function is based on the potential satisfaction functions of price ϕ_{price} and RE production ϕ_{RE}. These two functions are the ratios of the forecast instantaneous values of the price signal and the RE ratio to their minimum and maximum values over the day. They are therefore bounded between 0 and 1. By way of example, the maximum achievable satisfaction with the price is at the time of day when the price is minimum, and the maximum achievable satisfaction with RE consumption is at the time interval when the RE ratio is maximum. Indeed, the underlying assumption is that the actor cannot expect to obtain more satisfaction from these two factors over the day than from their respective daily extremes, which is in line with the initial assumption of working at constant energy. An example of this normalization is shown in Figure 5.9.

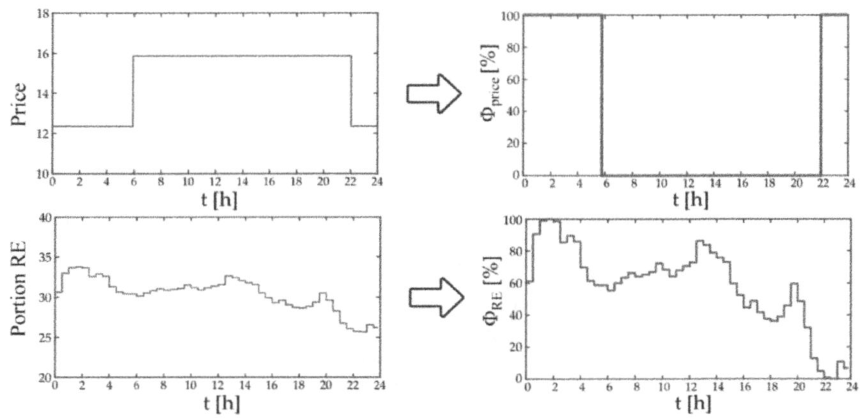

Figure 5.9. *Modeling potential satisfaction in terms of renewable energy production and price*

Note the inversion of the curve for price, reflecting the highest potential for satisfaction at the time when price is at its lowest, in contrast to RE production.

5.4.6.2. *Global algorithm*

The role of the aggregator in the optimization process is to circulate information on the state of the network and ensure that equilibrium is achieved. Price and RE production information can be sent at the first iteration only, assuming they remain unchanged during optimization. Indeed, if these two data streams were to change, the whole optimization process would have to be restarted.

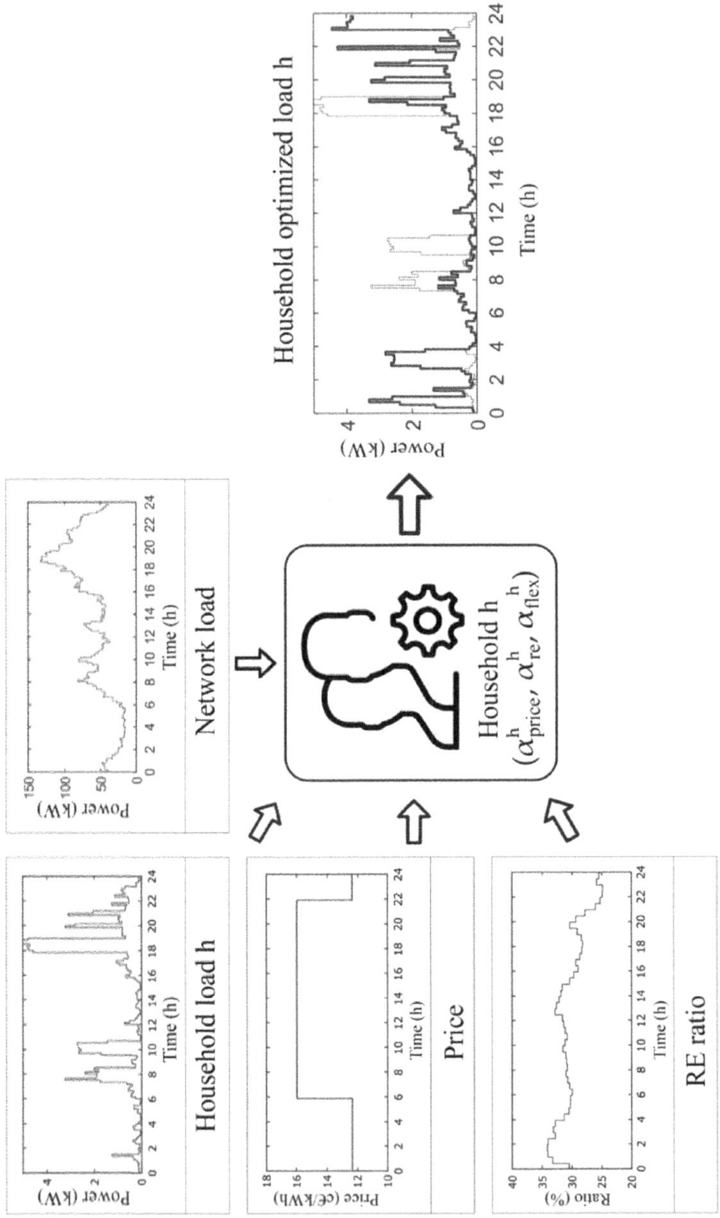

Figure 5.10. *Information processed locally to optimize the actor's consumption*

At the local level, when a household receives information on the total load on the network and if this has changed since its last optimization (and only if), the user recalculates its best consumption strategy using a dynamic programming algorithm presented in section 5.4.7 and sends it back to the aggregator. When a household is no longer interested in modifying its consumption (when the load on the network has not changed between two optimization rounds), it indicates this to the aggregator. Therefore, when all users maintain their strategy between two successive rounds, equilibrium is reached, and the algorithm stops. In fact, by optimizing consumption, a household's objective function U^h can only stagnate or decrease if the mathematical framework has been respected. Global optimization then converges on a fixed point, the Nash equilibrium.

5.4.6.3. *Global indicators*

To assess the value of the approach developed from a network point of view, two indicators are calculated to account for peak reduction and load smoothing:

– *peak to average ratio (PAR)*, the ratio between peak power and average power measured over the day on the network.

– *squared Euclidean distance (SED)*, the sum of the squared distances between the power at each time interval and the average power.

5.4.7. *Local optimization using dynamic programming*

5.4.7.1. *Information*

Locally, the consumer must integrate the information received from the aggregator with their own knowledge of their consumption, to calculate the consumption strategy that will increase their utility, given their constraints. A diagram summarizing this information is shown in Figure 5.10.

5.4.7.2. *Applying dynamic programming*

Dynamic programming is an optimization method that breaks down a problem into a multitude of sub-problems which, once solved, lead to the overall solution. The principle of this decomposition is illustrated in Figure 5.11: the period under consideration is divided into time intervals, for each of which the power to be consumed must be determined in order to consume the total energy forecast for the day. This energy is fixed, based on the assumption of constant energy.

As illustrated in Figure 5.11, the set of achievable energy levels over the whole day is defined by the following variables:

– the duration of the time interval τ;

– appliance power steps P_{step} and rated power;

– initial energy E_{min} and final energy E_{max}.

To these constraints, we can add time constraints for carrying out a particular activity (e.g. charging a vehicle before a specific time) by defining the maximum energy at the desired time interval. Similarly, the minimum energy is defined as zero for cycle-type appliances and corresponds to the initial state of charge for energy-storing appliances.

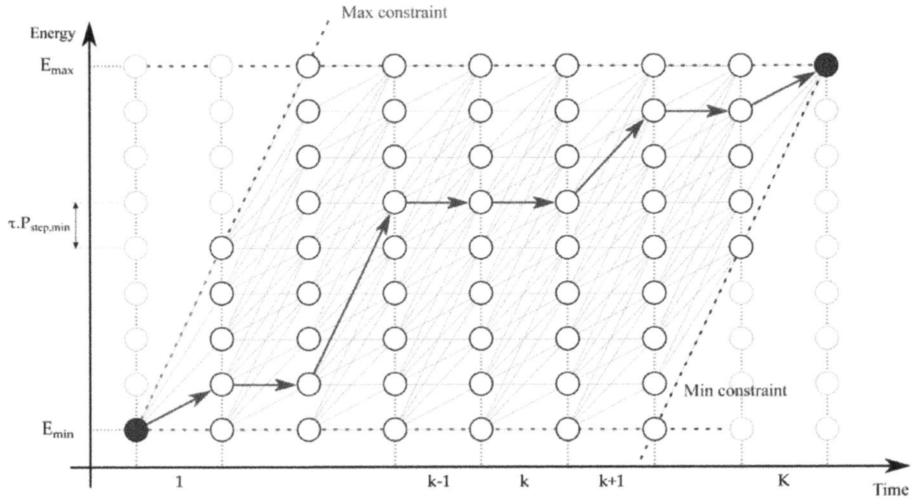

Figure 5.11. *Principle of dynamic programming applied to energy consumption, for an appliance with four power steps*

To understand how to solve the problem of reaching an energy level on a particular time interval, let us zoom in on Figure 5.11, as illustrated in Figure 5.12. Taking the level E^{p+5} reachable after time interval k, each possible power step x_k to reach this energy level is evaluated taking into account the cost (defined by the objective function U^h) needed to reach the starting energy level, that is, U_{k-1}. The path chosen is then that which minimizes the sum $U(x_k) + U_{k-1}$. Therefore, for each achievable energy level across all strategies, the previous power step leading to the minimum of the objective function is recorded. By proceeding in this way for each time interval, the minimum cost for reaching the last power step is therefore the last cost obtained. It then suffices to re-run the entire space from the last time interval to the first, selecting the best path recorded for each time interval. The strategy obtained is the one that minimizes the objective function U^h over all possible strategies.

5.4.7.3. *Local algorithm*

The local algorithm is the procedure that each consumer follows when they receive information about the load on the network. If this load is the same as the last information received, it has no interest in recalculating its strategy, and so it informs the aggregator. If it is different, the household recalculates its consumption for the day after tomorrow, using the dynamic programming shown, and sends it back to the aggregator, storing the updated total load information.

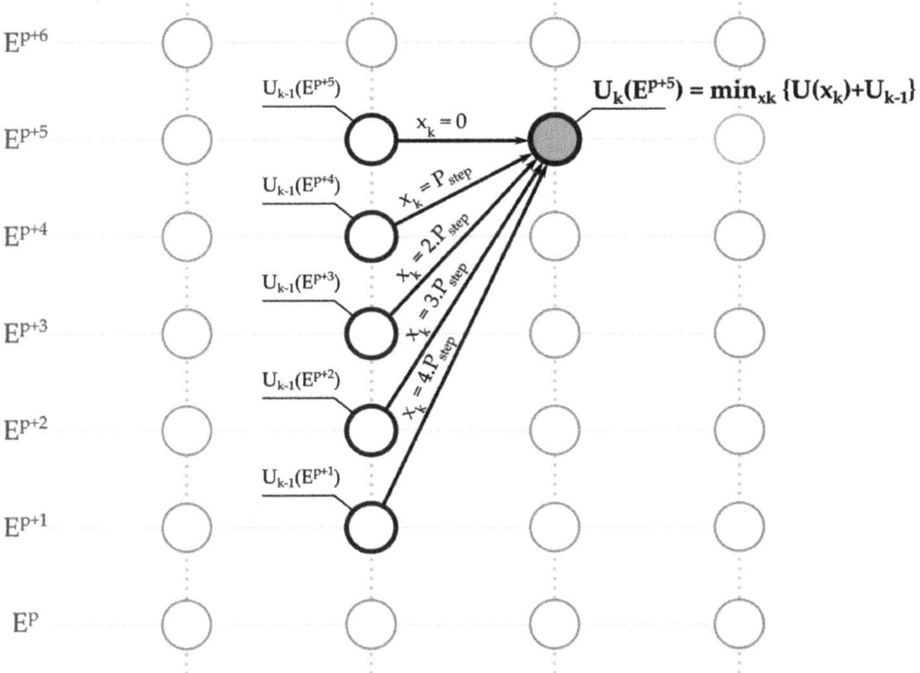

Figure 5.12. *Example of finding the minimum objective function on a time interval*

5.4.7.4. *Local indicators*

For each consumer, four indicators are introduced to account for:

– cost trends γ_{cost};

– its consumption of renewable energy γ_{re};

– the average time lag of cycle-type appliances γ_{dec};

– the overall evolution of satisfaction γ_{sat}.

For each of these indicators, the values before and after optimization will be comparable, with the convention that a positive evolution means a gain from the actor's point of view.

The final indicator is the actor's overall satisfaction, calculated according to their sensitivities. The time lag between appliances is not taken into account in this indicator, as this sensitivity is already included as a constraint, and is always respected during optimization. This shift in consumption is therefore always within the range accepted and defined by the user and made available to achieve their objectives. As the perceived benefit depends on the objectives sought, overall satisfaction γ_{sat} is calculated as the ratio between satisfied energy (defined by the function ρ^h) and total energy consumed over the day.

5.5. Real-time supervision

5.5.1. *Discussion of real-time supervision*

5.5.1.1. *Steps for real-time supervision*

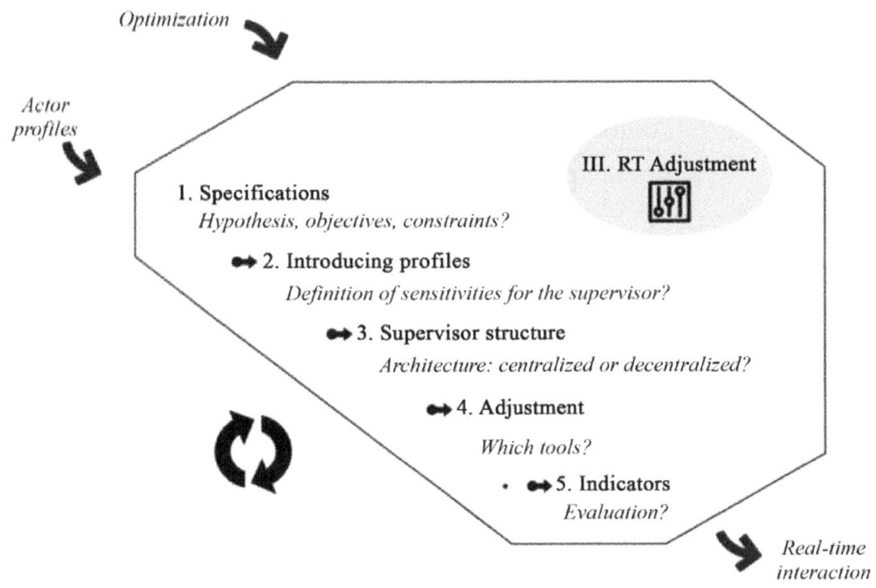

Figure 5.13. *The third step in the methodology, real-time supervision*

The final step of the methodology boils down to the following question: "Given the optimization already performed and the knowledge of the profiles, how can we adjust the energy flows in real time in response to the hazards observed?" The approach can be broken down into five steps, shown in Figure 5.13 and developed further below. As in the previous step, these different levels should be seen as an iterative process that must maintain overall coherence, returning to the previous levels when a choice requires us to revisit those made previously.

In a similar way to the optimization stage, the definition of technical scope for the real-time adjustment of energy flows must cover:

– discussion and compilation of assumptions and trade-offs, which can be adjusted at a later date;

– the objectives of the actors who will be taken into account in the chosen timeframe, and who must be integrated into the supervisor;

– constraints specific to real time, which will limit the supervisor's ability to adjust energy flows;

– the relevant indicators to be calculated and communicated in order to evaluate the approach and the gains made by the actors.

As with the prior optimization in the previous step, it is necessary to determine how the offsets will be controlled (manually or remotely), although the real-time constraint leads supervision at this stage to consider only the most easily controllable appliances.

Real time is to be understood in the broadest sense of the term, as the timeframe interval will be defined according to the flexibility available and the needs of the network. The shorter the time interval considered, the more limited the participation of actors will be, as only those with controllable resources will be able to react.

Consideration must also be given to the necessary arbitration of flexibilities: while it may be possible to readjust the production–consumption balance on D-1 by forecasting critical time intervals in advance, the real-time supervisor must react in the next time interval. It is therefore necessary to decide on the actors on whom the need to regulate the network will be based, and to what extent the actors increasing the imbalance will be called upon, even if this goes against their objectives.

5.5.1.2. *Continuity with D-1 optimization*

5.5.1.2.1. Introducing profiles

The optimization considerations remain the same for the part of the development happening in real time; only the timeframe is different. The essential difference with the previous stage is that profiles will only be taken into account for the flexibility that can be called up in the timeframe under consideration, bearing in mind that prior optimization has already taken place: we need to ensure that, all other things being equal, the actors do not modify their behavior. Sensitivities will no longer be based on nominal influence factor signals, but on their variations relative to the forecast. That said, the questions remain the same as in section 5.4:

– Introduction of influence signals: how are their relative deviations from the forecast made in the previous step to be introduced?

– Introduction of sensitivities: how will the sensitivities defined in the first step and used in the second step be taken into account, and will they weight the relative deviation of the influencing factors?

– Sensitivity of time lags: given the short-term timeframe of this stage, how can time lags be adjusted if they are linked to influencing factors? What trade-offs should be made between flexibilities?

At this stage, the same question arises as to the relative variation of signals in relation to each other, which needs to be studied in order to best model the reactions of actors and ensure that each actor sees their flexibility used according to their interests, when the balance of the network allows for this.

5.5.1.2.2. Supervisor structure

The two approaches to consider are centralized and decentralized management. As in the previous step, the supervisory structure is based on two factors: available infrastructure and data sensitivity. This leads us to consider the same structure as for the D-1 supervisor, where acceptance and technical factors have already been taken into account. In the case of decentralized pre-optimization, it may be possible to return to centralized adjustment if a very short timeframe is considered and if the infrastructure allows, but the flexibility called for must be 100% controllable, which greatly reduces the acceptance and involvement on which it will be possible to rely. The overall coherence of the two stages of the supervisor must be ensured by going back and forth between the two corresponding stages of the methodology to address all the issues that the choice of one or more approaches brings to the fore.

5.5.1.3. *Adjusting energy flows*

In the literature, the real-time approach can be implemented in one of two ways: either by optimization techniques, as discussed in the previous step, and if this is possible in the time step under consideration, optimization can be performed over a sliding time window; or by adjusting flows based on prior optimization. It is this second approach that is considered in this work, since the first is based solely on slaved flexibility and not on the implication that we are seeking to model here.

Here again, technique is not at the heart of the problem, as numerous approaches are discussed in the literature. There are two issues at stake when it comes to taking actors into account in the supervisor:

– The primary objective is to satisfy the actors by adapting flows to their objectives, but when the imbalance on the network increases, compensation is necessary. The threshold values defining the criticality of the imbalance therefore need to be carefully studied, as *restoring balance* may mean going against the actor's objectives and sensitivities. The decision to call on the flexibilities of a given actor must therefore be part of a more global reflection on the distribution of effort and the freedom left to each actor.

– Whether in a centralized or decentralized approach, the order in which flexibilities are called upon is also important to consider. Its influence on the state of the network, as well as on the effort made by the actors, needs to be assessed. The order in which flexibilities are called up is little studied in the literature, although it is critical, particularly for decentralized approaches. Evaluating this aspect does not allow us to judge but to become aware of the efforts required. For example, balancing can be based on actors with greater flexibility and whose objectives are in line with those of the network, or on the contrary, on the priority use of the flexibility of actors whose behavior runs counter to the network's objectives. The choice made must take into account acceptance issues and must be made in parallel with the model put in place to reward/manage the efforts made by the actors.

5.5.1.4. *Indicators*

The choice of indicators and the way in which they are calculated are essential in assessing the added value and relevance of the supervision developed. As in the optimization stage, the definition of indicators is a two-stage process:

– In the design phase: ensure that the indicators are relevant in assessing the supervisor's ability (real time or D-1 and real time) to include involvement profiles. Each actor's gains must therefore be assessed in the light of their objectives and sensitivities. If necessary, it may be necessary to return to the previous stages to assess the relevance of assumptions and choices concerning the introduction of profiles into the supervision system.

– In the operating phase: the challenges in the operating phase are communicating the gains achieved to each actor, developing the energy culture and skills of each of them and enabling them to ensure that their flexibility is being properly exploited. From a global point of view, indicators need to be put in place to assess the distribution and level of participation, particularly if a reward (or malus) system is put in place, or if the call for flexibility is restrictive for certain actors.

5.5.2. *Implementation of supervision in real time*

Like the observations made in section 5.4 for D-1 optimization, the literature on real-time network management does not take into account the actual profiles observed within a residential population and as described above. On the one hand, their different sensitivities are not integrated, and on the other hand, the satisfaction that can be derived from them is not evaluated either. To address these issues, the contributions of the real-time supervisor are in line with those of D-1:

– propose real-time management of residential consumption that is also decentralized, taking into account the different consumer profiles, their sensitivities and the appliances available to participate in this flexibility;

– keep the same method of sensitivity coefficients as that used for the D-1 optimization, in order to remain consistent and to be able to compare the results obtained;

– highlight the possible gains, in financial terms as well as in terms of satisfaction, for both consumers and the network, and according to the actor profiles observed;

– assess the impact of the proposed supervisor, by comparing the following four scenarios: "unsupervised", "only optimized at D-1", "only supervised in real time", and "coupling optimized at D-1 and supervised in real time".

5.5.3. *Continuity with D-1 supervisor*

5.5.3.1. *Supervision framework*

Before discussing the development of the supervisor, the network and actor objectives to be introduced, the coherence of the overall supervisor, including D-1 optimization and real-time management, needs to be discussed. The objectives and constraints presented above remain the basis of the real-time supervisor framework. The difference for real-time management is that we no longer speak explicitly of optimization, but of implicit optimization or adjustment. Indeed, the objectives sought can only be optimized with a global view of events (production, prices, consumption) over the day in question, which is done on D-1. Real-time management, on the other hand, only allows us to react to unforeseen events, while taking into account the objectives of each actor.

The framework for this real-time management remains the same as that described in section 5.4:

– The basic objectives of supervision are to smooth out the load and take into account the objectives of the different profiles.

– The constraints to be considered also remain the same.

– The assumption of constant energy is also retained for development purposes.

It is important that everyone's objectives are taken into account and optimized for D-1, and that real-time management allows us to react to unforeseen circumstances and to authorize or not authorize each user to shift their consumption in the quest for satisfaction. All other things being equal, the main aim is therefore to monitor D-1 since everything has been optimized according to the previous day's forecasts. The framework developed should then enable everyone to increase their satisfaction in the event of contingencies or unforeseen events (deviations in price forecasts or renewable energy production), within the limits of network balance. The information available and exchanged on the network remains the same as at D-1.

5.5.3.2. *Modeling profiles*

Continuing with the optimization at D-1, the sensitivities taken into account are the same as before: environmental sensitivity modeled as the share of RE in the actor's consumption, sensitivity to the energy price, and finally the actor's comfort considered as the acceptance or non-acceptance of an appliance shift over a given time range. While the sensitivity coefficients are the same for each profile, it should be noted that they are no longer applied to the raw value of the input signals (price and RE ratio), but to the value of the offset observed between the signals measured in real time and the forecast made the previous day.

5.5.4. *Fuzzy logic supervisor*

5.5.4.1. *Introduction to fuzzy logic*

By generalizing set theory, fuzzy logic was introduced by Lotfi Zadeh in 1965 [ZAD 65]. Boolean logic then became a special case of fuzzy logic, for which a proposition is no longer simply true or false but admits a degree of truth between 0 and 1. The interest of this artificial intelligence tool is being able to mathematically model the fuzziness of human representation in decision-making. In this way, imprecision and uncertainty can be formalized in the form of membership functions.

Used in many fields, this implicit optimization technique has been applied to electrical engineering for many years by the L2EP network team, which has developed an eight-stage methodology specifically for energy management [ROB 15]. These stages in the development of a fuzzy logic supervisor are as follows:

– Technical scope: this is the most important stage, defining objectives, constraints and means of action precisely, to facilitate implementation of the supervisor.

– Supervisor structure: all inputs/outputs and their processing must be listed, and at least one input must correspond to each objective. The first step is therefore crucial to the development of the supervisor, which can then be used to limit the number of inputs, making the next steps more complex.

– Functional graphs: the various operating states, the actions to be performed, and the transitions between each of them are specified.

– Graphs for each of the characteristics.

– Operational graphs: these graphs are the translation of the functional graphs, based on the membership functions defined above.

– Fuzzy rules: derived directly from the previous step, fuzzy rules are then listed, and simplifications can be made.

– Indicators: used to measure whether objectives have been achieved or not.

– Optimization: once the supervisor has been defined, it is possible to optimize its parameters, particularly the definition of membership functions.

The main steps are presented below, with the aim of integrating the different actor profiles and showing how the preferences of each can be introduced.

5.5.4.2. *Specifications*

Within the framework defined above, the technical scope (objectives/constraints/means of action) is defined and summarized in Table 5.4. Network equilibrium is considered here as respecting the planning at D-1. The supervisor's main challenge is therefore to allow consumers to shift their search for maximum satisfaction according to contingencies, while ensuring this balance.

5.5.4.3. *Supervisor structure*

5.5.4.3.1. Based on D-1

The supervisor architecture is shown in Figure 5.14. It is based on the same decentralization principle as the D-1 supervisor: the aggregator observes the total load on the network, compares it with the forecast and sends information to all households on the need for adjustment (consumption or load shedding), at each time step. Each household then integrates this information into its fuzzy logic supervisor to decide whether or not to act on its consumption. The notations used are as follows:

– for energy prices:

- ψ_{D-1} represents the forecast made on D-1;

- ψ represents its observed value in real time;

– for the ratio of renewable energies to total production:

- ξ_{D-1} represents the forecast made at D-1;

- ξ represents its observed value in real time.

Objectives	Ensure network balance based on D-1 forecasts
	Respect actors' objectives as long as the network is balanced
	Maximizing the availability of flexibility
Constraints	Total power less than the sum of the subscribed powers ($P_{\text{Totale}} \leq P_{\text{PS}}$)
	Electric vehicle (EV) storage capacity
	Time slots accepted by the user
	Household power lower than subscribed power ($P_{\text{actor}} \leq P_{\text{Subscribed}}$)
	Constant daily energy
Means of action	Charging power of electric vehicle: P_{EV}
	Power of cycle-type appliances: P_{cycles}
	Charging power of hot water tank: P_{boiler}

Table 5.4. *Technical scope for supervision in real time*

Reducing the number of inputs to the supervisor, which is necessary to limit the number of decision rules, involves combining several variables, as shown in Figure 5.14. Since an initial optimization was carried out at D-1, the inputs must take into account the time lag with the forecast used for the first stage of the supervisor. The inputs are therefore as follows:

– Adjustment of overall consumption $\Delta P_{\text{Adjustment}}$: the delta of total consumption on the network is normalized by the maximum attainable value, which is theoretically the sum of the subscribed powers.

– Consumer's interest in shifting or not $\Delta P_{\text{Interest}}$: the real-time values of the capped price and RE ratio (as we do not know the actual extreme values over the day) are compared with their predicted values, then normalized and weighted by the corresponding sensitivity coefficient.

– The local consumption offset ΔP_{Consum}: the offset between consumption defined on D-1 and that observed in real time is calculated and then normalized by the household's subscribed power.

– The last three entries are the flexibility indicators of the appliances participating in flexibility: $\text{Flex}_{\text{cycle}}$, $\text{Flex}_{\text{boiler}}$ and Flex_{EV}.

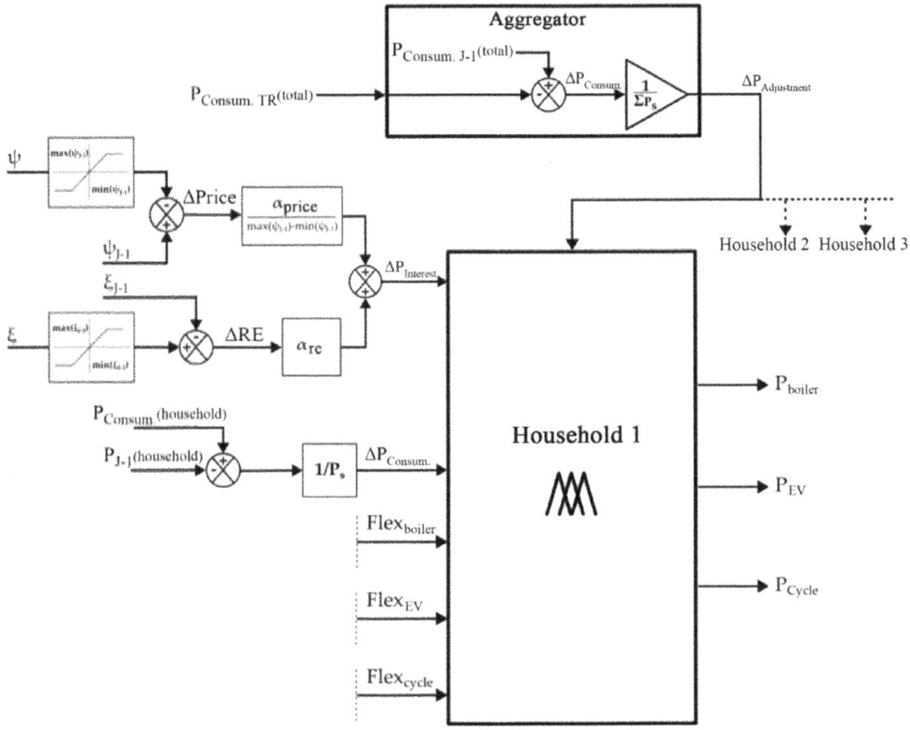

Figure 5.14. *Structure of the decentralized supervisor in real time*

For the first three entries, the following formalism will be employed:

– A $\Delta P > 0$ indicates a demand to consume more ($\Delta P_{\text{adjustment}}$), an interest in consuming more ($\Delta P_{\text{interest}}$) or consumption higher than the D-1 forecast (ΔP_{consum}).

– A $\Delta P = 0$ indicates no specific need on the part of the network ($\Delta P_{\text{adjustment}}$), an interest in not changing consumption ($\Delta P_{\text{interest}}$) or consumption equal to the D-1 forecast (ΔP_{consum}).

– A $\Delta P < 0$ indicates a request to consume less ($\Delta P_{adjustment}$), an interest in consuming less (($\Delta P_{interest}$) or consumption lower than the D-1 forecast (ΔP_{consum}).

Depending on the type of appliance, the *flexibility* variables Flex_a are calculated as a function of the appliance power, the energy consumed at the time interval under consideration and the total energy to be consumed over the day, using the time required to consume the total energy planned for this appliance. When the flexibility reaches zero, the appliance is compulsorily started. An example of how to calculate this variable is shown in Figure 5.15, for a cycle-type appliance.

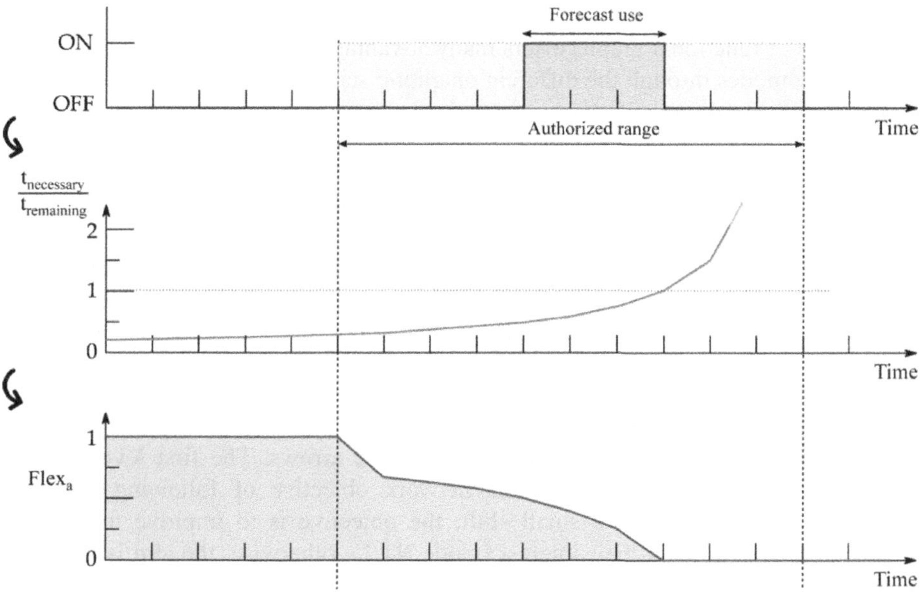

Figure 5.15. *Modeling flexibility in real time: how to transform the planned and authorized period for an appliance (top) into a $Flex_a$ decision variable (below)*

5.5.4.3.2. Adaptation when only real-time supervision is applied

From a methodological point of view, a scenario to test the relevance of real-time supervision is that of supervision without prior D-1 optimization. Following the same reasoning as above, this real-time-only supervision relies solely on the forecast (which exists for traditional network management), which we seek to monitor, as the daily production plan has been designed for this scenario.

The major difference in this context is that consumers have no prior visibility of the day and react only to the absolute signal in real time. In this case, the inputs are:

– $\Delta P_{\text{Adjustment}}$: the calculation remains the same, but the comparison is with the forecast and not with optimized consumption.

– $\Delta P_{\text{Interest}}$: clipping is performed as before, but normalization is achieved by refocusing the signals around zero, based on the mean forecast.

– ΔP_{Consum}: the forecast and not the optimized consumption value.

5.5.4.4. *Functional graphs*

The use of functional graphs offers many advantages for representing the various supervisory modes through the different operating states, thanks, on the one hand, to the graphical and linguistic expression of steps and transitions, and, on the other hand, to the simplified but complete synthesis of the overall operation.

The approach used for the supervisor is to accept consumption shifts, whatever they may be, as long as the total consumption on the network does not deviate from the reference consumption. The further away from the reference consumption, the less appliance shifts that run counter to the network objective are rejected. If the offset is too great, the flexibility of users is used to return to an acceptable offset.

The functional graphs for the supervisor developed in this chapter are shown in Figure 5.16. The transitions between the different levels are conditioned by the states of the system inputs and symbolized by the arrows. The first level, N1, in Figure 5.16, is the translation of the network objective of following the total optimized load at D-1: for a small shift, the objective is to improve the actor's satisfaction by following their interest (mode N1.1); otherwise, the aim is to return to equilibrium. For a medium shift in total load (modes N1.2 and N1.3), the consumer's flexibility is called upon if their interest and consumption go against the network objective (example of mode N1.2.2 in Figure 5.17). When the total load is significantly shifted, the network objective takes over and all flexibilities are called upon (modes N1.4 and N1.5). Appliance control, for example, to increase the actor's load (mode N1.1.2 in Figure 5.17), is carried out with the greatest possible flexibility. To achieve this, cycle-type appliances are used first, followed by other appliances in order of flexibility via the $Flex_a$ input.

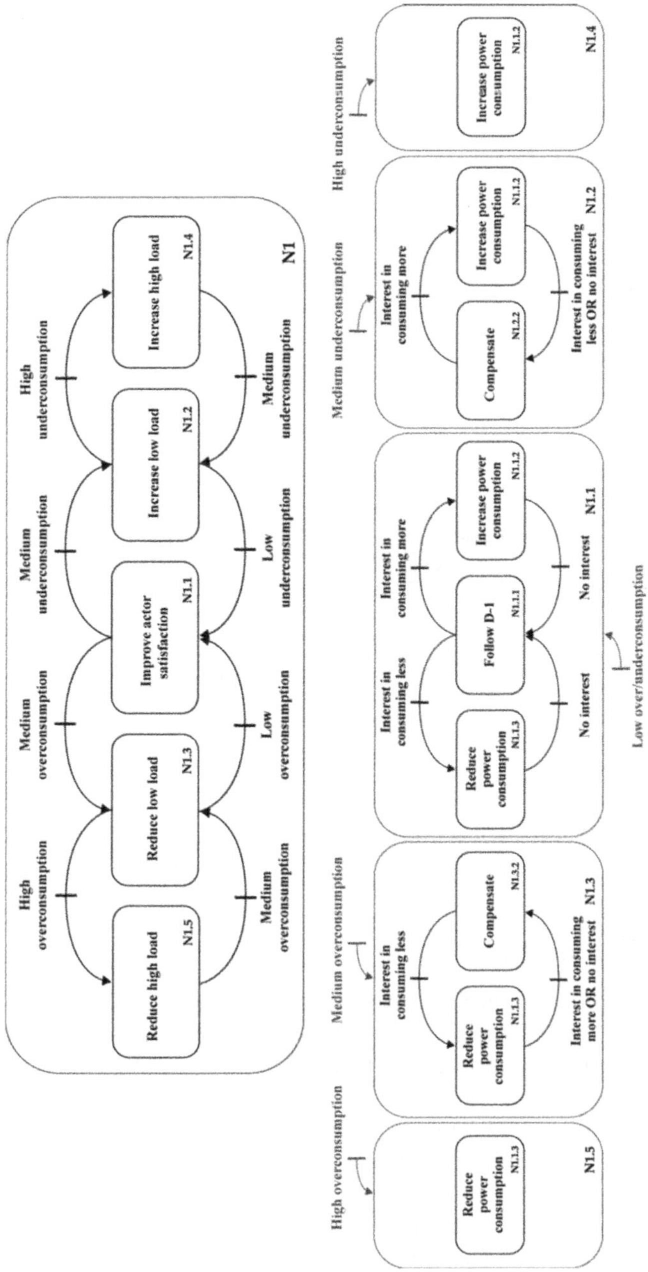

Figure 5.16. *Functional graphs of the supervisor in real time for levels 1 and 2*

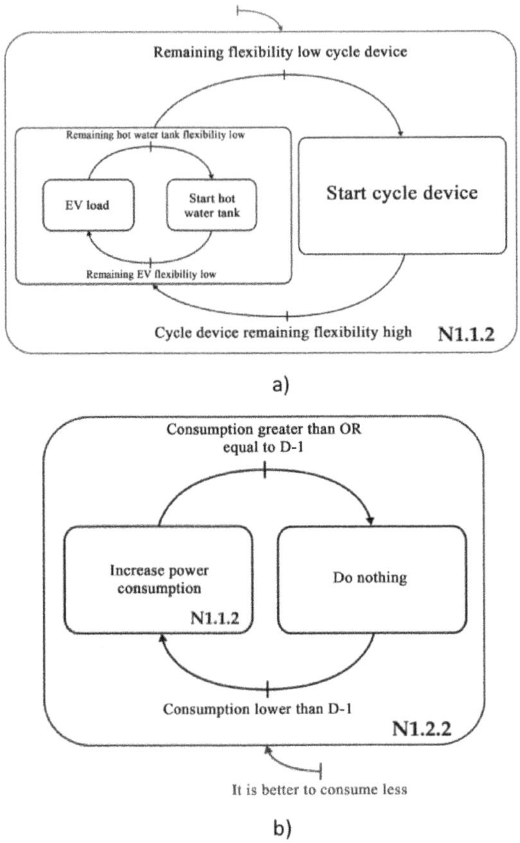

Figure 5.17. *Functional graphs of the supervisor in real time, level 3 examples for modes: a) N1.1.2 and b) N1.2.2*

5.5.4.5. *Membership functions*

Modes of operation and transitions have already been defined in the functional graphs. The possible system states represented by the inputs and outputs must then be translated numerically, in the so-called fuzzification step [ROB 19]. This is based on the normalized input values determined for the function graphs. This numerical translation is done empirically by means of membership functions, which associate the variable value with a defined level, based on knowledge of the system.

For $\Delta P_{\text{Adjustment}}$, five membership functions are considered: zero, negative medium, negative large, positive medium and positive large. For household-specific inputs, the variables $\Delta P_{\text{Interest}}$ and ΔP_{Consum} are each divided into three membership functions: negative, zero and positive. For appliance flexibility, only two membership functions are considered: small and large. All these inputs and their membership functions are shown in Figure 5.18. For the outputs, the levels are binary (on or off), and only two non-overlapping membership functions are required: ON and OFF.

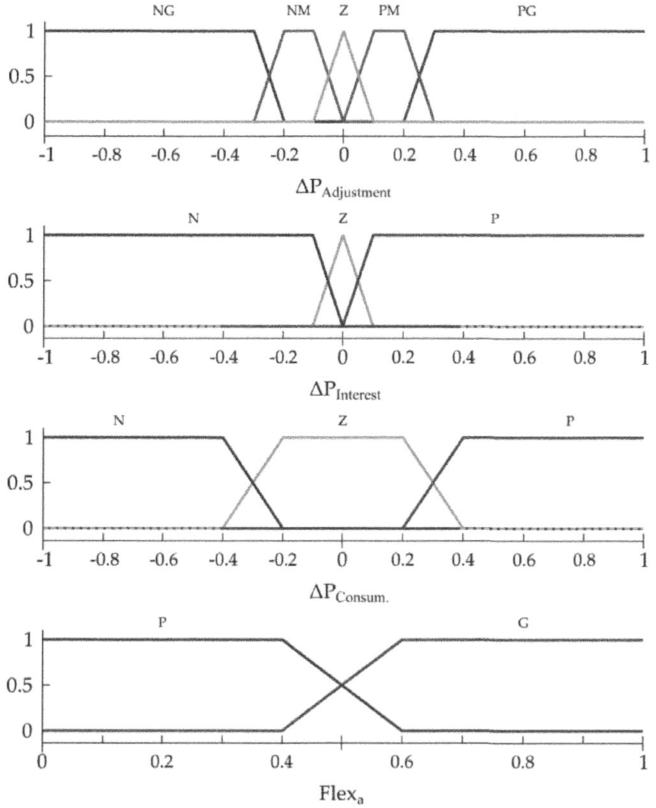

Figure 5.18. *Fuzzy logic supervisor input membership functions*

5.5.4.6. *Fuzzy rules*

The fuzzy rules are deduced directly from the operational graphs and synthesize all possible management cases. The interest of the different steps of the methodology is then apparent, as the system could theoretically reach 360 rules ($5 \times 3 \times 3 \times 2 \times 2 \times 2$, due to the six inputs and their respective number of membership functions). Applying the various steps simplifies the search for rules and reduces the number to 30 by concatenation. The set of rules is presented in Table 5.5

N1	N1.1	N1.1.1	If $\Delta P_{adjustment}$ is Z	and	$\Delta P_{interest}$ is Z	and	ΔP_{consum} is N	and	$Flex_{cycle}$ is P					then	P_{cycle} is ON
			If $\Delta P_{adjustment}$ is Z	and	$\Delta P_{interest}$ is Z	and	ΔP_{consum} is N	and	$Flex_{cycle}$ is G	and	$Flex_{Boiler}$ is P			then	P_{Boiler} is ON
			If $\Delta P_{adjustment}$ is Z	and	$\Delta P_{interest}$ is Z	and	ΔP_{consum} is N	and	$Flex_{cycle}$ is G			and	$Flex_{EV}$ is P	then	P_{EV} is ON
			If $\Delta P_{adjustment}$ is Z	and	$\Delta P_{interest}$ is Z	and	ΔP_{consum} is P			and	$Flex_{Boiler}$ is G			then	P_{Boiler} is OFF
			If $\Delta P_{adjustment}$ is Z	and	$\Delta P_{interest}$ is Z	and	ΔP_{consum} is P					and	$Flex_{EV}$ is G	then	P_{EV} is OFF
		N1.1.2	If $\Delta P_{adjustment}$ is Z	and	$\Delta P_{interest}$ is P	and		and	$Flex_{cycle}$ is P					then	P_{cycle} is ON
			If $\Delta P_{adjustment}$ is Z	and	$\Delta P_{interest}$ is P	and		and	$Flex_{cycle}$ is G	and	$Flex_{Boiler}$ is P			then	P_{Boiler} is ON
			If $\Delta P_{adjustment}$ is Z	and	$\Delta P_{interest}$ is P	and		and	$Flex_{cycle}$ is G			and	$Flex_{EV}$ is P	then	P_{EV} is ON
		N1.1.3	If $\Delta P_{adjustment}$ is Z	and	$\Delta P_{interest}$ is N	and		and		and	$Flex_{Boiler}$ is G			then	P_{Boiler} is OFF
			If $\Delta P_{adjustment}$ is Z	and	$\Delta P_{interest}$ is N	and		and				and	$Flex_{EV}$ is G	then	P_{EV} is OFF
	N1.2	N1.2.1	If $\Delta P_{adjustment}$ is PM	and	$\Delta P_{interest}$ is P	and			$Flex_{cycle}$ is P					then	P_{cycle} is ON
			If $\Delta P_{adjustment}$ is PM	and	$\Delta P_{interest}$ is P	and			$Flex_{cycle}$ is G	and	$Flex_{Boiler}$ is P			then	P_{Boiler} is ON
			If $\Delta P_{adjustment}$ is PM	and	$\Delta P_{interest}$ is P	and			$Flex_{cycle}$ is G			and	$Flex_{EV}$ is P	then	P_{EV} is ON
			If $\Delta P_{adjustment}$ is PM	and	$\Delta P_{interest}$ is Z	and			$Flex_{cycle}$ is P					then	P_{cycle} is ON
			If $\Delta P_{adjustment}$ is PM	and	$\Delta P_{interest}$ is Z	and			$Flex_{cycle}$ is G	and	$Flex_{Boiler}$ is P			then	P_{Boiler} is ON
			If $\Delta P_{adjustment}$ is PM	and	$\Delta P_{interest}$ is Z	and			$Flex_{cycle}$ is G			and	$Flex_{EV}$ is P	then	P_{EV} is ON
		N1.2.2	If $\Delta P_{adjustment}$ is PM	and	$\Delta P_{interest}$ is N	and	ΔP_{consum} is N	and	$Flex_{cycle}$ is P					then	P_{cycle} is ON
			If $\Delta P_{adjustment}$ is PM	and	$\Delta P_{interest}$ is N	and	ΔP_{consum} is N	and	$Flex_{cycle}$ is G	and	$Flex_{Boiler}$ is P			then	P_{Boiler} is ON
			If $\Delta P_{adjustment}$ is PM	and	$\Delta P_{interest}$ is N	and	ΔP_{consum} is N	and	$Flex_{cycle}$ is G			and	$Flex_{EV}$ is P	then	P_{EV} is ON
	N1.3	N1.3.1	If $\Delta P_{adjustment}$ is NM	and	$\Delta P_{interest}$ is N					and	$Flex_{Boiler}$ is G			then	P_{Boiler} is OFF
			If $\Delta P_{adjustment}$ is NM	and	$\Delta P_{interest}$ is N							and	$Flex_{EV}$ is G	then	P_{EV} is OFF
			If $\Delta P_{adjustment}$ is NM	and	$\Delta P_{interest}$ is Z					and	$Flex_{Boiler}$ is G			then	P_{Boiler} is OFF
			If $\Delta P_{adjustment}$ is NM	and	$\Delta P_{interest}$ is Z							and	$Flex_{EV}$ is G	then	P_{EV} is OFF
		N1.3.2	If $\Delta P_{adjustment}$ is NM	and	$\Delta P_{interest}$ is P	and	ΔP_{consum} is P			and	$Flex_{Boiler}$ is G			then	P_{Boiler} is OFF
			If $\Delta P_{adjustment}$ is NM	and	$\Delta P_{interest}$ is P	and	ΔP_{consum} is P					and	$Flex_{EV}$ is G	then	P_{EV} is OFF
	N1.4		If $\Delta P_{adjustment}$ is PG					and	$Flex_{cycle}$ is P					then	P_{cycle} is ON
			If $\Delta P_{adjustment}$ is PG					and	$Flex_{cycle}$ is G	and	$Flex_{Boiler}$ is P			then	P_{Boiler} is ON
			If $\Delta P_{adjustment}$ is PG					and	$Flex_{cycle}$ is G			and	$Flex_{EV}$ is P	then	P_{EV} is ON
	N1.5		If $\Delta P_{adjustment}$ is NG							and	$Flex_{Boiler}$ is G			then	P_{Boiler} is OFF
			If $\Delta P_{adjustment}$ is NG									and	$Flex_{EV}$ is G	then	P_{EV} is OFF

Table 5.5. *Fuzzy rules for supervisors in real-time.*

5.5.5. *Indicators*

In order to evaluate and compare the results obtained by the real-time supervisor with regard to the defined objectives, and to compare them with unsupervised, supervised-only-on-D-1 or supervised-only-in-real-time scenarios, the indicators used in this chapter are constructed as before. These indicators are compared directly with the unsupervised scenario (before D-1), based on the data actually observed in real time, that is, taking into account the hazards actually observed.

For the network, the two indicators measuring the peaks and fluctuations we aim to reduce are:

– *peak to average ratio (PAR)*, the ratio between peak power and average power measured over the day on the network, here compared to its unsupervised value r;

– *squared Euclidean distance (SED)*, the sum of the squared distances between the power at each time interval and the average power, here compared to its unsupervised value r.

From the consumer's point of view, three indicators are calculated to reflect the satisfaction obtained according to the different profiles:

– overall satisfaction, measured as the ratio of satisfied energy (defined by the function ρ^h, equation [5.2]) to total household consumption: β^h_{sat};

– cost indicator, calculated as the household bill, and compared to the unsupervised cost: β^h_{cost};

– RE consumption indicator compared to unsupervised RE consumption: β^h_{re}.

5.6. Two-week prospective simulations of the global supervisor

5.6.1. *Scenarios*

To observe the contribution of the supervisor developed for energy management, a realistic two-week context is simulated. The relevance of such a supervisor is in fact to react to fluctuations (hazards) on the three main input data, that is, grid consumption, price and RE production. The current French price (mostly fixed or at two levels: peak/off-peak) does not allow for sufficiently interesting dynamics. it will have to be more dynamic in the future to take full advantage of the flexibility offered by the actors' sensitivity to it.

To introduce a dynamic price that reflects the real fluctuations in its level for the grid operator, we rely, as a first approximation, on daily market data from the

European grid. Similarly, differences in consumption and RE production forecasts must be realistically included.

To do this, the following simulations are based on data from the German electricity grid (which allows the three corresponding hazards to occur simultaneously) over the first two weeks of January 2018:

– Forecast (to D-1) and actual consumption and RE production data are from Amprion, one of the four German transmission system operators and a member of the European Network of Transmission System Operators (ENTSO-E) [AMP 18]. These data, available at 15-minute intervals, are shown graphically in Figure 5.19 for the deviation of consumption from the forecast, and in Figure 5.20 for the RE forecast profile (in blue) and actual generation (in orange) for the two weeks considered in the simulation.

– For price, the data used are the German forecast (at D-1) and actual data observed on the EPEX SPOT European power exchange over the period under consideration [EPE 18], available for a time interval of one hour. These data are represented graphically in Figure 5.20: the forecast price profile on the electricity market (in blue) and actual price (in orange) for the two weeks considered in the simulation.

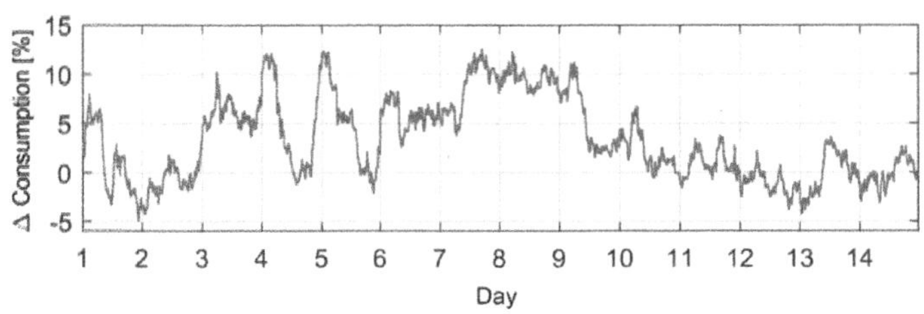

Figure 5.19. *Difference between the predicted and the actual consumption for the two weeks considered in the simulation*

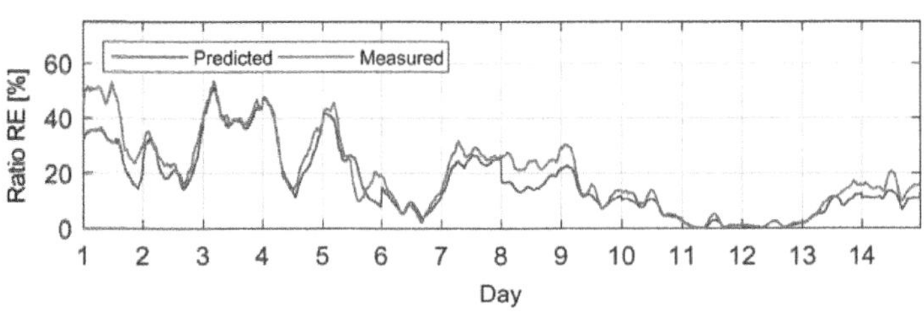

Figure 5.20. *RE forecast profile (in blue) and actual production (in orange) for the two weeks considered in the simulation*

In this simulated case study, we tested the population of 140 households (see section 5.4) with the five homogeneously distributed profiles.

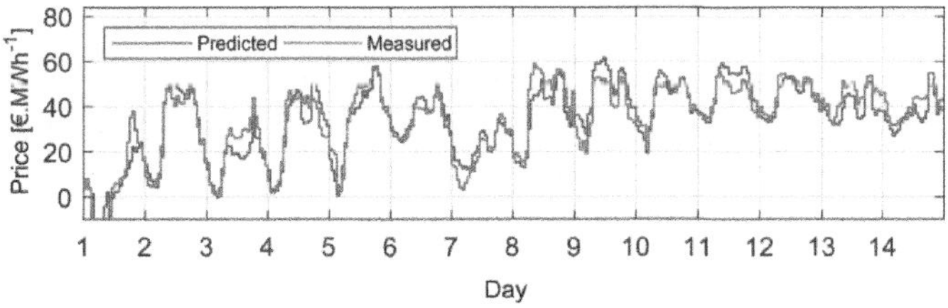

Figure 5.21. *Forecast price profile for the electricity market (in blue) and the actual price (in orange) for the two weeks considered in the simulation*

5.6.2. *Results and discussion*

The indicators for this scenario are shown in Figure 5.22 for the network, and the overall indicators for users are shown in Figure 5.23. Details of the results by objective and by profile group for overall satisfaction are shown in Figure 5.23; for price in Figure 5.24; and for RE consumption in Figure 5.25. The results for real-time supervision alone clearly show the value of a D-1 coupling based on forecasts followed by real-time supervision adjusting consumption. It should be borne in mind that the real-time supervisor can still be optimized, but the difficulty here is that they have no vision of the day to help them, hence the importance of prior D-1 supervision. What is more, as already mentioned, real-time adjustment is not optimization in itself. Adjusting its membership functions is tantamount to arbitrarily deciding on the desired balance between the network objective and user satisfaction.

The results are conclusive: Figure 5.22 shows a balance between these two aspects, which real time alone fails to achieve. Indeed, the full supervisor achieves a 68% reduction in fluctuations (decrease in the SED indicator between Init and TR+D-1), while increasing average satisfaction by 34%, as shown in Figure 5.23 (increase in the satisfaction indicator between Init and TR+D-1). With the exception of serious cases of network imbalance, the question of whether or not real time is necessary in relation to D-1 can then be considered, since all three supervised cases show an improvement over the unsupervised case: a decrease in both SED and PAR indicators for the network, whatever the level of supervision (Figure 5.22), and an increase in the satisfaction indicator (Figure 5.22).

This balance is also reflected in the results by the profile group in Figures 5.23–5.26, which show the evolution of each indicator for each profile group: the satisfaction indicator in Figure 5.23, the cost indicator in Figure 5.24 and the RE consumption indicator in Figure 5.25. What emerges from the study of these different figures is that each profile group does indeed gain in satisfaction according to its sensitivities, and the trade-off in the event of a shift is indeed observed on the weakest sensitivity: this phenomenon is visible, for example, for the economic profiles, which do indeed achieve a reduction in cost (Figure 5.24) but which, to achieve this, lose out on the RE factor to which they are only slightly sensitive (Figure 5.25). The relatively high gains observed for *indifferent* profiles in the three figures (5.23–5.25) can be explained by the greater contribution of hot water tanks to flexibility, ahead of other appliances, especially for real-time optimization.

A sensitivity study shows that, for low sensitivities, the gain is very random, but that this disparity diminishes with increasing sensitivity, which is consistent with the objective of taking profiles into account: *a guarantee of gain is only observed for the most sensitive and committed consumers*.

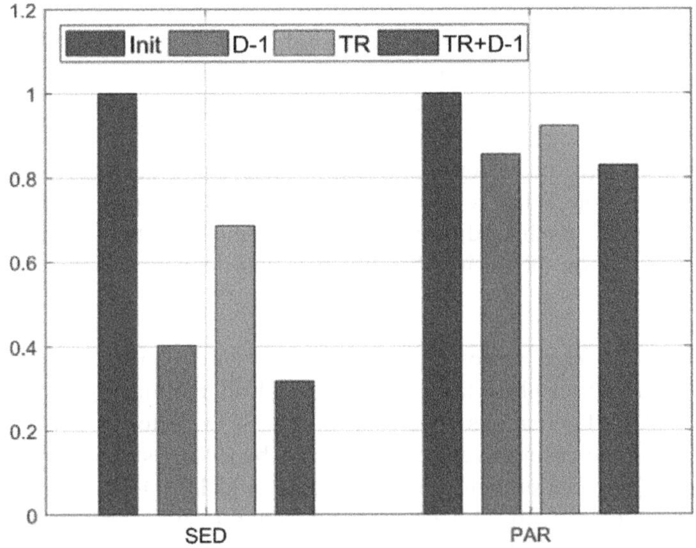

Figure 5.22. *Overall supervisor output indicators in response to two weeks of contingencies, from the network point of view (fluctuation measurements, SED – peak measurements, PAR) and for each supervision level (without, with D-1, with real time, with both). For a color version of this figure, see www.iste.co.uk/robyns/smartgrids.zip*

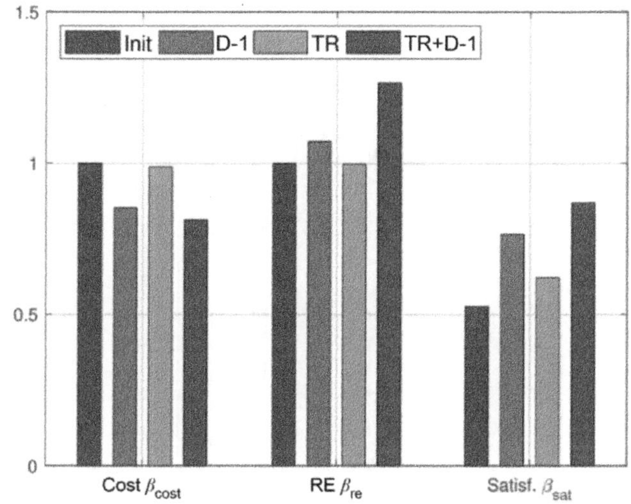

Figure 5.23. *Global indicators at the supervisor's output in response to two weeks of hazards, from the consumer's point of view (measures of costs, RE consumption, overall satisfaction) and for each level of supervision (without, with D-1, with real time, with both). For a color version of this figure, see www.iste.co.uk/robyns/smartgrids.zip*

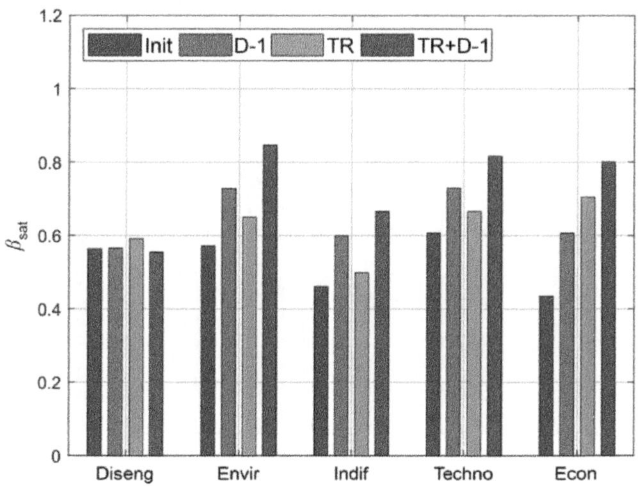

Figure 5.24. *Satisfaction indicator by profile group at the real-time supervisor output in response to two hazards and for each supervision level (without, with D-1, with real time, with both). For a color version of this figure, see www.iste.co.uk/robyns/smartgrids.zip*

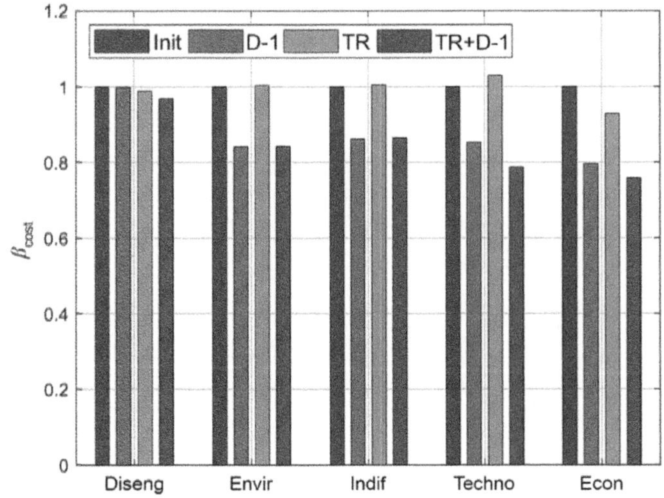

Figure 5.25. *Cost indicator by profile group at the real-time supervisor output in response to two contingencies and for each supervisory level (without, with D-1, with real time, with both). For a color version of this figure, see www.iste.co.uk/robyns/smartgrids.zip*

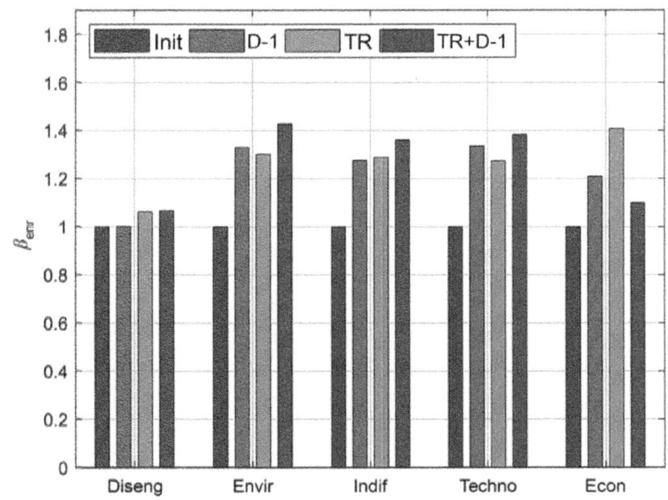

Figure 5.26. *RE consumption indicator by profile group at the real-time supervisor output in response to two contingencies and for each supervisory level (without, with D-1, with real time, with both). For a color version of this figure, see www.iste.co.uk/robyns/smartgrids.zip*

The interest of the results presented lies more in the orders of magnitude and the signs of the evolution of the various spin-offs for the network and consumers, than in the absolute values. Indeed, the main aim of this study is to propose a methodology for taking into account actor profiles, and to show the possible gains. Once feasibility has been demonstrated, it is a matter of fine-tuning and arbitration. These adjustments can only be made on a real population in a functional, observable network. With this in mind, the various tests and simulations carried out showed several results:

– The advantages of real-time supervision coupled with D-1 optimization compared with the unsupervised case or the case where only D-1 optimization is performed.

– Real-time supervision alone (without prior optimization) does not deliver as good results as the global supervisor but does achieve an interesting balance of objectives compared to the unsupervised case.

If we look at the various appliances involved in flexibility, the BECs are the ones that participate most in real time in response to observed contingencies. Since EVs are not available during the day, the assumption of constant energy leads them to make up for any discrepancies during real-time optimization in the last hours of the day. A non-residential case study (tertiary sector, for example) could demonstrate their value in a more diversified network. Cycle appliances, on the other hand, play a greater role in flexibility during D-1 optimization, as their constraints are greater than the real-time limit.

The flexibility of cycle machines is in fact more complicated to use in real time, because if a hazard occurs, it is necessary, on the one hand, to be within an operating range authorized by the user, and, on the other hand, to accept the fact that there is no medium-term vision if a reverse hazard occurs during the cycle, which can then not be stopped. The way in which this flexibility is taken into account needs to be improved in this model: on the one hand, integrating this effect, for example, by limiting it through price contractualization (a fixed price could be guaranteed once the machine is up and running); on the other hand, taking into account the more restrictive aspect of this flexibility for users when calculating satisfaction. *An economic model could also be introduced to compensate for this participation.*

More generally, for the satisfaction indicator, an interesting prospect would be to introduce the degree of flexibility into its calculation, since for the average indicator, uninvolved profiles restrict the overall rise in the population. Similarly, satisfaction with environmental sensitivity can be questioned on the basis of the RE ratio alone. Indeed, as society evolves towards a predominantly *eco-sensitive* segmentation, the

RE criterion is no longer necessarily sufficient: other impacts need to be taken into account, such as network support, which can also be beneficial in this respect.

The results show that the different levels of supervision offer an interesting balance for both the network and consumers, managing to satisfy the different profiles. Once we have demonstrated the relevance of energy management that takes into account different involvement profiles, we need to bear in mind that everything is a matter of compromise, and that there is no perfect choice, but rather a limit to be arbitrarily set in the freedom left to each individual. The various simulations show that real time coupled with D-1 offers the best balance between the different profiles, their satisfaction, their gains and the flexibility brought to the network. If, for certain points, real-time optimization alone achieves better results, this is to the detriment of other factors. It is therefore a factor to be considered, although the question of arbitration is more of a strategic and societal choice, taking into account the network's ability to withstand fluctuations and deviations.

The performance of the real-time supervisor can be improved by optimizing the membership functions, which have been arbitrarily constructed on the basis of global knowledge of the system. This optimization can relate to the form of the functions, their definition values or the gain of each input. It can be performed by evolutionary algorithms and has been shown to be relevant in [BRE 13]. However, the same problem of balancing objectives will arise: *what priority should be given to the network objective over the user objectives?* This may limit the value of optimization.

In any case, as the three levels of supervision provide added value for the various actors involved, an interesting point to consider is the date on which to carry out the D-1 optimization. As the latter achieves good results, it might be appropriate to run it again on the same day if a more accurate forecast is available. This choice must, of course, be a *compromise between the best possible accuracy* (which can only be achieved as close as possible to the optimized day) and *the physical capacity of the actors to integrate the information*: a fully automated system can certainly withstand being rescheduled at any time, although the preferences of the actors may change, but a system based on the manual participation of users' needs to be able to schedule equipment early enough without changing at the last moment.

In addition to thinking in terms of the supervisor, it may be worth looking at different timeframes. Raising awareness may lead to greater participation in the short term, which is wise, and, in the long term, to the purchase of more flexible appliances (remote-controlled or more finely programmable). Paradoxically, this can also lead to the purchase of less energy-intensive appliances, which in turn reduces this flexibility in terms of energy volume.

5.7. Conclusion

The answer to the problem of how to manage these networks, that is, how to integrate the involvement of different actors in order to achieve greater flexibility, lies in proposing a methodology that is open to interdisciplinarity. Taking these actors into account, with their specific characteristics, objectives and constraints, is not a purely technical issue, but requires us to consider all aspects of flexibility with each of them. The three stages of this methodology are:

– defining the profiles of the actors involved;

– D-1 energy management (optimization stage);

– real-time energy management (adjustment stage).

This approach is illustrated by the application of the methodology to a residential case study, questioning the ability and willingness of consumers to participate in network balance while increasing their satisfaction. One of the key points to emerge from this methodology and these simulations is *the junction between the engineering sciences and the human and social sciences in terms of introducing profiles into optimization*.

Optimization of the proposed supervisor is always possible to improve the observed results, but it requires an arbitrary decision between the gains of the network and the gains of the network's users. The big question here is indeed that of compromise, which cannot be answered without a real case study. Discussions on the involvement of actors, particularly residential ones, show that many non-technical parameters come into play, and if the optimization of this supervision can be done, *optimizing for the network to the detriment of consumers can lead to their disengagement*. The technical aspects proposed, if they bring a gain for both the network and consumers, are based precisely on this involvement linked to the implementation of these technical aspects, which cannot be completely anticipated. This work therefore questions the complexity of the task, and the prospective results, while encouraging, should be viewed with caution. What is more, the assumption that the network's objective will be integrated into the behavior of each actor may be due either to a collective awareness (if no network, no service), or to a pricing incentive. The study of the translation of actors' objectives is therefore important.

The paradigm can be either *manual* operation (like ADEME's S2), allowing for the maximum number of loads to participate in flexibility, or *automated* operation (like ADEME's S3), raising the question of user control. With further development of the model, and in a context of advanced and adequate information, it may also be possible to forecast loads and their shifting. As discussed in this work, the user's sense of control is an important factor in acceptance and involvement. The paradigm of the total automation of flexibility is therefore unrealistic, and the effectiveness of such a system relies instead on a mix of the two, adapted to the context and the profiles of the actors involved.

The economic model to be implemented thus becomes an important question at the retail market level. Of course, a static model does not allow for any flexibility, but by making it more dynamic, the question becomes: should we punish negative behavior (in the sense of worsening network equilibrium) or reward virtuous behavior? The answer is surely a compromise between the two, given the differing sensitivities of the actors involved. However, the scale of reflection must not be confined to the micro-economic framework, but must also take place on a macroscopic scale, as this touches on issues of public policy and economics. The discussions in this work thus highlight a crucial point in the definition of new viable models: they *must not place the burden on the weakest*, all the more so given that beyond the commitment, the impact of the most affluent is the most significant. To this end, and as discussed below, a paradigm shift in electricity consumption might be to see access to electricity as a vital right, with everyone able to consume a defined minimum, and the rest subject to higher constraints (tariff, limitation). The definition of the minimum then requires particular attention but could bring new flexibility to the network.

Here again, the challenge is not just a technical one since the actors' perception of the benefits and constraints involved influences their acceptance and involvement. The point of this work is to define what can be done and put in place, as well as to ask the questions necessary to take into account the actors in order to help the network function. What emerges is that it is technically possible to play on flexibility, but that to make a system truly effective, it is necessary to get the actors involved, and this requires a real, more global project for society. The current crises are naturally leading us towards this paradigm shift, with greater involvement of network actors, as we move from a paradigm of *infinite energy* to a world wherein energy supply is no longer automatically guaranteed, unless efforts are made at the national level.

Indeed, taking into account the sensitivities and objectives of each individual, even when these run counter to the collective interest (over-consumption, network imbalance), will in no way solve the problem, and may even make it worse. For example, can we equate an *actor's desire to reduce their environmental impact* with their desire to consume as they see fit, even if this goes against the common good? Or the interest of an RE producer vis-à-vis a fossil fuel producer, in full knowledge of the issues we face today and in the not-so-distant future presented at the beginning of this book?

From an overall point of view, the final conclusion of Chapters 4 and 5 is that managing the network with flexible, involved and satisfied actors is technically interesting and feasible, and possibly rewarding for the actors involved in view of the results presented. However, the discussions show that the realization of such a scenario goes far beyond the simple point of view of the network and the simple approach of getting solutions accepted: it requires the construction of a real global project, drawing on all disciplines, and demands a significant and sustained energy culture among all network actors. The two possible directions are either the imposition of a restrictive model justified by environmental urgency, but which would be difficult to win the support of all actors, or the reconciliation of everyone's objectives, but for this model to work, *a majority of actors must be involved and convinced.* The methodology and the feasibility of respecting (to a certain extent) everyone's objectives are a first step in this direction, so that instead of *having to do something,* we instead *work together* to achieve the objective.

5.8. Acknowledgments

The authors would like to thank the Hauts-de-France region for its contribution to the funding of this research.

5.9. References

[ALQ 18] ALQUTHAMI T., SAKIS MELIOPOULOS A.P., "Smart house management and control without customer inconvenience", *IEEE Transactions on Smart Grid*, vol. 9, no. 4, pp. 2553–2562, 2018.

[AMP 18] AMPRION, Grid data, available at: https://www.amprion.net/Grid-Data/, 2018.

[BRE 13] BREBAN S., SAUDEMONT C., VIEILLARD S. et al., "Experimental design and genetic algorithm optimization of a fuzzy-logic supervisor for embedded electrical power systems", *Mathematics and Computers in Simulation*, vol. 91, pp. 91–107, 2013.

[CAP 14] CAPRINO D., DELLA VEDOVA M.L., FACCHINETTI T., "Peak shaving through real-time scheduling of household appliances", *Energy and Buildings*, vol. 75, pp. 133–148, 2014.

[CHA 14] CHAI B., CHEN J., YANG Z. et al., "Demand response management with multiple utility companies: A two-level game approach", *IEEE Transactions on Smart Grid*, vol. 5, no. 2, pp. 722–731, 2014.

[DEC 19] DE CHRISTO T.M., PERRON S., FARDIN J.F. et al., "Demand-side energy management by cooperative combination of plans: A multi-objective method applicable to isolated communities", *Applied Energy*, vol. 240, pp. 453–472, 2019.

[DLA 12] DLAMINI N.G., CROMIERES F., "Implementing peak load reduction algorithms for household electrical appliances", *Energy Policy*, pp. 280–290, 2012.

[DUR 19] DURILLON B., ModAICSS : modélisation de l'acceptation et de l'implication de producteurs et consommateurs d'énergie électrique en vue d'une stratégie de supervision énergétique multi-acteurs, PhD thesis, University of Lille 1, Lille, 2019.

[DUR 20] DURILLON B., SALOMEZ F., DAVIGNY A. et al., "Consumers' sensitivities and preferences modelling and integration in a decentralised two levels energy supervisor", *Mathematics and Computers in Simulation*, doi: 10.1016/j.matcom.2020.02.003, 2020.

[DUR 21] DURILLON B., DAVIGNY A., KAZMIERCZAK S. et al., "Decentralized neighbourhood energy management considering residential profiles and welfare for grid load smoothing", *Sustainable Cities and Society*, vol. 63, pp. 102464, doi: 10.1016/j.scs.2020.102464, 2021.

[EPE 18] EPEX SPOT, Données de marché, available at: https://www.epexspot.com/fr/donnees_de_marche/, 2018.

[FAH 18] FAHAD ZIA M., ELBOUCHIKHI E., BENBOUZID M., "Microgrids energy management systems : A critical review on methods, solutions, and prospects", *Applied Energy*, vol. 222, pp. 1033–1055, doi: 10.1016/j.apenergy.2018.04.103, 2018.

[GEE 13] GEELEN D., REINDERS A., KEYSON D., "Empowering the end-user in smart grids: Recommendations for the design of products and services", *Energy Policy*, vol. 61, pp. 151–161, 2013.

[HAR 10] HARGREAVES T., NYE M., BURGESS J., "Making energy visible: A qualitative field study of how householders interact with feedback from smart energy monitors", *Energy Policy*, vol. 38, no. 10, pp. 6111–6119, doi: 10.1016/j.enpol.2010.05.068, 2010.

[LIU 18] LIU Y., GENG G., GAO S. et al., "Non-intrusive energy use monitoring for a group of electrical appliances", *IEEE Transactions on Smart Grid*, vol. 9, no. 4, pp. 3801–3810, 2018.

[LOG 12] LOGENTHIRAN T., SRINIVASAN D., SHUN T.Z., "Demand side management in smart grid using heuristic optimization", *IEEE Transactions on Smart Grid*, vol. 3, pp. 1244–1252, 2012.

[LON 17] LONI A., PARAND F.-A., "A survey of game theory approach in smart grid with emphasis on cooperative games", *2017 IEEE International Conference on Smart Grid and Smart Cities (ICSGSC)*, pp. 237–242, 2017.

[LUT 16] LUTHANDER R., WIDEN J., MUNKHAMMAR J. et al. "Self-consumption enhancement and peak shaving of residential photovoltaics using storage and curtailment", *Energy*, vol. 112, pp. 221–231, 2016.

[MAR 10] MARLER R.T., ARORA J.S., "The weighted sum method for multi-objective optimization: New insights", *Structural and Multidisciplinary Optimization*, vol. 41, no. 6, pp. 853–862, 2010.

[MCK 16] MCKENNA E., THOMSON M., "High-resolution stochastic integrated thermal–electrical domestic demand model", *Applied Energy*, vol. 165, pp. 445–461, 2016.

[MCK 20] MCKENNA E., THOMSON M., CREST demand model, available at: https://figshare.com/articles/CREST_Demand_Model_v2_0/2001129 [Accessed 15 January 2023], 2020.

[MEY 17] MEYABADI A.F., DEIHIMI M.H., "A review of demand-side management: Reconsidering theoretical framework", *Renewable and Sustainable Energy Reviews*, vol. 80, pp. 367–379, doi: 10.1016/j.rser.2017.05.207, 2017.

[REI 16] REIHANI E., MOTALLEB M., GHORBANI R. et al., "Load peak shaving and power smoothing of a distribution grid with high renewable energy penetration", *Renewable Energy*, vol. 86, pp. 1372–1379, 2016.

[RÉS 16] RÉSEAU DE TRANSPORT D'ÉLECTRICITÉ, Écowatt Bretagne, le bon geste énergie, Report, available at : http://www.ecowatt-bretagne.fr, 2016.

[RIC 10] RICHARDSON I., THOMSON M., INFIELD D. et al., "Domestic electricity use: A high-resolution energy demand model", *Energy and Buildings*, vol. 42, pp. 1878–1887, 2010.

[ROB 15] ROBYNS B., FRANÇOIS B., DELILLE G., et al., *Energy Storage in Electric Power Grids*. ISTE Ltd, London and John Wiley & Sons, New York, 2015.

[ROB 19] ROBYNS B., DAVIGNY A., BARRY H. et al., *Electrical Energy Storage for Buildings in Smart Grids*, ISTE Ltd, London and John Wiley & Sons, New York, 2019.

[SAA 12] SAAD W., HAN Z., POOR H.V. et al., "Game-theoretic methods for the smart grid: An overview of microgrid systems, demand-side management, and smart grid communications", *IEEE Signal Processing Magazine*, vol. 29, no. 5, pp. 86–105, 2012.

[SHA 16] SHAHEEN N., JAVAID N., NISA N. et al., "Appliance scheduling for energy management with user preferences", *International Conference on Innovative Mobile and Internet Services in Ubiquitous Computing Appliance, Proceedings*, no. 10, 2016.

[ZAD 65] ZADEH L.A., "Fuzzy sets", *Information and Control*, vol. 8, pp. 338–353, doi: 10.1016/S0019-9958(65)90241-X, 1965.

[ZHE 14] ZHENG M., MEINRENKEN C.J., LACKNER K.S., "Agent-based model for electricity consumption and storage to evaluate economic viability of tariff arbitrage for residential sector demand response", *Applied Energy*, pp. 297–306, 2014.

[ZHE 15] ZHENG M., MEINRENKEN C.J., LACKNER K.S., "Smart households: Dispatch strategies and economic analysis of distributed energy storage for residential peak shaving", *Applied Energy*, vol. 147, pp. 246–257, 2015.

6

Self-Consumption within a Local Renewable Energy Community

6.1. Introduction

The development of self-generation and self-consumption of renewable electrical energy (see section 1.7.3 of Chapter 1) encourages better energy management and lower electricity bills (lower costs of connection to the electrical distribution network, lower subscriber power and therefore lower subscription fees, and potentially lower taxes). When collective, it allows for better optimization by grouping together consumers (buildings, businesses, transport systems such as electric vehicles, etc.) with different consumption time profiles. Added to these potential gains is the possibility of making the most of energy shedding by modulating consumption, acting on loads or through self-consumption of local production, with or without a storage system. Finally, in addition to these financial gains, there are also gains for the community when the energy consumed is of renewable origin (which is a major objective of self-consumption), as it is distributed over multiple potential production sites (especially rooftop photovoltaics) and consuming locally produced energy avoids or limits the losses incurred by transporting this energy over long distances.

This development will call into question the role of suppliers: faced with a drop in marketing revenue due to the development of self-consumption, they could increasingly take on the role of aggregator, which consists of piloting production facilities, loads and distributed storage, or even that of installer of these infrastructures.

The development of self-generation and self-consumption, the interest in energy flexibility and the quest for energy sufficiency are thus leading each actor to become active in their own production and/or consumption, with a variety of objectives: economic, energy-related, environmental, etc. To enable these new actors to act within the energy system, new methods and tools, such as blockchain, could be implemented.

A blockchain is a collaborative system for managing a distributed ledger that can provide various data-related services: storage, data exchange, code execution (smart contracts). The registry in question takes the form of a "linked list" of interconnected blocks of information. The challenge is to define a management method that prevents a single actor from holding, reading, writing and validating data on the blockchain. The blockchain can be used as the IT solution for managing the transactions of a collective self-consumption operation [GIO 21].

In France, most distribution network operators have a local monopoly on metering, and a single operator manages 95% of the distribution networks. As these operators, playing the role of trusted third parties, are the only ones able to provide the "raw" metering data needed to contractually manage the energy exchanges of a collective self-consumption operation, there is no disintermediation value to blockchain.

However, blockchain could present disintermediation value for collective self-consumption [ECU 17]:

– In France, in the event of derogations from the metering monopoly; the case of Voltalis (consumption aggregator), which has de facto taken on this role on an experimental basis, is an illustration of this.

– In other countries, where the use of blockchain could replace an ad hoc trusted third party for transaction management. This is the case when:

- metering is not subject to a local monopoly, as in the UK;
- several local monopolies are adjacent, as in Germany and Switzerland;
- consumers form consumer groups, as in Switzerland.

The writing and validation of transaction data on a blockchain will generally be managed by the metering operators, or by service providers. In the latter case, operators transmit "raw" metering data to service providers, who return the results integrating transactions (smart contracts that automatically execute autonomously) between collective self-consumption participants.

This chapter proposes a techno-economic energy exchange model using a game-theoretic approach combined with a (distributed, secure and decentralized) blockchain, to maximize self-consumption within the community [STE 20, STE 21a, STE 21b]:

– Actors exchange energy without intermediaries (decentralized market).

– Exchanges are automated according to photovoltaic (PV) production and consumption via smart contracts, taking into account cost, PV self-consumption and actor comfort.

– A wide range of actors can be taken into account: permanent consumers, transient consumers such as electric vehicles (EVs), consumer/producer, producer, storage operator, public distribution network operator, etc.

Section 6.2 introduces the notion of a local renewable energy community.

Section 6.3 presents the electrical network of the Université Catholique de Lille's historical island demonstration site, developed through the Live TREE program [ROB 23], which serves as a case study for the application of the energy management methods developed in this chapter.

Section 6.4 proposes a distributed optimization based on game theory and the Alternating Direction Method of Multipliers (ADMM) algorithm. Through utility functions (inspired by the developments presented in Chapter 4), actor profiles are taken into account. Simulations on the digital twin of the demonstration site's electrical network illustrate the performance of the proposed methodology.

Section 6.5 deals with the management of energy exchanges via blockchain technology. An implementation example is developed. The energy consumption of different blockchain technologies is evaluated and compared with more conventional centralized management.

The development of renewable energy communities clearly contributes to SDG 7 by promoting and optimizing the use of clean energy. Taking into account electric mobility, in addition to seeking renewable energy consumption by buildings, contributes to SDG 11 "Sustainable cities and communities".

Renewable energy communities are part of ADEME's S2 scenario, "Territorial Cooperation".

6.2. Local renewable energy communities

With the falling cost of renewable energies, and photovoltaics in particular, as well as the financial support provided by governments, some consumers have chosen to install their own photovoltaic systems to generate their own electricity. In France, there are two possible situations: either they sell all this electricity to Électricité de France (EDF) at advantageous rates, or they consume it themselves. The latter case corresponds to individual self-consumption, in which the consumer reduces their bill to their energy supplier [ADE 19]. To take full advantage of this situation, the challenge is to increase the rate of self-consumption, that is, the proportion of local production that is consumed locally. As photovoltaic production is intermittent, this means adapting consumption to production periods [LUT 15].

For a single consumer, the possibilities of making consumption coincide with production are limited: a residential consumer, for example, has little electrical equipment that can effectively increase this flexibility. Therefore, to make the most of local production facilities and encourage their development, collective self-consumption was authorized in France in 2017 for producers and consumers on the low-voltage grid [LEG 17]. Gradually, the maximum radius of self-consumption operations has been agreed and it can now be up to 20 kilometers [ECO 17]. The idea behind collective self-consumption is to take advantage of the overlapping nature of consumption, that is, the differences in consumption profiles between different users, and to combine their flexibility to bring consumption into line with local production, thereby maximizing the self-consumption rate of an installation.

Figure 6.1 shows both situations of individual and collective self-consumption on the power grid.

In Europe, other countries have gradually followed the same trajectory, authorizing and then gradually increasing the radius of collective self-consumption operations, first within a building, and then on a neighborhood scale. However, the development of collective self-consumption is being driven above all by the European Union, which has introduced the concept of energy communities into its Clean Energy Package [EUR 17]. This text was adopted by the European Parliament in 2019. It defines renewable energy communities and citizen energy communities. The first term refers to a situation in which users join forces to share production from renewable sources on a local scale. The similar term "citizen energy community" has a broader definition and does not refer only to renewable energy sources [FRI 19]. Through legislating on such concepts, the European Union hopes to facilitate the insertion of renewable energies into electricity grids and also better integrate consumers into the electricity system, giving them a more active role, and thus fostering the acceptability of renewable energy sources [AZA 19]. The stated aim is for such communities to own 17% and 21% of solar and wind generation capacity, respectively, by 2030 [EUR 17].

The 2019 European directive was transposed into French law in March 2021, with the legal framework for renewable and citizen energy communities gradually becoming clearer. The aim is to be able to go beyond collective self-consumption as defined by the French legislator, which is rather restrictive, limiting the development of consumption modes [STE 21b].

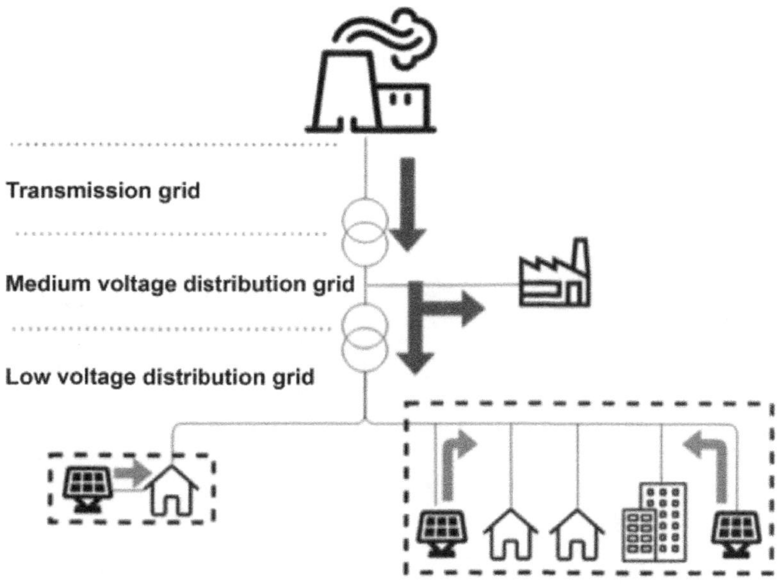

Figure 6.1. *Two situations of individual self-consumption (bottom left) and collective self-consumption (bottom right) on the electricity grid [STE 21b]*

6.3. Modeling a tertiary-sector case study

6.3.1. *Historic block at the Université Catholique de Lille*

The electrical network of the Université Catholique de Lille's historic block demonstration site serves as a case study for the application of the energy management methods developed in this chapter. The Université Catholique de Lille, through the Live TREE program [ROB 23], has embarked on an energy renovation of the campus, promoting energy sufficiency, energy efficiency and the installation of renewable energy sources. Objectives include encouraging energy-saving behavior and significantly increasing self-consumption and self-generation.

The power grid in question, shown in Figure 6.2, has the following infrastructure:

– two photovoltaic power plants with outputs of 185 kWp and 28 kWp;

– a 203 kWh Li-ion battery, with a charging capacity of 40 kW, a discharging capacity of 80 kW and an efficiency of 97%;

– six charging points for electric vehicles, rated at 22 kW;

– four buildings for different types of loads (air conditioning, lighting, heating, etc.);

– a connection point to the medium voltage (15 kV) distribution grid.

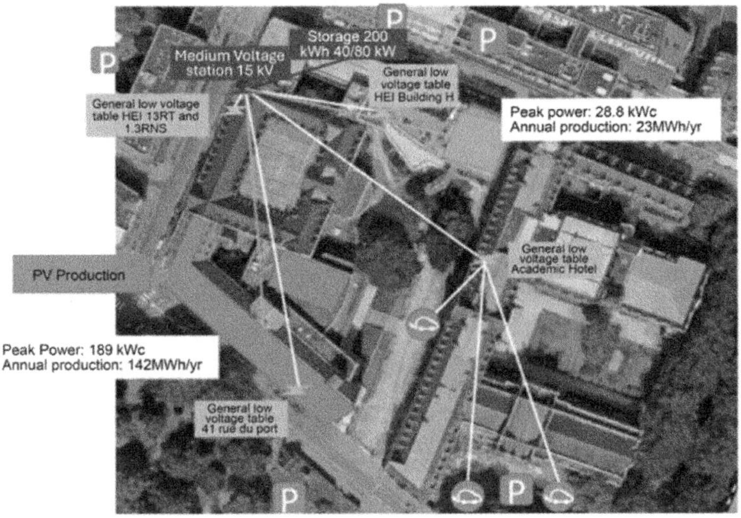

Figure 6.2. *The historic block at the Université Catholique de Lille demonstration site [STE 21b]*

In addition, a central monitoring system has been developed [ROB 23], which collects electrical data from the network (consumption and production power, voltages) and controls storage. It also provides consumption and photovoltaic production forecasts (see Chapter 3).

With all this equipment, the network forms a real local energy community, with a wide range of actors. It is therefore a relevant model for studying the optimization of energy exchanges between users. The model presented in this section has been developed with this case study in mind. However, it can be adapted to any other energy community.

6.3.2. *Modeling the electrical network*

The PowerFactory software package, published by DIgSILENT, is used to develop a physical model of the network and perform power flow calculations. It is used in particular by distribution and transmission system operators [KUL 10]. In this software, a network model of the Université Catholique de Lille's historic block is created. The model is shown in Figure 6.3, which distinguishes between loads (red), battery storage (blue), photovoltaic generators (green) and electric vehicle charging points (orange).

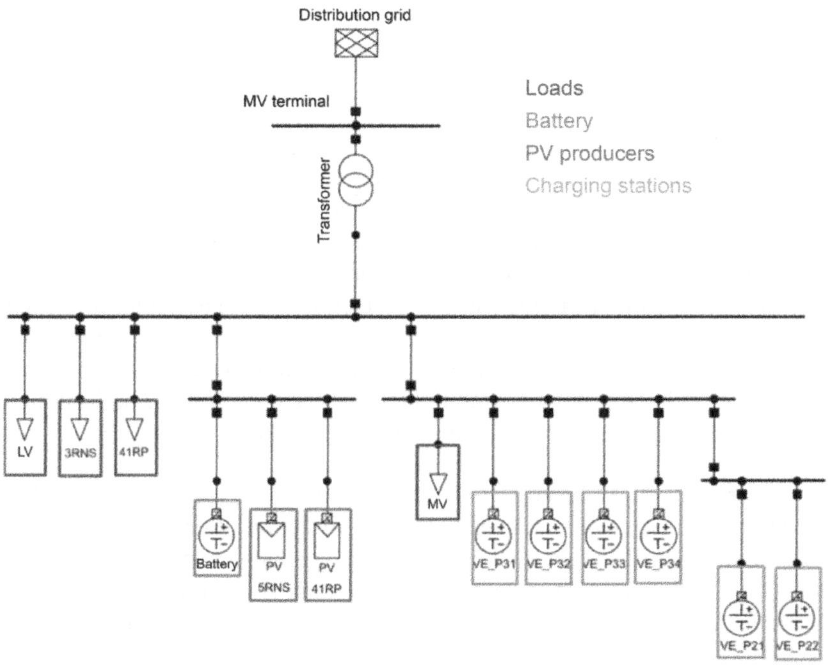

Figure 6.3. *Electrical model of the Université Catholique de Lille historical block: framed loads in red, storage in blue, photovoltaic generators in green, electric vehicle charging stations in orange [STE 21b]*

In this model, the connection to the external network is modeled by a distribution network element called the external network. In concrete terms, this is the reference node (slack bus), which defines the reference phase and for which the voltage is imposed. Loads and electric vehicles are modeled by PQ nodes (active and reactive power), for which consumption profiles are given as input.

Following the models proposed by PowerFactory [GON 18], a PQ node model for the battery and photovoltaic generators is implemented, with production profiles from optimization as input data.

Lines are modeled on the basis of technical documentation, which specifies the type of cable used and its length. PowerFactory offers a cable library, which ensures correct line modeling [STE 21b].

6.4. Distributed energy optimization

6.4.1. *Introduction*

Energy communities bring together a variety of actors, whether energy producers or consumers, with different habits and behaviors, who exchange electrical energy on a local scale. As mentioned in section 6.2, the interest of these communities lies in the coordination of their participants to optimize the use of locally produced energy, promote self-consumption and thus facilitate the integration of renewable energy sources as close as possible to points of consumption. Cooperation is therefore at the very heart of how these communities operate: actors can influence their behavior and reach a consensus to achieve a common goal, be it minimizing electricity costs or increasing the rate of self-consumption.

More formally, the objective of such a community can be described by an optimization problem, which defines the means of action of each participant, on the one hand, and the goal sought by the community as a whole, on the other. Therefore, the selection and development of an appropriate optimization method is a real challenge for ensuring the coordination of community participants. Indeed, over and above performance, in terms of achieving the desired objective and convergence, more pragmatic considerations must also be taken into account: "Is the chosen method relevant to an energy community? Is it applicable in a real-life situation? What does it imply for the participants?"

Indeed, the chosen formulation of the optimization problem reflects a certain representation of the energy community and is based on precise hypotheses concerning the behavior of the agents. The choice of mathematical model must therefore be carefully considered and answer the questions mentioned above.

6.4.2. *Energy exchanges within communities*

6.4.2.1. *Game theory*

With the emergence of distributed approaches, the scientific community is also increasingly interested in game theory. This mathematical framework, originally developed for micro-economics, is an analytical tool for studying interactions between several supposedly rational actors [ROB 19]. It is therefore used to describe a system in which several agents interact, and in which each seeks to fulfill its own objectives.

In this way, game theory can be used to build distributed approaches, with a vision that first focuses on the behavior of each agent in the system [LI 13]. More precisely, starting from a global optimization objective, game theory consists of defining a sub-problem for each agent in the system, in which the agent rationally optimizes its individual situation. Then, in a second step, an algorithm called gradient play is employed so that the agents can iteratively converge towards an equilibrium [YAN 10]. Game theory thus provides a framework for predicting and analyzing a situation in which different agents react individually and rationally to limited information from their environment [BOU 14].

Game theory divides games into two categories: cooperative and non-cooperative, depending on whether actors can coordinate or compete with each other. Generally speaking, the behavior of each user is described by a utility function, that is, a mathematical function representing the objectives of the actor in question, and a strategy, that is, the set of possible actions. Since each actor is assumed to be rational, they act on their strategy to maximize their utility, which is sometimes expressed in terms of well-being. In the context of distributed control, writing utility functions is a system modeling choice, but the advantage of this notion is that it can take into account the socio-economic aspects of agents' behavior. In this way, game theory can be used to describe a variety of situations, with actors having a wide range of individual behaviors and objectives. The mathematical framework of game theory guarantees, under certain conditions of utility functions, that the system arrives at a state of equilibrium, called a Nash equilibrium: in this situation, no actor can unilaterally increase their utility, in other words, no actor has any interest in changing their strategy [LIU 17].

Current uses of game theory for the power grid include microgrid management, demand side management and load optimization for a fleet of electric vehicles [MOR 16].

Game theory is therefore a relevant tool for the power grid, to build distributed optimization approaches and promote grid decentralization. Indeed, since utility functions can account for a variety of behaviors and objectives, game theory is able to capture the heterogeneity of the network. Numerous situations can thus be modeled, and distributed optimization techniques applied.

6.4.2.2. *Community energy-sharing markets*

Game theory and distributed optimization are also applied to the problem of energy exchanges within energy communities. In this context, users (consumers, producers, prosumers) of a local electricity network exchange electrical energy directly with each other, thus forming a local electricity market. Three main market categories are generally distinguished: community structures, peer-to-peer markets and mixed approaches [SOU 19].

6.4.2.2.1. Community-based markets

In this type of market, an agent coordinates exchanges within the energy community. In concrete terms, this role can be played either by the grid operator or by an energy manager, who is responsible for energy exchanges. In [TUS 21], the authors use the term "community manager" to describe this role in general terms, as the intermediary between the community and the distribution network. This community market is illustrated in Figure 6.4.

The community manager decides on the price of electricity or applies a price mechanism previously accepted by all community users. This pricing mechanism reflects the general interests of the community, such as encouraging the sharing of locally produced energy, or minimizing costs. In this structure, agents locally optimize their behavior according to this price mechanism, deciding, for example, how much energy they wish to buy or sell. The community manager collects this information and then organizes energy and financial exchanges between the users of the energy community [TUS 21].

Thanks to the role played by the energy manager, this community market structure has the particular advantage of involving all community participants on an equal footing. They are treated in the same way by the energy manager and have the same responsibilities. In addition, this energy manager can act as an aggregator at the interface between the community's producers and consumers, on the one hand, and the distribution system operator (DSO), on the other, and can therefore offer services to the operator, such as reducing peak load over a certain period.

On the other hand, the main problem with a market of this type is protecting the privacy of users, who must communicate their data to the energy manager, who thus reintroduces a form of centralization.

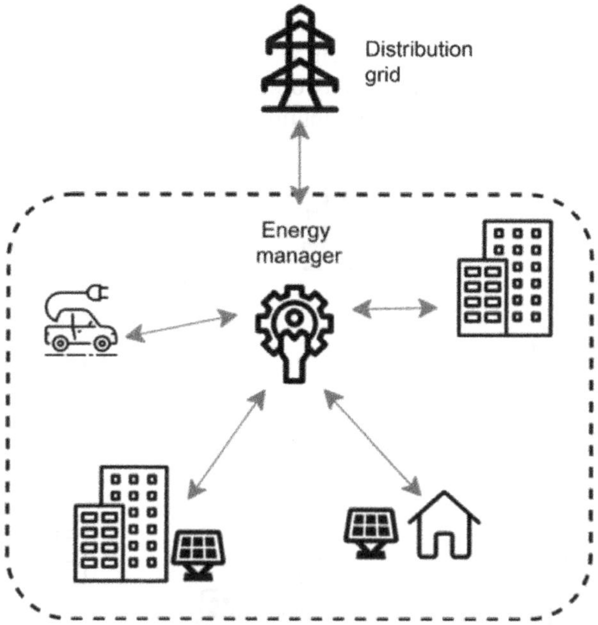

Figure 6.4. *Diagram of a community market [STE 21b]*

6.4.2.2.2. Peer-to-peer market

Peer-to-peer (P2P) markets represent a more decentralized approach to managing energy exchanges within a community. In this situation, community users exchange electricity directly, without going through an aggregator or energy manager. They therefore directly set the price of the electricity they buy or sell, negotiate among themselves and carry out transactions directly [SOR 19]. In general, the negotiation phase leads to a balancing of offers to buy and sell electricity around an equilibrium price. This process is known as *market clearing*. Figure 6.5 illustrates the peer-to-peer market.

The P2P approach is interesting because it gives community actors total freedom in the exchanges they make. In theory, they are free to set the costs and quantities of electricity they wish to exchange. This structure allows for completely decentralized management.

However, to be effective, a P2P market requires a significant communication infrastructure, as each participant in the community must be in contact with all the other participants, so as not to bias the market mechanism. On the other hand, convergence towards an equilibrium may take longer to achieve, as participants have more freedom. Finally, with this approach, we lose the possibility of the community providing a service to the distribution network. For the moment, this is a developing concept of electricity sharing on the distribution network.

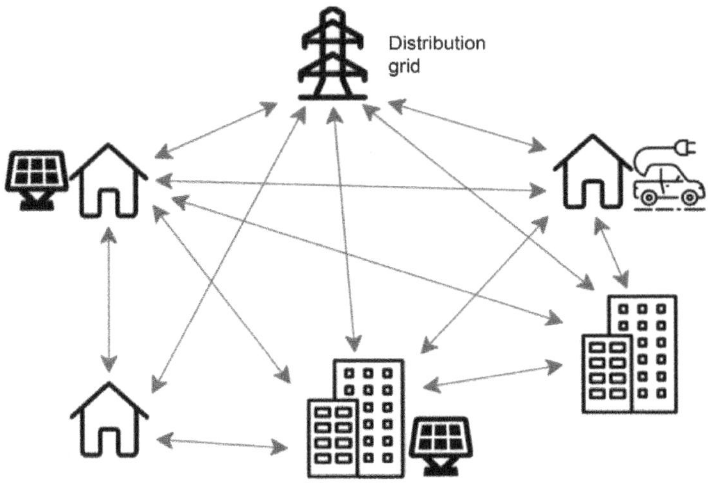

Figure 6.5. *Diagram of a peer-to-peer (P2P) market [STE 21b]*

6.4.2.2.3. Hybrid market

Finally, the last category corresponds to a hybrid structure drawing on the previous two, containing two levels. Firstly, there is a P2P exchange level in which individual agents and community managers exchange energy directly. On a second level, users within communities exchange energy using community market mechanisms.

This mixed approach between a community market and a pure P2P market is interesting because it combines the advantages of both categories. It gives actors freedom, as well as preserving the services that communities can provide to the distribution network. Moreover, in theory, it allows actors to combine exchanges regulated by a community manager within a community, and deregulated exchanges with a P2P market [SOR 19]. However, this approach is currently very theoretical, and in practice requires coordination that is difficult to implement to make both individual agents and renewable communities interact within the same mechanism [TUS 21].

6.4.3. *Distributed optimization of energy exchanges with game theory*

6.4.3.1. *Objectives*

In this section, an energy management method adapted to an energy community is developed, taking as a case study the historical block of the Université Catholique de Lille power grid and considering the following objectives [STE 21a, STE 21b]:

– The strategy must be distributed, to encourage decentralization of the power grid and promote the inclusion of end-users in electricity management mechanisms. What is more, the distributed nature of the system limits the exchange of information between users, and therefore better protects their privacy.

– To be applicable to real-life energy communities and acceptable to the actors involved, the strategy must consider the diversity of actors, as well as their consumption (or production) habits and preferences. With this in mind, game theory is used to model the behavior of these actors and include aspects that are not solely economic.

– A community-based approach is favored over a purely P2P approach. Indeed, even if P2P approaches allow for more complete decentralization, they remain rather theoretical and, in practice, are subject to legislation to authorize the buying and selling of electricity directly between users. However, the aim is to develop a management strategy that can be applied in real cases, notably to the Université Catholique's energy community, and for this reason a community approach is favored. Concerning the role of the community manager, section 6.5 explains how a blockchain can be used specifically to fulfill this role and avoid recentralization through the presence of such an agent.

6.4.3.2. *Formulating the optimization problem*

This section describes the modeling of the mathematical problem for optimizing energy exchanges within an energy community. For the reasons explained in section 6.4.2.1, game theory is used to model the diversity of actors and take into account their preferences. Furthermore, it is assumed that actors are not naturally inclined to cooperate with each other, but instead act with purely individual reasoning. Indeed, assuming cooperation between actors is a strong assumption, and not always a realistic one, given that participants are primarily guided by personal considerations [MAR 10]. The problem is then modeled by a non-cooperative game, which therefore applies to more general situations than a cooperative game [STE 21b].

The non-cooperative game G defined by $G = \{N, (S_i)_{i \in N}, (U_i)_{i \in N}\}$, where N represents the set of N actors, that is, the participants in the energy community, is therefore considered. Each actor is associated with a utility function U_i, sometimes also called the objective function in optimization. This function is a representation of the actor's satisfaction or well-being. In other words, it mathematically models the individual goals of the actor in question. In the economic theory that gave rise to game theory, this utility function is used to model objectives that are not just financial, but more social or behavioral, such as comfort, or satiation in the consumption of a good [LIU 17]. The use of game theory thus enables us to take more detailed account of the behavior of actors, without limiting ourselves to purely financial interests.

The set $(S_i)_{i \in N}$ represents the set of actors' strategies, that is, the set of their possible actions. Game theory assumes that actors are rational agents, that is, they adapt their strategy to optimize their utility. Here, the strategy of an actor i is their consumption profile x_i, or production profile, depending on their role (consumer or producer), defined with a time interval Δt. Average powers are therefore considered over time intervals of the type $[t; t + \Delta t]$. Adapting the model defined by [TUS 21] and used in particular in [MOR 19a], the profile x_i is defined more precisely by $x_i = (p_i, q_i, \lambda_i, \mu_i)$ with the following variables:

– p_i is the net power output for each actor. Therefore, $p_i > 0$ corresponds to production, and $p_i < 0$ to consumption.

– q_i represents the local photovoltaic power imported by agent i. For photovoltaic generators, $q_i < 0$ because they export their production.

– λ_i is the power imported from the distribution network. For all agents, $\lambda_i \geq 0$. This is because it is assumed that the photovoltaic generators and the battery cannot send energy back to the distribution grid. This assumption is consistent with the objective of maximizing self-consumption within the energy community, and therefore obligatorily consuming local generation in the neighborhood.

– r_i represents, for actors who are consumers, the power imported from the battery (in this case $r_i > 0$), and for the storage system, the power exported to consumers (in this case $r_i < 0$).

A power balance imposes the following relationship between the preceding variables:

$$p_i(t) + q_i(t) + \lambda_i(t) + r_i(t) = 0, \forall i \in N, \forall t \in [t; t + \Delta t] \quad [6.1]$$

Each actor optimizes its profile x_i in advance, for the period between instants t_i and t_f, in time intervals Δt. It is assumed that all users know the total local photovoltaic generation forecast $P_{PVtotal}^{predicted}$, the price π_{imp} of electricity imported from the grid as well as the price π_{PV} of locally generated photovoltaic electricity and the price π_s of stored electricity.

In the rest of the chapter, $S = \{S_1, \ldots, S_N\}$ represents the set of strategies and $X = [x_1, \ldots, x_N]$ the set of profiles of all actors. Also, so as not to weigh down the equations, time dependence is not systematically specified: unless otherwise stated, all the equations that follow in this chapter are true for any t in $[t_i; t_f]$ [STE 21a, STE 21b].

6.4.3.2.1. Individual objectives

The utility functions must transcribe the objectives of each actor. In [JIA 20], for example, the authors carried out a sociological analysis of the most important factors for consumer participation in energy management mechanisms. Based on the findings of this study, we can distinguish three main objectives pursued by each actor:

– electricity cost minimization (or profit maximization for producers);

– maximizing comfort, defined as the fact of having a production or consumption profile identical to the one initially desired by the user;

– maximizing the use of locally produced renewable energy.

Utility functions therefore combine these objectives. To account for the variety of actor profiles and behaviors, weighting coefficients are introduced to prioritize the sensitivities of each user. Each actor is assigned a utility function according to their type (consumer, producer, electric vehicle user, storage system). The actor themself then defines their profile and preferences by prioritizing the three objectives specified above, via these coefficients.

The use of such weighting coefficients is a classic feature of multi-objective optimization. In the remainder of this chapter, these coefficients are denoted with the Greek letters α, β, γ and δ. To be able to compare the different preferences of each actor on the same scale, these coefficients must lie in the interval [0;1], and for each user, the sum of these coefficients must be 1 [KAN 17].

In practice, the blockchain system presented in section 6.5 constitutes a distributed platform in which each user can directly indicate the value they choose for these coefficients for the upcoming optimization period, according to their personal preferences [STE 21a, STE 21b].

6.4.3.2.2. Electric vehicles

It is assumed that an electric vehicle user, when arriving at the charging station, specifies their arrival time and their departure time (respectively denoted t_{arrive} and t_{depart}), as well as the minimum state-of-charge (SOC) of their vehicle on departure (denoted $SOC_{EV}^{required}$). The initial state-of-charge SOC_{EV} ($t = t_{\text{arrive}}$) is assumed to be automatically recovered by the terminal when the vehicle arrives.

The following *utility function* is proposed for electric vehicle users:

$$U_{EV}\left(x_{EV}\right)=-\alpha_1\left(\pi_{PV}\cdot q_{EV}+\pi_{imp}\cdot \lambda_{EV}+\pi_s\cdot r_{EV}\right)$$
$$\cdot\Delta t-\alpha_2\left(p_{PV,total}^{predicted}\cdot\Delta t-q_{EV}\cdot\Delta t\right)^2+\alpha_3\ln\left(1-p_{EV}\cdot\Delta t\right) \quad [6.2]$$

The first term (weighted by α_1) of this function is akin to the cost of electricity paid for the load, depending on where the electricity comes from. The second term $\alpha_2(P_{PVtotal}^{predicted}\cdot\Delta t - q_{EV}\cdot\Delta t)^2$ corresponds to the willingness to consume local photovoltaic generation. Finally, the term $\alpha_3\ln(1-p_{EV}\cdot\Delta t)$ represents comfort, which is defined as the willingness to have an electric vehicle as fully charged as possible at the end of the charging period. For this term, a logarithm, which is classically used in the scientific literature to model energy consumers, is implemented. The logarithm function accounts for consumption satiation [DEH 18].

As previously stated, the actor prioritizes their preferences with the coefficients α_1, α_2 and α_3, all within the interval [0;1] and which verify: $\alpha_1+\alpha_2+\alpha_3=1$. What is more, each of the three terms of the utility function can take on very different values from the other terms, which therefore prevents a comparison. So, to be able to rank these three terms on the same scale, in practice they must be normalized, so that they take their values equally in the set [0;1]. This applies to all the utility functions presented in this section. However, in order not to detract from the clarity of the equations, only the non-normalized equations are presented.

The following *constraints* apply to the profile $x_{EV}=(p_{EV}, q_{EV}, \lambda_{EV}, r_{EV})$:

$$\begin{cases} p_{EV}+q_{EV}+\lambda_{EV}+r_{EV}=0 & [6.3] \\ p_{EV}(t)=0,\forall t\notin\left[t_{arrive};t_{depart}\right] & [6.4] \\ 0\leq -p_{EV}\leq -P_{EV}^{\max} & [6.5] \\ SOC_{EV}^{\min}\leq SOC_{EV}\leq SOC_{EV}^{\max} & [6.6] \\ SOC_{EV}\left(t=t_{depart}\right)\geq SOC_{EV}^{required} & [6.7] \end{cases}$$

Equation [6.3] is simply the power balance for the electric vehicle. Constraint [6.4] dictates that the charging power is zero when the vehicle is not connected to a charging point. Constraint [6.5] requires charging power to be negative and less than the maximum charging power of the terminal to which the vehicle is connected, denoted P_{EV}^{max} This is because the electric vehicles considered in this study are not reversible, so they cannot send energy back to the community.

Secondly, constraints apply to the electric vehicle's state-of-charge. According to [WOU 15], in line with the sign conventions adopted here, the state-of-charge is defined by:

$$SOC_{EV}(t+\Delta t) = SOC_{EV}(t) - \frac{\eta_{EV} \cdot \Delta t}{E_{EV}^{\max}} p_{EV}(t) \qquad [6.8]$$

with η_{EV} being the efficiency of the electric vehicle under load and E_{EV}^{max} being the vehicle's battery capacity.

Constraint [6.6] requires the state-of-charge to be between the maximum and minimum levels defined by the manufacturer. Furthermore, equation [6.7] guarantees that when the user returns to the vehicle at the end of the charging period, they will find it again with a charge level at least equal to the one they required, thus ensuring a certain level of satisfaction [STE 21a, STE 21b].

6.4.3.2.3. Electrical energy storage system

In this study, the storage system is considered as a full-fledged actor in the energy community. Consequently, a utility function specific to this user is defined [STE 21a, STE 21b].

For the storage system, the notion of comfort has no meaning. Consequently, the *utility function* has only two terms, representing the price and consumption of local photovoltaic production:

$$U_s(x_s) = -\beta_1\left(\pi_{PV} \cdot q_s + \pi_{imp} \cdot \lambda_s - \pi_s \cdot r_s\right) \cdot \Delta t - \beta_2\left(p_{PV,total}^{predicted} \cdot \Delta t - q_s \cdot \Delta t\right)^2 \qquad [6.9]$$

As for electric vehicles, the weighting coefficients β_1 and β_2 are between 0 and 1 and verify $\beta_1 + \beta_2 = 1$.

The following *constraints* apply to the storage system:

$$\begin{cases} p_s + q_s + \lambda_s + r_s = 0 & [6.10] \\ -P_s^{charge} \leq -p_s \leq -P_s^{discharge} & [6.11] \\ SOC_s^{\min} \leq SOC_s \leq SOC_s^{\max} & [6.12] \\ SOC_s(t_f) = SOC_s(t_i) & [6.13] \end{cases}$$

Equation [6.10] is always the power balance applied to the storage system. Constraint [6.11] bounds the charge and discharge power between the limit values

P_s^{charge} and $P_s^{discharge}$, specified by the battery manufacturer (and assumed positive in this study).

Constraint [6.12] limits the state-of-charge between the operating limits SOC_s^{min} and SOC_s^{max}, and equation [6.13] requires the battery to return to its initial state-of-charge at the end of the optimization period, to cycle. Similar to electric vehicles, with the sign conventions adopted, the battery's state-of-charge is defined by:

$$SOC_s(t+\Delta t) = SOC_s(t) - \frac{\eta_s \cdot \Delta t}{E_s^{max}} p_s(t) \qquad [6.14]$$

6.4.3.2.4. Photovoltaic generators

Photovoltaic generators aim to maximize their income, maximize exchanges within the community and have a production profile equal to their forecasts. The latter can be likened to a notion of comfort: the photovoltaic generator wishes to optimize its production as far as forecasts allow, without having to resort to clipping. As there is a delay between optimization and the application of the setpoints, fluctuations in real-time production are not taken into account. In other words, for a photovoltaic generator, a difference between the optimized profile and the forecast production profile, denoted $p_{PV}^{predicted}$, represents a loss of comfort for this actor (hence the importance of good forecasts as explained in Chapter 3). The following *utility function* is therefore proposed:

$$U_{PV}(x_{PV}) = -\gamma_1 \pi_{PV} \cdot q_{PV} \cdot \Delta t - \gamma_2 \left(p_{PV}^{predicted} \cdot \Delta t - q_{PV \cdot \Delta t}\right)^2 - \gamma_3 \left(p_{PV}^{predicted} \cdot \Delta t - p_{PV} \cdot \Delta t\right)^2 \qquad [6.15]$$

For the comfort term, a quadratic function is used, as in [MA 14, ROS 65]. In practice, these three terms all aim to maximize photovoltaic production as much as possible and are redundant here. However, the advantage of distinguishing three different terms, even if they have the same effects, is in constructing a generic model, which can be adapted to other situations and other types of generators, for example. The three coefficients γ_1, γ_2 and γ_3 are positive and such that $\gamma_1 + \gamma_2 + \gamma_3 = 1$.

For photovoltaic generators, $\lambda_{PV} = 0$ and $r_{PV} = 0$, as they do not import electricity from the grid, nor from the battery. Consequently, the following constraints apply:

$$\begin{cases} p_{PV} + q_{PV} = 0 & [6.16] \\ 0 \le p_{PV} \le p_{PV}^{predicted} & [6.17] \end{cases}$$

Equation [6.16] specifies the power balance for a photovoltaic generator. Since $\lambda_{PV} = 0$ and $r_{PV} = 0$, this relationship simply indicates that all photovoltaic production is exported to the community, and not to the distribution grid.

As already mentioned, optimization does not take place in real time, but in advance, and therefore $p_{PV}^{predicted}$ represents the maximum production profile for the optimization period, which is reflected in the constraint [6.17], according to [STE 21a, STE 21b].

6.4.3.2.5. Consumers

Within an energy community, the consumers are the buildings, whether tertiary or residential. In the approach considered in this study, consumers can generally be residents, or the managers of tertiary buildings. More pragmatically, the actors considered here are those who actually have their own electricity bill, in other words, those who are responsible for their own consumption. In this study, the management of consumption within a home or a tertiary building is considered, the modeling stopping at the electricity meter of these consumers.

The three main objectives are to reduce electricity costs, maximize comfort and consume local photovoltaic production. It is assumed in this model that each consumer knows their predicted consumption $p_c^{predicted}$ (denoted positive) for the next optimization period. The following *utility function* is proposed, defining comfort in the same way as for photovoltaic generators, by a quadratic function:

$$U_c(x_c) = -\delta_1\left(\pi_{PV} \cdot q_c + \pi_{imp} \cdot \lambda_c + \pi_S \cdot r_c\right)\Delta t - \delta_2\left(p_{PV,total}^{predicted} \cdot \Delta t - q_{load} \cdot \Delta t\right)^2 - \delta_3\left(p_c^{predicted} \cdot \Delta t - p_c \cdot \Delta t\right)^2 \quad [6.18]$$

The weighting coefficients δ_1, δ_2 and δ_3 are also positive and verify $\delta 1 + \delta_2 + \delta_3 = 1$.

It is assumed that consumers have a certain flexibility $f \in [0;1]$, which corresponds to the fraction of consumption that can be shifted in time. It is this flexibility that allows consumers to play on their consumption profile.

For consumers, the following constraints apply:

$$\begin{cases} p_c + q_c + \lambda_c + r_c = 0 & [6.19] \\ (1-f)\, p_c^{predicted} \leq -p_c \leq (1+f)\, p_c^{predicted} & [6.20] \\ -\sum_{t=t_i}^{t_f} p_c(t) = \sum_{t=t_i}^{t_f} p_c^{predicted}(t) & [6.21] \end{cases}$$

Equation [6.19] expresses the power balance for a consumer. Constraint [6.20] defines the flexibility, that is, the bounds within which the optimized consumption profile can vary from the planned profile. Furthermore, to guarantee a minimum level of comfort, consumers can simply shift their consumption, but not decrease the energy consumed over the entire optimization period. In other words, the amount of energy consumed remains the same overall, to ensure a minimum level of comfort for consumers. This is indicated by constraint [6.21], according to [STE 21a, STE 21b].

6.4.3.2.6. Global optimization problem

In the model developed, each stakeholder in the community optimizes their utility function (which depends on their type), according to their preferences, which they specify with weighting coefficients. At the level of the energy community, as a whole, this amounts to solving the following problem:

$$\max_{x_i \in \mathcal{S}_i} \sum_{i \in \mathcal{N}} U_i(x_i) \quad [6.22]$$

under local constraints [6.3]–[6.7], [6.10]–[6.13], [6.16], [6.17] and [6.19]–[6.21].

Moreover, an additional constraint is imposed on the energy community: a maximum power limit (denoted P_{max}) that the community can draw from the distribution network. In the case of the Université Catholique network, this constraint corresponds to the withdrawal limit authorized by the network manager.

This constraint is therefore written as:

$$\sum_{i \in \mathcal{N}} \lambda_i \leq P_{max} \quad [6.23]$$

Furthermore, exchanges within the community, between photovoltaic generators, the storage system and consumers, must always be balanced. Consequently, the following two global constraints also apply:

$$\begin{cases} \sum_{i \in \mathcal{N}} q_i = 0 & [6.24] \\ \sum_{i \in \mathcal{N}} r_i = 0 & [6.25] \end{cases}$$

The problem of optimizing energy exchanges within the energy community is then completely defined. As equation [6.22] shows, in reality, the self-consumption rate is not optimized as such, but the sum of the actors' utilities, that is, the overall well-being of the actors in the energy community. In the distributed approach considered in this study, the optimization of the self-consumption rate results from the actions of the actors, and therefore from their preferences [STE 21a, STE 21b].

6.4.3.2.7. Nash equilibrium

An important result of game theory is the concept of Nash equilibrium. This term defines a situation in which no actor can increase their utility if they are the only one to change their strategy. Thus, in this situation, the system reaches equilibrium [LIU 17]. Mathematically, the Nash equilibrium, denoted $X^* = [x_1^*, \ldots, x_N^*]$, is defined by:

$$U(x_i^*, X_{-i}^*) \geq U(x_i, X_{-i}^*), \forall i \in \mathcal{N}, \forall x_i \in S_i \qquad [6.26]$$

with: $X_{-i} = [x_1, \ldots, x_{i-1}, x_{i+1}, \ldots, x_N]$.

In the system modeling developed in this chapter, all utility functions U_i are concave with respect to each of the $x_i = (p_i, q_i, \lambda_i, r_i)$ variables, for all actors i. On the other hand, for each type of actor, the constraints define strategy sets S_i that are closed sets of $\mathbb{R}$. Consequently, these two conditions guarantee the existence of at least one Nash equilibrium, according to [BOU 14].

Moreover, for an actor i, its utility function U_i depends solely on the variables $(p_i, q_i, \lambda_i, r_i)$) specific to the actor in question, and not on the variables of the other actors.

In other words, the Jacobian matrix of all system functions is diagonal. Since the utility functions are concave, this condition proves the uniqueness of the Nash equilibrium, according to [GOO 80, MÜN 17].

For the model developed in this chapter, the existence and uniqueness of the Nash equilibrium shows that, if each actor individually optimizes their utility function, then the system as a whole arrives at an equilibrium situation [STE 21a, STE 21b].

6.4.3.3. *Optimization algorithm: ADMM*

6.4.3.3.1. General principle of the algorithm

Once the system model has been developed, the question to be addressed is: How can we achieve a situation of system equilibrium in practice? This requires the use of a distributed optimization algorithm to solve the global problem of equation [6.22].

Among the distributed algorithms available, ADMM offers interesting features for optimizing energy in a community or microgrid [TUS 21]. This algorithm has also been used for energy sharing in [BOY 11, ZHE 18, LIU 19].

This algorithm, based on Lagrangian decomposition, is particularly suitable when a global constraint applies to all actors. The constraint is shared by all actors by progressively imposing a penalty parameter to force actors to respect the constraint. A full description of the algorithm is available in [LEE 20], whose methodology for implementing the algorithm is applied in this study.

The global problem is defined by [6.22], with the global power constraint [6.23], the global equality constraints [6.24] and [6.25] and the local constraints [6.3]–[6.7], [6.10]–[6.13], [6.16], [6.17] and [6.19]–[6.21].

This problem can be rewritten by translating the global constraints with a convex g function to have a problem of the following form:

$$\min_{x_i \in \mathcal{S}_i} \sum_{i \in \mathcal{N}} (-U_i(x_i)) + g(\sum_{i \in \mathcal{N}} x_i) \qquad [6.27]$$

always under the set of local constraints.

Since U_i utility functions are concave, -U_i functions are convex. Similarly, g must be convex for the algorithm to converge. To translate the global constraints into the function g, we use the following form:

$$g\left(\sum_{i \in \mathcal{N}} x_i\right) = \left(\exp \sum_{i \in \mathcal{N}} \left(\lambda_i - P_{\max}^{global}\right)\right)^6 + \left(\exp\left(\sum_{i \in \mathcal{N}} q_i\right)\right)^6 + \left(\exp\left(\sum_{i \in \mathcal{N}} r_i\right)\right)^6 \qquad [6.28]$$

The idea behind writing g in this way is that, as soon as one of the global constraints is violated, the g function takes on very large values with respect to the first term of the global optimization problem $\Sigma_{i \in \mathcal{N}}(-U_i(x_i))$. In practice, then, minimization requires each of the three constraints to be satisfied [STE 21a, STE 21b].

6.4.3.3.2. Implementation

In the remainder of this section, let us note that $\bar{x}$ is the average of x_i, that is, $\bar{x} = (1/N)\sum_{i=1}^{N} x_i$. Two new variables z and u are introduced, defined on the same set as $\bar{x}$. [LEE 20] also defines the primal and dual residue at iteration k+1, respectively, by: $r^{k+1} = x^{k+1} - z^{k+1}$ and $s^{k+1} = \rho(z^{k+1} - z^k)$. The algorithm stops when these two quantities become smaller than the stopping criteria ϵ^{primal} and ϵ^{dual}. In practice, these criteria ensure that the value of z is almost equal to that of x, and that z no longer evolves between two successive iterations. The steps of the algorithm are shown in Table 6.1 (noting ||.|| the Euclidean norm).

Algorithm 1 ADMM algorithm
1. **while** $\left\Vert r^k \right\Vert > \in^{primal}$ and $\left\Vert s^k \right\Vert > \in^{dual}$ **do**
2. $x_i^{k+1} \leftarrow argmin_{x_i}\left(-U_i(x_i) + \frac{\rho}{2}\left\Vert x_i - x_i^k + \bar{x}^k - z^k + u^k \right\Vert^2\right)$
3. $z^{k+1} \leftarrow argmin_z\left(g(Nz) + \frac{N\rho}{2}\left\Vert z - u^k - \bar{x}^{k+1}\right\Vert^2\right)$
4. $u^{k+1} \leftarrow u^k + \bar{x}^{k+1} - z^{k+1}$
5. **end while**
6. **return** $X = [x_1, \ldots, x_N]$

Table 6.1. *ADMM algorithm [STE 21b]*

In the first step (line 2 of Table 6.1), each actor optimizes their own utility function locally, according to their preferences, and under their individual constraints. The term $(\rho/2)\Vert x_i - x^k_i + \bar{x}^k - z^k + u^k\Vert^2$ is the penalty term, applied equally to all actors, to force compliance with global constraints.

Lines 3 and 4 of Table 6.1 correspond to an aggregation step: all the solutions *xi* of the local problems are collected, then the variable z is calculated, taking into account the global constraints through the g function. Finally, the u variable is also calculated.

Following the iteration, the penalty term $L^k = \bar{x}^k - z^k + u^k$ is passed on to all actors for the new iteration.

The variable ρ is the penalty parameter. In practice, the greater the value of this parameter, the faster actors are encouraged to respect the global constraints. This improves convergence, albeit to the detriment of actors' utility values. The choice of this parameter therefore has a major influence on the algorithm's final results. [LEE 20] points out that often, the best method for choosing the value of ρ is empirical. However, to reduce the importance of the initial choice of this parameter and have a balanced value between convergence and local utility optimization, as the authors of [LEE 20] suggest, the value of ρ is adapted at each iteration, as follows:

$$\rho^{k+1} = \begin{cases} t^{+}\rho^{k} & \text{if} \quad \left\|r^{k}\right\|_2 > \mu\left\|s^{k}\right\|_2 \\ \rho^{k}/t^{-} & \text{if} \quad \left\|s^{k}\right\|_2 > \mu\left\|r^{k}\right\|_2 \\ \rho^{k} & \text{if not} \end{cases} \qquad [6.29]$$

The parameters t^{+} and t^{-} quantify the speed at which the parameter ρ increases or decreases, and μ is the criterion for deciding under which condition the value of this parameter is updated. Again, the choice of these parameters affects the speed of convergence, but after a few tests, the values $t^{+} = 5$, $t^{-} = 5$ and $\mu = 20$ are chosen, values very close to those suggested by the authors of [LEE 20] in the general case.

In practice, the ADMM algorithm has a drawback in that, working with $\bar{x}$, during successive iterations it progressively averages x_i. However, in this study, with actors who are producers (and for whom $p_i > 0$) and others who are consumers (with $p_i < 0$), the variables have different signs and, as a result, the algorithm tends to decrease the absolute value of all p_i, thus reducing photovoltaic production and consumption. However, this consequence is undesirable, as it reduces energy exchanges, even if all constraints are respected, simply for mathematical reasons. The self-consumption rate is then artificially improved, as the energy produced locally is reduced.

To overcome this shortcoming, the ADMM is improved as follows: after the first iteration, when the penalty term is negligible, that is, when users have a profile that is optimized to the maximum according to their personal preferences only, the quantity $\Sigma_{i\in N}\, q_i$ is calculated. In this case, there is no point in reducing photovoltaic production. Consequently, the magnitude L^k is adapted for photovoltaic generators only and for all iterations, by not applying a penalty to them, so that they do not suffer from the averaging effect of the algorithm. Only buyers of local photovoltaic electricity will have to reduce their import q_i in successive iterations.

On the contrary, if $\Sigma_{i\in N}\, q_i$ is negative after the first iteration, this means that photovoltaic production is greater than the overall quantity that buyers can or want to import. In this case, it makes sense to decrease photovoltaic production if necessary; however, L^k is modified so that buyers are not also driven to decrease the amount of local electricity they want to import [STE 21a, STE 21b].

6.4.3.3.3. Global vision

With these improvements, ADMM is adapted more specifically to the case of a local energy community. By building a game-theoretic model of the network, a distributed optimization algorithm has been developed which lets actors specify their preferences (via weighting coefficients) and then locally optimize their consumption or production profile. An aggregation step then verifies that the global constraints applying to the community are respected, and if necessary, imposes a penalty, equally on all actors, until the algorithm converges. Figure 6.6 gives an overall representation of this algorithm, for an example with three actors.

Even if each actors optimize their utility locally, an aggregation step is nevertheless necessary to coordinate all the actors, seemingly reintroducing centralization by a controller who would be responsible for collecting all the solutions of local optimizations. The distributed nature of blockchain avoids this phenomenon. This aspect is dealt with in section 6.5.

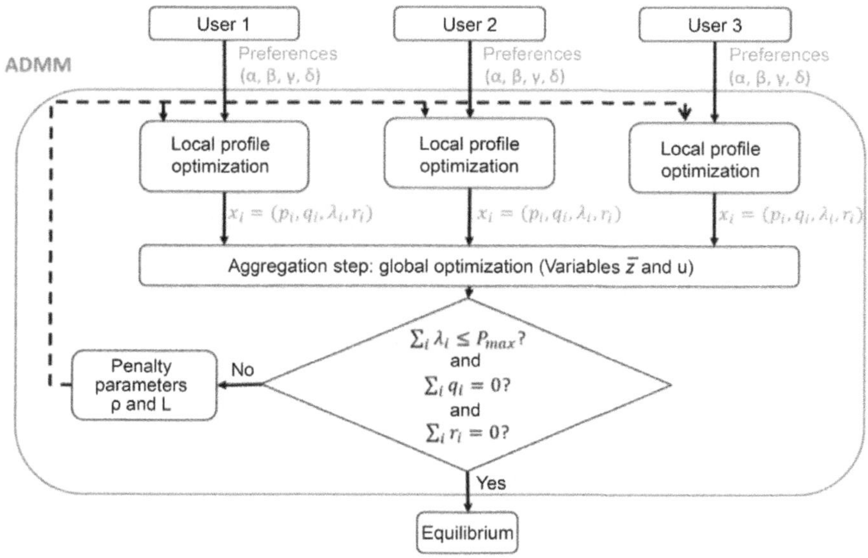

Figure 6.6. *Schematic diagram of the ADMM algorithm [STE 21b]*

6.4.4. *Simulation results*

6.4.4.1. *Hypotheses*

This section presents the results of simulations carried out with data from the Université Catholique de Lille demonstration network presented in section 6.3, in order to analyze the performance of the management strategy developed. Several cases were studied in summer and winter, with different user behaviors. It was assumed that the price of electricity is fixed in advance (PV, storage and distribution network), that production and consumption are known and that optimization will take place on D-1 (that is the day before), with time intervals of 15 minutes [STE 21b]:

a) Simulation parameters. the optimization was simulated for two distinct days, in summer (June 10th, 2019) and winter (December 16th, 2019), with t_i = 00 h 00 and $t_f = t_i$ + 24 h 00 and in time intervals Δt = 15 minutes.

b) Price of imported electricity. The price of electricity imported from the πimp network for the Université Catholique de Lille network is of the peak/off-peak type, with a seasonal variation between summer and winter. For summer 2019, it is 11.15 or 6.05 cEUR/kWh, and in winter 18.04 or 12.26 cEUR/kWh, respectively, in peak and off-peak hours.

c) Price of photovoltaic electricity. The proposed model assumes that locally produced photovoltaic electricity is traded at a single price, regardless of the number of photovoltaic generators in the community, that is, all photovoltaic generators sell at the same price. This selling price is estimated at πPV = 10.0 cEUR/kWh.

d) Price of storage. Considering that a battery is at the end of its life when degradation reaches 80%, the life of a battery, under the considered assumptions, can be evaluated at 8.11 years. However, the installed battery would have an initial capacity of 203 kWh, a total charge/discharge efficiency of approximately 85% (taking auxiliaries into account), an installation cost (CAPEX) of EUR 165,500 and a maintenance cost (OPEX) of approximately 0.01% of the CAPEX. So, following [SCH 19], this gives us an average levelized cost of storage (LCOS) of around 21.90 cEUR/kWh. This price reflects only the cost of electricity storage, but to determine the cost of selling the stored electricity, the cost of charging the electricity must also be taken into account. With a high assumption, we set: $\pi s = \max(\pi_{PV}, \pi_{imp}) + LCOS$.

e) Physical parameters of different users. This section presents the physical parameters used to model the different actors. In total, there are N = 11 actors in the community, with four electric vehicles, one battery, two photovoltaic generators and four tertiary buildings.

It is assumed that four electric vehicles come to charge during the day. The characteristics of the electric vehicles, presented in Table 6.2, concern the Renault Zoé and Hyundai Ioniq models.

The battery characteristics, taken from the manufacturer's data, are summarized in Table 6.3.

Property	EV_1	EV_2	EV_3	EV_4
t_{arrive}	9:15	10:30	16:30	10:00
t_{depart}	12:30	19:00	11:00	18:00
$SOC_{EV}(t = t_{arrive})$	20%	45%	40%	25%
$SOC_{EV}^{required}$	60%	80%	80%	85%
SOC_{EV}^{min}	20%	20%	10%	20%
SOC_{EV}^{max}	100%	100%	100%	100%
$E_{EV}^{max}(\text{kWh})$	41	41	39	41
$P_{EV}^{max}(\text{kW})$	22	22	7.2	22
η_{EV}	80%	80%	80%	80%

Table 6.2. *Physical parameters of electric vehicles [STE 21b]*

Consumers are represented in the Université Catholique de Lille network by four tertiary buildings. For their $P_c^{predicted}$, the actual consumption data are used. Flexibility is essentially based on hot water tanks, heating, lighting, ventilation and air conditioning. According to a study by ADEME [ADE 17], approximately 50% of heating, ventilation and air-conditioning consumption can be shifted in less than an hour. To illustrate the effects of the algorithm when loads have some leeway, this f_c flexibility is set at 25% for all buildings.

Property	Battery
SOC_s^{min}	100%
SOC_s^{max}	10%
$SOC_s(t_i)$	35%
$E_s^{max}(\text{kWh})$	203
$P_s^{charge}(\text{kW})$	40
$P_s^{discharge}(\text{kW})$	80

Table 6.3. *Physical parameters of storage systems [STE 21b]*

For the power limit imported by the whole community from the distribution network, this value is set at Pmax = 580 kW, according to the energy contract.

6.4.4.2. *Scenarios*

In order to test the algorithm in different situations, several scenarios were considered, reflecting different behaviors of the actors. Five scenarios can be distinguished [STE 21b]:

– an economic scenario in which actors want to reduce their costs or maximize their profits (denoted I);

– a scenario in which users favor comfort (denoted II);

– a scenario in which actors primarily want to exchange energy within the community (denoted III);

– a scenario providing a balance of the first three (denoted IV);

– a scenario in which random coefficients are used for each user (denoted V).

It should be remembered that, in a concrete situation, the preference coefficients corresponding to each of these scenarios will be directly indicated by the users themselves, in a local blockchain. The aim of this study was simply to simulate different scenarios in order to analyze the algorithm's results. Therefore, the simulated scenarios do not claim to represent a real situation, but serve as tests for the ADMM algorithm. In particular, scenarios I, II and III correspond to borderline situations in which all actors have broadly the same objectives. Scenario IV represents a balanced situation in which all participants have proportionate objectives. Finally, in scenario IV, random coefficients are chosen.

6.4.4.3. *Results*

In this section, simulation results for a summer day (June 10th, 2019) are presented. Figure 6.7 illustrates the price of electricity coming from the public grid in blue (π_{imp}), of photovoltaic electricity in green (π_{PV}) and of storage electricity in orange (π_s).

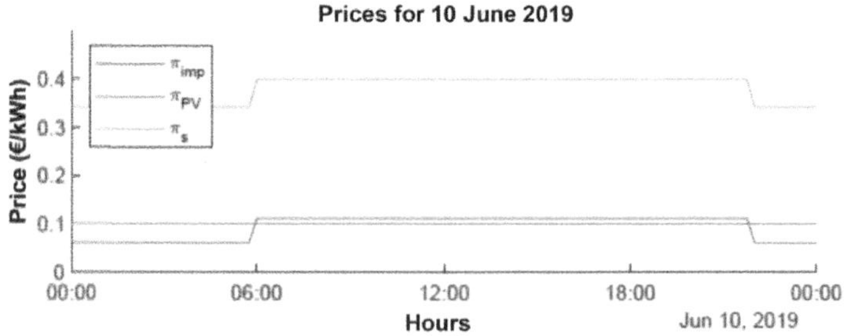

Figure 6.7. *Electricity prices by source [STE 21b]. For a color version of this figure, see www.iste.co.uk/robyns/smartgrids.zip*

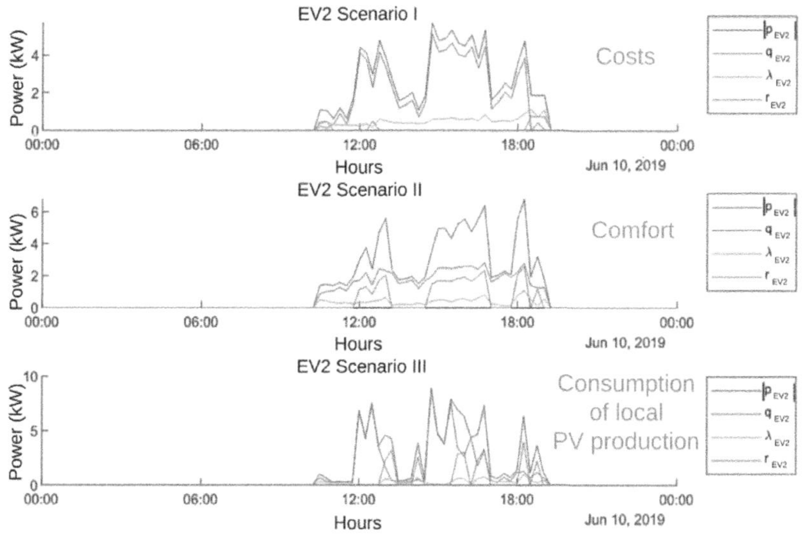

Figure 6.8. *Power profiles of an electric vehicle's load and various sources supplying this load for scenarios I–III (summer day). $|P_{VE2}|$: total power consumed by vehicle EV2; q_{EV2}: photovoltaic power consumed by vehicle EV2; λ_{EV2} : power imported from the distribution grid by vehicle EV2; r_{EV2}: power imported from the battery by vehicle EV2 [STE 21b]. For a color version of this figure, see www.iste.co.uk/robyns/smartgrids.zip*

For the case of electric vehicle charging, the curves in Figure 6.8, showing the total charging profile and the different charging sources in scenarios I–III, clearly indicate that when the actor wants to maximize the use of local photovoltaic production (scenario III), then indeed the q_{EV} term represents the major contribution to total power. This is also true for economic scenario I, as photovoltaic electricity is cheapest in the middle of the day. Furthermore, the state-of-charge (SOC) curves presented in Figure 6.9 for the five scenarios show that in scenario III, where the vehicle wants to optimize its comfort and therefore maximize its state-of-charge, the SOC actually reaches 100% at the end of charging. In contrast, in the economic scenario (scenario I), the SOC is not maximized, in order to reduce charging costs.

Similarly, for buildings, Figure 6.10 shows the total predicted and consumed power in the five scenarios, as well as the different sources used. This figure shows that when people maximize their comfort (scenario II), their optimized profile p_c coincides with the predicted consumption $P_c^{predicted}$ (shown as a black dotted line). Furthermore, in the economic scenario (scenario I), as the price of local photovoltaic electricity becomes lower than the price of off-peak grid electricity for the day

(Figure 6.7), buildings tend to shift their load to consume a maximum of this photovoltaic electricity in the middle of the day. What is more, since the price of electricity from the battery is the highest, they tend to consume as little as possible. More results, including a comparison of scenarios using indicators, can be found in [STE 21b].

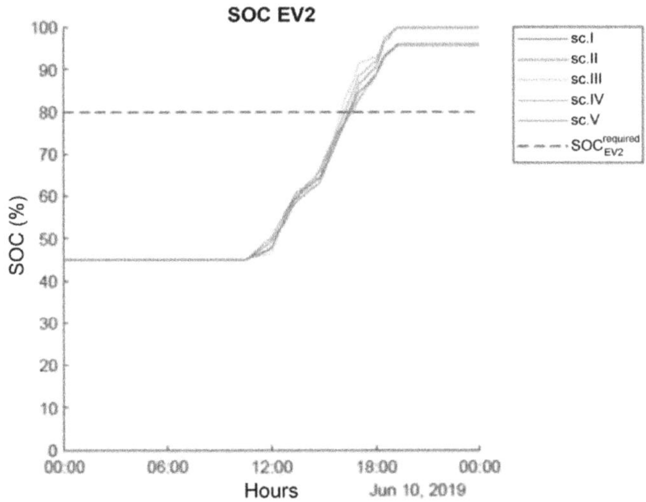

Figure 6.9. *State of charge of an electric vehicle for different scenarios (summer day) [STE 21b]. For a color version of this figure, see www.iste.co.uk/robyns/smartgrids.zip*

To judge the performance of the ADMM algorithm, the results were compared with a completely centralized approach, in which the global problem of equation [6.22] (with the same set of local and global constraints) is solved by a central algorithm, which therefore collects all the necessary information and aggregates the utility functions into a single objective function. Specifically, sequential quadratic programming (SQP) is implemented; it is widely used for solving convex nonlinear problems and is renowned for its good performance. Reference [NOC 06] gives details of this algorithm. The results of various simulations show that:

– In terms of the energy community's costs, comfort and self-production rate, the distributed approach with ADMM gives results close to those of a centralized method with SQP, and often even better.

– In terms of the community's overall utility, the centralized method outperforms the distributed method, which suffers from a lack of cooperation between actors.

– However, the centralized approach favors certain actors who carry more weight to the detriment of others, whereas with ADMM, optimizing utilities locally guarantees a minimum level of satisfaction for all actors. The centralized method therefore results in some actors adopting profiles that do not individually maximize their utility (e.g. clipping photovoltaic output from generators in scenario III), but which ultimately increase the community's total utility, as the utility of others compensates for this loss. When photovoltaic production is sufficient, the distributed approach is therefore actually more equitable, as illustrated by the distribution of locally produced photovoltaic energy among the community's consumers [STE 21a, STE 21b].

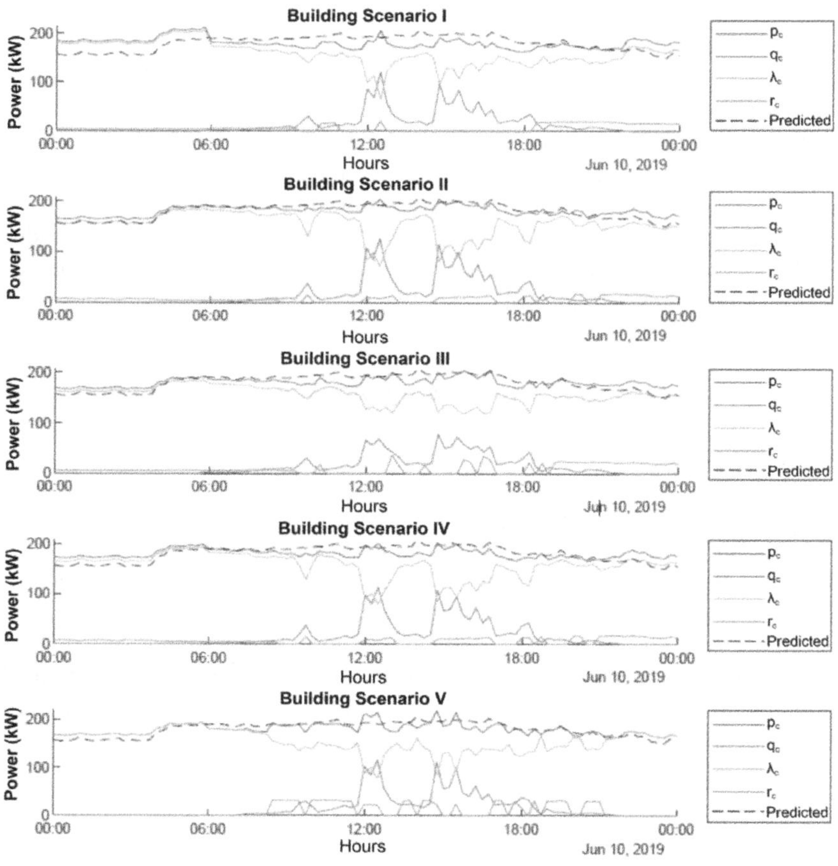

Figure 6.10. *Power profiles of tertiary buildings for different scenarios (summer day). p_c: total power consumed by buildings; q_c: photovoltaic power consumed by buildings; λ_c: power imported by buildings from the grid; r_c: power imported by buildings from the battery. Dashed line: production forecasts [STE 21b]*

6.4.4.4. *Interpretations*

These results illustrate the relevance of a distributed approach using ADMM to optimize energy exchanges within an energy community. Even if the sum of users' utilities is slightly lower than that obtained with a centralized method, the distribution of the problem guarantees greater fairness between actors. Indeed, by optimizing their utility individually, no actor is harmed to the detriment of others with greater weight. This result demonstrates the value of distributed methods for managing energy communities. By developing approaches centered on the users themselves, and letting them optimize their behavior, it is possible to fairly increase the satisfaction of all users.

As a result, these approaches are more likely to meet user expectations and therefore have greater acceptability. In the proposed model, actors do not cooperate, and the penalty term is applied equally to all actors. To improve results, a negotiation phase could be introduced, so that when a global constraint is not respected, users can agree on how to adapt their profiles. For example, when the overall power constraint is exceeded, some actors may be more inclined to lower their consumption than others. This working perspective would therefore introduce a phase of cooperation between users and would also help to strengthen the cohesion of the energy community [STE 21a, STE 21b].

6.5. Managing energy exchanges using blockchain technology

6.5.1. *Introduction*

The term "blockchain" refers to a computer technology that connects several users and enables them to interact, that is, to exchange or store information or carry out transactions. It is one of the technologies known as digital ledger technologies (DLTs): the database is said to be distributed, as it is replicated on all network nodes, unlike centralized databases, which are hosted on a single server. Indeed, DLTs were initially created with the aim of ensuring decentralization and enabling direct interaction between users, without the need for a trusted third party [BUR 18]. DLTs encompass several technologies, which differ in the way data is verified and added to the database. Blockchain technology is currently the most widespread and most widely developed [STE 20, STE 21b].

6.5.2. *The principle of blockchain*

6.5.2.1. *Hypotheses*

Generally speaking, blockchain defines a peer-to-peer distributed system, secured by cryptographic tools, in which data can only be added to and accessed, but not deleted. This system is immutable and updated by a consensus mechanism between users [BAS 17].

In more practical terms, blockchain refers to a P2P platform through which users exchange data, via transactions registered on the platform, without recourse to a trusted third party [NAK 08].

Therefore, in general, blockchain technology refers to the above-mentioned characteristics common to all blockchains. If a particular blockchain is being considered, it would be a specific platform, that is, the P2P system in question, defined by its nodes and consensus mechanism, with which users actually interact. There are many different blockchains, all with their own architecture and varying consensus mechanisms.

6.5.2.2. *The organization of a blockchain*

A blockchain[1] is a technology that enables information to be shared between multiple users, whether financial, informational or material, in a fully automated and secure way. It can be defined as a peer-to-peer network containing a shared ledger, the contents of which are validated and secured by a calculation algorithm known as "mining", or by cryptographic keys. Mining is an operation used to validate a transaction on a blockchain network by means of a mathematical calculation. Proof-of-work mining secures the blockchain. To validate a transaction on the blockchain network, a miner must provide proof of work.

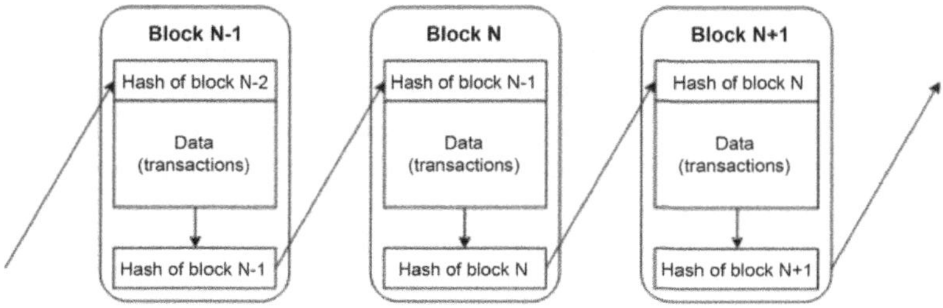

Figure 6.11. *Illustration of a block chain*

1 See: https://www.inria.fr/fr/comment-fonctionne-une-blockchain.

The blockchain is made up of three important concepts: blocks, nodes and miners. When a transaction is shared on the peer-to-peer network, it is collected by miners and, for reasons of efficiency, processed in batches (called blocks).

Each block is made up of four components: the hash that references the previous block, all the transactions recorded in it, an arbitrary number given in cryptography called the nonce[2], and finally the block's unique "hash". The hash is a key element of the blockchain. It is used to link and condense groups of transactions into blocks, and to produce cryptographic links between each block, creating a blockchain. Hashing is a type of cryptography that creates a coded string of letters and numbers permanently unique to each individual block in the chained sequence. The hash is in effect a "digital fingerprint".

Miners create new blocks following on from existing ones, in a process known as "mining". This makes it virtually impossible to extract an existing block to modify it, or to falsify it. Once verified by thousands of computers, the block is added to a node, called a blockchain, and the miner can typically be rewarded financially; hence, the idea of "blockchain".

6.5.2.3. *The properties of blockchains*

The structure of blockchain technology has the following properties [STE 21b]:

– *Decentralization*: data exchanges within the system are carried out via P2P transactions, that is, without recourse to a trusted third party. As a result, the system no longer relies on a central body, but solely on its users.

– *Resilience*: the blockchain is a database replicated across all the nodes in the system. This means that, should a node malfunction, the system remains operational thanks to the other nodes that keep the blockchain running. This data replication structure offers greater resistance to failure than a centralized system with a single point of failure.

– *Transparency:* before being considered, each transaction must be validated by all users, using a consensus mechanism. In this way, all users are aware of what has been added to the database, although the information contained in the transaction may be encrypted and only readable by some. Data replication on each node, moreover, makes it impossible for all data to be controlled by a single actor, who would then have the ability to restrict access to this data and validate transactions alone.

---

2 In cryptography, a "nonce" is a random number that is used only once in a specific context, such as when creating a new block on a blockchain network.

– *Immutability*: once a transaction has been validated and recorded in the blockchain, it can no longer be modified. Cryptographic tools guarantee that the data stored in the blockchain cannot be subsequently modified or deleted.

– *Automation*: this property is not common to all blockchains, but only to those that support the execution of computer programs (such as Ethereum). These computer programs, known as smart contracts, enable transactions to be automatically triggered according to predefined conditions. In this way, it is possible to automate certain tasks with a blockchain [THE 18].

6.5.3. *Development of a local blockchain for managing energy exchanges in the renewable energy community*

In this section, an implementation of a local blockchain that translates the distributed optimization problem of energy exchanges with the ADMM algorithm presented in section 6.4.3.3 is proposed. The aim is to test the smart contract under realistic conditions, and thus provide clues as to how to put blockchain technology into practice within a local energy community. The smart contract is tested on an Ethereum test blockchain, while taking into account the constraints of the local power grid via co-simulation with PowerFactory software [STE 20, STE 21b].

6.5.3.1. *The type of blockchain used*

A private blockchain is chosen, in which each participant in the energy community owns a node in the communication network. In this case study, the blockchain is not deployed on the actual community, that is, the blockchain is not actually deployed among all actors as illustrated in Figure 6.3, but a blockchain representative of the community is built, with as many nodes as users, all of whose nodes are hosted on a single computer. As a result, aspects of remote communication between multiple actors are not addressed. From the perspective of real deployment, this point will need to be further explored, since each node will be hosted on a server or processor specific to each user. On this blockchain, each actor has an account that is both a public and a private key pair, in order to be able to interact with other users and with the smart contract.

6.5.3.2. *The choice of platform*

There are several options when it comes to choosing the blockchain platform used for this application, including Ethereum, Hyperledger Fabric and Multichain. These open-source technologies all offer tools for developing private blockchains. Their operation is described in detail in [CHO 19].

Among these platforms, Ethereum was chosen. There are several reasons for this choice. Firstly, Ethereum is used by a large community, and therefore benefits from significant online development. It is maintained by numerous developers, with very regular updates. What is more, numerous pages of documentation are available and regularly improved. This aspect should not be overlooked, as blockchain technology is so recent and there are, as of yet, no standard tools for developing a decentralized application. On the other hand, the Ethereum blockchain[3] can easily be combined with other IT tools, as numerous application programming interfaces (APIs) are available to communicate with programs in Python or JavaScript.

This possibility, only offered by some of the blockchains presented in [CHO 19], enables the blockchain to be integrated into a complete environment and thus used in broader processes, while automating operations on the blockchain. Finally, Ethereum technology has been designed for the development of decentralized applications and smart contracts using the Solidity programming language. It combines tools for deploying blockchain nodes with tools for writing and testing smart contracts. These tools make it possible to quickly set up a few nodes on a private blockchain and concentrate on writing the smart contract. From these aspects, it appears that Ethereum is currently the most mature and easily accessible platform for the development of a local blockchain and smart contracts [MAC 17]. These considerations lead us to choose the Ethereum platform. In particular, the Geth client, a Go-language implementation of Ethereum, was implemented. In concrete terms, Geth enables nodes to be created either to join a public blockchain, or to create a private blockchain. A Geth node can host several user accounts.

6.5.3.3. *The architecture*

As mentioned above, in the proposed approach, each user has a blockchain node and an account. In practice, as many nodes as users are created, and these are linked together to create a private blockchain. The creation of such a blockchain requires identification of the nodes wishing to join the system.

As a reminder, in this approach, the ADMM algorithm contains two main steps. After having indicated their preferences (costs, comfort or willingness to exchange energy locally), the actors must carry out a local optimization of their power profile. Next, an aggregation stage collects these profiles, checks that the global constraint is respected and, if not, calculates the penalty to be sent to users for a new iteration.

To make this approach completely distributed and dispense with the need for a single entity to receive profiles from all actors, the aggregation step is performed by a smart contract. This program then runs simultaneously on all the nodes in the network, guaranteeing a completely distributed operation.

3 Ethereum.org, "Backend API libraries". Available at: https://ethereum.org.

To collect user preferences, the power profile os iptimized locally according to these preferences and interactions are automated with the smart contract, Python scripts have been developed. These scripts model the different classes of actors in the community (consumers, photovoltaic producers, electric vehicles and storage systems). When the blockchain is initiated, each user declares their type in the Python script, which then assigns them the corresponding utility function.

The global environment of the developed blockchain is shown in Figure 6.12. The blockchain nodes communicate with each other (shown in blue). Each actor has a Python client associated with a blockchain node. In practice, this assumes that all users of the system have a computer or microprocessor, such as a Raspberry Pi, to host a node and run optimization calculations in Python. In [HEC 20] the feasibility of such an approach was shown for a local blockchain. The smart contract translating the ADMM aggregation step is deployed on the blockchain and executed simultaneously on all nodes.

In addition to the energy community actors, an agent is added that is not a community actor. This agent simply deploys the smart contract by defining the following parameters:

– the number of users;

– the overall constraint, that is, the maximum power that can be drawn from the distribution network;

– the time interval for the optimization period;

– the price of electricity from the distribution network for this period.

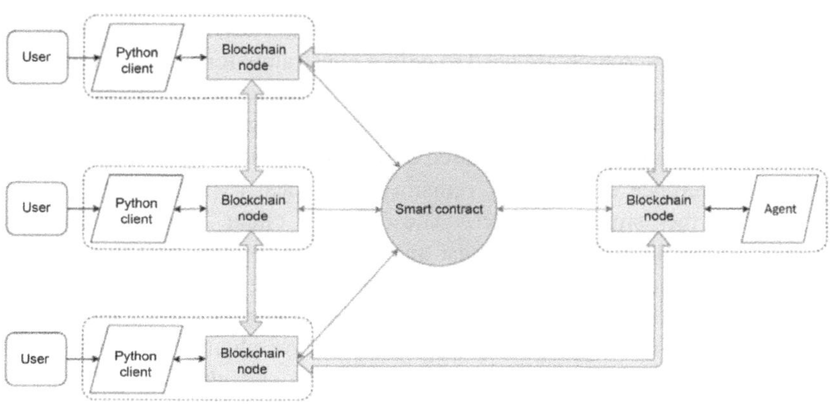

Figure 6.12. *Diagram of the blockchain environment developed [STE 21b]*

This agent does not take part in the optimization process and does not read any data from the smart contract. As such, it does not recentralize the process. The agent's sole responsibility is to specify the parameters required for optimization, and to deploy the smart contract. In practice, this role can be played by any node. However, the data to be indicated in the smart contract presupposes that this agent knows the composition of the energy community (notably the number of users) and is in contact with the grid operator, to indicate the power limit and electricity prices. In reality, this role could be played by an energy manager from the energy community, who would precisely perform these communication functions with the distribution network operator.

The entire smart contract optimization process from a user's point of view is shown in Figure 6.13.

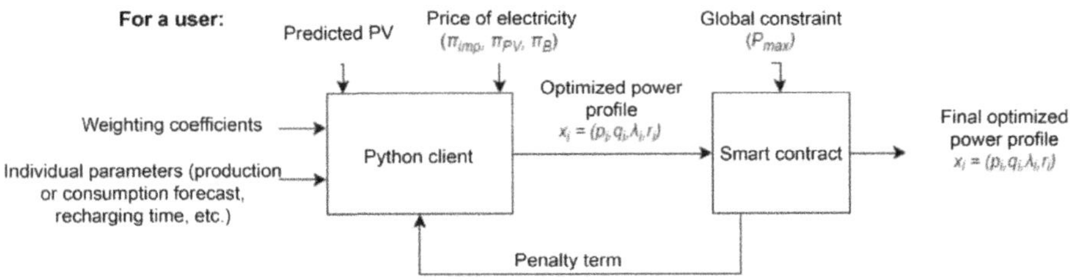

Figure 6.13. *Optimization on the blockchain with ADMM from a user's point of view [STE 21b]*

The user must specify their type (consumer, photovoltaic producer, electric vehicle, storage) and preferences in the Python client. They must also specify their individual parameters, which are used to define constraints for the local optimization of the power profile. For example, for an electric vehicle, these parameters are the charging power limit, the vehicle capacity, the charging time, the state-of-charge terminals and the state-of-charge required at the end of charging. These private data, such as the preference coefficients, remain on the local Python client and are not transferred to the smart contract. The Python client retrieves the energy community's total photovoltaic forecast and electricity prices from the blockchain. All these data are used to locally optimize the user's power profile according to their preferences. This profile is then sent by the Python client to the smart contract. In return, after the global aggregation stage, the Python client receives the penalty parameter to be applied for the new iteration. In the end, the user gets back their power profile, which optimizes their utility while ensuring that the entire community respects the global constraint.

6.5.3.4. *The consensus-building mechanism*

The consensus mechanism is an important blockchain parameter since, as explained above, it impacts the way the blockchain is collectively managed and has consequences for the energy consumption of the nodes.

Proof-of-work (PoW), since it generates significant energy consumption to mine new blocks, is not relevant for energy management applications [THE 18].

For this application, a private blockchain with a limited number of users (approximately 10) could be developed. It is legitimate to assume that these users, being neighbors within the same local energy community, trust each other because they know each other. Under these conditions, a proof-of-authority (PoA) mechanism proves more relevant. It eliminates the need for PoW calculations. On the other hand, unlike PoW, the time between each new block can be explicitly specified when the blockchain is created. With this feature, new blocks are created only when transactions are awaiting validation. This limits the size of the database. For these reasons, PoA is suitable for a local blockchain with a limited number of actors [CHR 16].

In practice, when the blockchain is initialized, it must be declared which nodes are authorized to vote for the creation of new blocks (sealers). In the proposed approach, for the optimization process to be decentralized and for all nodes to have equal importance, all are sealers. By allowing only one sealer, a central agent would be reintroduced.

In this study, both PoW and PoA consensus mechanisms were tested, in order to compare and quantify the differences, particularly in terms of energy consumption.

6.5.3.5. *The smart contract*

This section explains in detail how the smart contract that implements the ADMM algorithm works, and how it interacts with users, describing and illustrating the sequence of each step. The code is available in [STE 21b].

As previously mentioned, once all the nodes of the blockchain have been created and linked, the agent deploys the smart contract, specifying the total number of users N, the power constraint P_{max} not to be exceeded for the community, and the more specific optimization parameters, that is, the initial penalty parameter ρ and the stopping criterion ϵ (step 1).

Once the contract has been deployed, each user creates a class in the Python script, corresponding to its type (electric vehicle, PV generator, storage or consumer). They also explicitly indicate the value of their preference coefficients to prioritize their objectives, as well as their own physical parameters, which specify the local constraints that apply to their power profile. These data enable us to fully define each actor's utility function.

In the proposed model, all these data are entered directly into the Python script. However, this model could easily be made more interactive and user-friendly for concrete implementation, by developing an ergonomic interface for the user.

When the user class is initialized in the Python client, the latter automatically creates a corresponding structure in the smart contract. A structure, in Solidity[4], is a storage tool that associates several variables. Therefore, each user is represented by a structure in the smart contract, with which its type and power profile are associated. However, these data cannot be read by other users, nor by the agent deploying the smart contract. When N user structures have been created, an event is triggered, informing the agent that all users have logged in. In Solidity, an event is associated with a transaction and signals a defined change of state in the blockchain. Users can listen to the blockchain and detect events by inspecting new transactions.

These two steps are illustrated in Figure 6.14. In the following figures, the orange blocks represent the information sent by the blockchain nodes to the smart contract. The green blocks correspond to data processed locally on the Python clients, and the blue blocks are data received from the smart contract by the nodes.

Then, in steps 3 and 4 shown in Figure 6.15, users receive the data they need to optimize their profile locally. In particular, the agent indicates in the smart contract the time interval for the next optimization period and the price of electricity over this period. Here too, an event is triggered to warn users that these data are unavailable, and the Python clients import them automatically (step 3 in Figure 6.15(a)).

For local optimization, users also need to know the forecast of total photovoltaic production on the community's power grid, in order to define local constraints and the initial vector for local optimization. To do this, participants who are PV generators send their production forecasts to the smart contract. The smart contract

4 See: https://soliditylang.org/.

adds up all the predictions, and when all the PV generators have sent their data, it triggers an event. Other users can then import the global photovoltaic forecasts (step 4 in Figure 6.15(b)). Users do not know the details of the forecast for each PV generator; they only have access to the total production for the community.

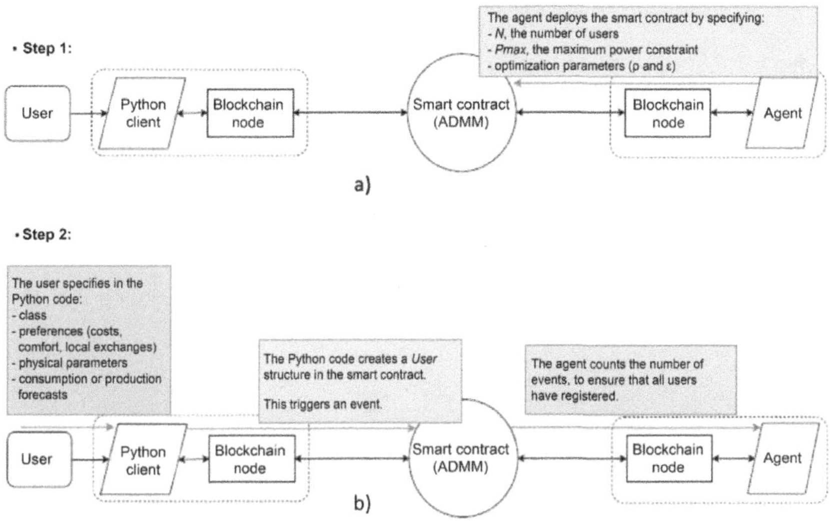

Figure 6.14. *Smart contract deployment and initialization: a) step 1 and b) step 2 [STE 21b]*

Finally, the optimization steps can begin, as illustrated in Figure 6.16. At each iteration, the Python client retrieves the penalty parameters to be applied, that is, the vector L and the parameter ρ (step 5 in Figure 6.16(a)). In the initial step, L = 0, so no penalty is applied. These data are used for local optimization in the Python client. Once the user profile has been optimized, it is sent to the smart contract. When all parties have updated their profiles, an event is triggered, warning the agent that the local calculations are complete (step 6 in Figure 6.16(b)). The agent then triggers the aggregation step. In concrete terms, it only calls the corresponding function in the smart contract but receives no data. The calculation takes place directly in the smart contract and is therefore completely distributed. At the end of the calculation, an event is triggered to inform users whether the final state has been reached (step 7 in Figure 6.16(c)). If not, a new iteration is started.

The smart contract thus enables the ADMM algorithm to be implemented in a completely distributed way. This algorithm combines local optimization with Python clients and global optimization with the smart contract. Events are used to synchronize all users and ensure step continuity. The detailed operation of the smart contract shows that the agent in charge of deploying the smart contract does not interfere in the optimization calculation, but only serves to guarantee its correct execution. This shows that the process is not centralized.

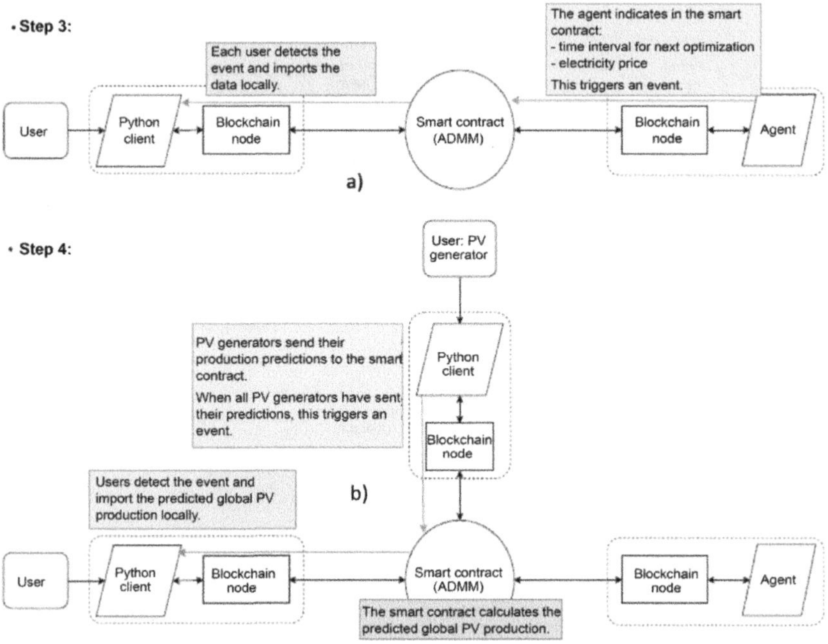

Figure 6.15. *Data exchange prior to optimization: a) step 3 and b) step 4 [STE 21b]*

On the other hand, during the entire optimization process, the participants' personal data remain on the local Python clients and are therefore not shared. Only the predicted PV production profiles and the profiles optimized after each iteration are sent to the blockchain. However, these data are read-only and accessible only to the owner. Other users only have access to the sum of the community's PV production and the global optimization results. This approach therefore guarantees the protection of personal data, that is, the individual preferences of participants and their consumption or production profiles [STE 20, STE 20b].

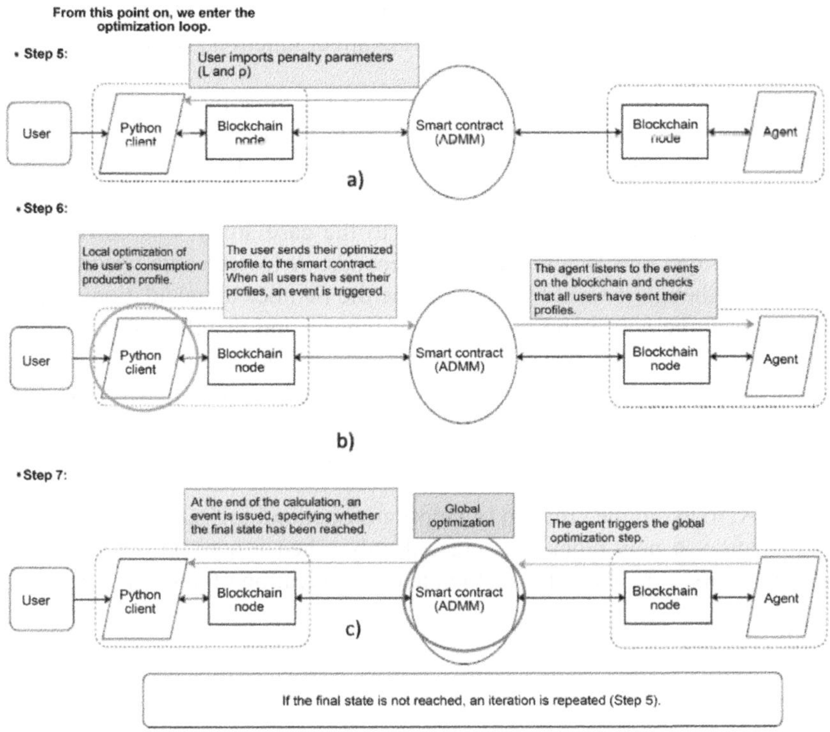

Figure 6.16. *Optimization steps: a) step 5, b) step 6 and c) step 7 [STE 21b]*

6.5.4. *Simulations and results*

6.5.4.1. *Simulation parameters*

In order to verify the functioning of the developed environment, simulations with a private blockchain were carried out. This blockchain was hosted on an Ubuntu server (Version 20.04.1 LTS) with a Linux kernel (5.4.0-52) and a 2.20 GHz Intel Xeon E5-2660 processor with 16 GB RAM. To carry out these tests, Geth software (Version 1.9.23) was used to create Ethereum nodes, Solidity (Version 0.6.0) for smart contracts and Python (Version 3.8.5) for local optimization.

The purpose of the simulations was to evaluate the performance of the blockchain, depending on the number of nodes and the consensus mechanism used (PoA or PoW), particularly with regard to energy consumption. This parameter was calculated by means of the gas expended, which translates the number of elementary operations triggered by each call of a smart contract function [WOO 19].

For PoW, in order to limit the energy required to mine blocks, a difficulty of 0 × 1,000 (i.e. 4,096 in decimal) was set. This value determines the probability of a calculated hash being correct to validate the new block. In practical terms, this means that miners need to calculate an average of 4,096 hashes to generate a new block. In practice, on the server allocated to the case study, this corresponded to approximately one new block every two seconds. A lower value was not relevant, because even though it would further reduce the number of calculations required, it would result in the very rapid generation of blocks, and therefore system instability (creation of numerous forks)[5]. As the present application required precise synchronization of transactions to achieve algorithm convergence, this situation would prove problematic. In the present blockchain, to guarantee the decentralization of the process, all nodes played the role of miners.

For PoA, to avoid the creation of empty transaction blocks, which are useless for our application, a new block was created only when a new transaction had to be mined. All users were also allowed to vote on the creation of new blocks.

6.5.4.2. *Results*

6.5.4.2.1. Gas consumption

Tables 6.4 and 6.5 show the gas results of the simulations for different numbers of nodes, for PoA and PoW, respectively. The number of iterations increases with the number of users, as convergence to the final state logically requires more steps with more users.

Number of nodes	Number of iterations	Total gas	Average gas by iteration and by node (outside agent)	Average gas by agent iteration
5	43	2.006×10^9	5.522×10^6	2.456×10^7
7	53	2.773×10^9	4.635×10^6	2.450×10^7
9	59	3.717×10^9	4.820×10^6	2.444×10^7
Average	–	–	4.990×10^6	2.450×10^7

Table 6.4. *Blockchain gas consumption with PoA [STE 21b]*

5 In the context of blockchain, a fork is a division of the blockchain into two distinct chains. A fork can be triggered by a divergence in consensus rules, in which case it is referred to as a hard fork, which designates a separation of the chain that is generally permanent. A fork is essentially a change in the blockchain protocol. See: https://fr.wikipedia.org/wiki/Fork_(blockchain).

Number of nodes	Number of iterations	Total gas	Average gas by iteration and by node (outside agent)	Average gas by agent iteration
5	43	2.146×10^9	5.401×10^6	2.830×10^7
7	53	3.879×10^9	6.655×10^6	3.327×10^7
9	59	4.455×10^9	5.278×10^6	3.329×10^7
Average	–	–	5.778×10^6	3.162×10^7

Table 6.5. *Blockchain gas consumption with PoW [STE 21b]*

A comparison of PoA and PoW mechanisms showed that, in general, PoW resulted in higher gas consumption, averaging 22% of total gas consumption. Indeed, even if the difficulty of PoW is reduced to a minimum, block mining is still energy-intensive, as it involves a calculus including all miners on the network. These results are in line with the articles [YAN 19, SED 20] and especially [JAV 20], which experimentally observed greater gas consumption with PoW than with PoA and found an increase of approximately 15% in gas consumption for PoW compared with PoA in the case of smart contract deployment. This value does not represent a very significant difference, but it does correspond to a private blockchain, with a reduced number of users. Therefore, the value of the difficulty of PoW is low, and the computational expenses much more limited than with a public blockchain, whereas it is precisely the greater difficulty of PoW that guarantees the security of the blockchain.

6.5.4.2.2. Energy consumption

By converting gas into CPU usage time and estimating CPU power, it is possible to calculate the energy consumption specific to the blockchain. The results are shown in Tables 6.6 and 6.7. These results correspond to the energy consumption of the blockchain nodes.

Number of nodes	Total consumption	Average consumption by iteration and by node, outside agent (Wh)	Average consumption by agent iteration (Wh)
5	2.79	7.67×10^{-3}	3.41×10^{-2}
7	3.85	6.44×10^{-3}	3.40×10^{-2}
9	5.16	6.69×10^{-3}	3.39×10^{-2}
Average	–	6.93×10^{-3}	3.40×10^{-2}

Table 6.6. *Energy consumption of blockchain nodes with PoA [STE 21a]*

Number of nodes	Total consumption	Average consumption by iteration and by node, outside agent (Wh)	Average consumption by agent iteration (Wh)
5	2.98	7.50×10^{-3}	3.93×10^{-2}
7	5.39	9.24×10^{-3}	4.62×10^{-2}
9	619	7.33×10^{-3}	4.62×10^{-2}
Average	–	8.03×10^{-3}	4.39×10^{-2}

Table 6.7. *Energy consumption of blockchain nodes with PoW [STE 21b]*

In addition, consumption due to optimization calculations in Python clients must be taken into account. In this way, the total consumption of the blockchain system for PoA and PoW can be calculated by adding together the component due to node operation and the component due to optimization calculations in Python. These results are compared with a centralized solution, where the global optimization problem defined in section 6.4.3 is not decomposed but solved directly by a single SQP algorithm [NOC 06]. In practice, Matlab is used for this calculation, and CPU time is measured. These results are re-grouped in Table 6.8 and for greater clarity are illustrated in Figure 6.15, where trend curves sketch the evolution of consumption as a function of the number of users. These results should be treated with caution, however, given the small number of measuring points, and need to be confirmed. What limited the number of simulations is detailed below. On the curves in Figure 6.17, the share of consumption due to blockchain nodes and the share due to Python clients for the blockchain system with PoA are differentiated.

Energy consumption (Wh)			
Number of users	**PoA**	**PoW**	**Centralized solution**
4	21.9	22.1	4.8
6	36.6	38.1	16.6
8	52.8	52.8	35.4

Table 6.8. *Overall consumption of the blockchain system and a centralized solution [STE 21b]*

Analysis of these results shows two main things. Firstly, for a reduced number of nodes, the difference in consumption between PoA and PoW is negligible. This is because the main source of power consumption comes from Python clients and not from the operation of the blockchain nodes. However, it should be noted that this is only true for a small number of nodes. Indeed, when this number increases, to guarantee the security of the blockchain, it is imperative to increase the difficulty of the PoW. The PoW system's power consumption would then be much higher. This is particularly true of public blockchains such as bitcoin.

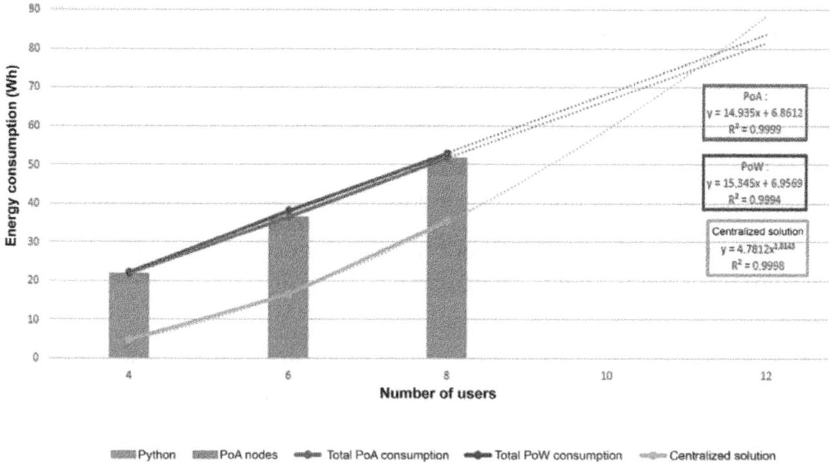

Figure 6.17. *Comparison of the energy consumption of the blockchain and a centralized solution [STE 21b]*

On the other hand, it appears that the consumption with a centralized solution and that with a blockchain solution are of the same order of magnitude, even if our results show that the centralized solution is more energy efficient. However, the trend curves suggest that the consumption of this centralized solution grows almost as a function of the square of the number of users. This is because, as the number of users increases, so does the complexity of the overall optimization problem. The blockchain system uses the ADMM algorithm to decompose this problem and parallelize the calculation, by giving each user a sub-problem to solve, the complexity of which is independent of the total number of users. In this way, the consumption of Python code in the blockchain system simply increases linearly with the number of users. Consequently, the results suggest that there is a critical number of users for whom the complexity of the overall problem makes the centralized solution more energy-intensive than a blockchain solution with PoA.

These findings corroborate the results presented in [LIC 20]. By comparing a centralized and a distributed multi-agent algorithm for the optimization of a heat network, the authors conclude that the distributed algorithm consumes less computational resources, whereas the centralized algorithm is more vulnerable to the lack of computational capacity when the difficulty of the optimization increases, that is, when the number of users increases.

The results presented here are also in line with [SED 20], from which Figure 6.18 is taken, which analyzed the energy consumption of different blockchain architectures but was limited to a theoretical study.

Figure 6.18 shows the approximate energy consumption (per transaction) per blockchain type (the vertical Y axis is logarithmic). The following cases are compared[6]:

– simple server: transactions processed with low energy consumption;

– typical centralized system without blockchain: uses a more complex database and backups, consuming slightly more energy;

– small-scale authorized blockchain (e.g. business-to-business or private): similar redundancy, but additional overheads and power consumption per transaction (e.g. more complex PoA consensus and cryptography);

– an authorization-free blockchain without PoW with a large number of nodes: high power consumption;

– bitcoin-type proof-of-work blockchain: intrinsically extremely energy-intensive.

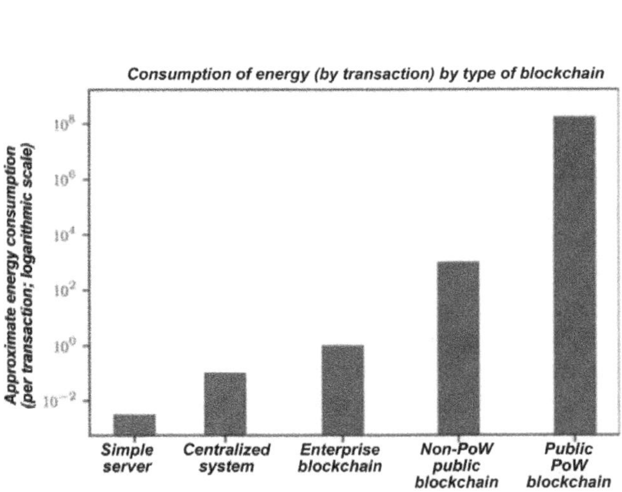

Figure 6.18. *Comparison of the energy consumption of different architectures (sources: [SED 20] and Wikipedia)*

6 See: https://en.wikipedia.org/wiki/Blockchain.

6.6. Interpretations and experience feedback

The tests carried out with blockchain also highlighted some of the technology's limitations. Firstly, the complexity of setting up a blockchain is certainly a factor. Nodes have to be initialized individually and then connected to each other. There is no procedure for automating blockchain setup, and the process is time-consuming and fairly complex. What is more, for the resolution of a distributed optimization algorithm, synchronization is essential, so failure to detect an event or initiate a transaction is critical for the whole system. This scaling problem is one of the main shortcomings of the technology [ZHO 20].

On the other hand, the present study is not exhaustive and needs to be more detailed. The results are simply estimates based on server power consumption and conversion of gas to CPU time. To consolidate the results, it would be interesting to measure gas consumption and energy consumption more precisely.

Furthermore, the results of this study were obtained using a limited number of nodes. To extend the conclusions, particularly concerning the comparison between centralized and distributed solutions, a more in-depth study with a larger number of nodes is required. Furthermore, the conclusions drawn here did not take into account the aspect of communication between nodes, as they were hosted on the same server. In real applications, however, the nodes would be hosted on different computers or controllers. Therefore, the communication aspect is crucial in distributed systems in general and in blockchain in particular, as all nodes participate in the consensus mechanism. Following [HEC 20], the present work can be complemented by a blockchain implementation using microprocessors to host blockchain nodes. By doing so, the energy consumption of the nodes could be more easily measured, while addressing the communication aspect.

6.7. Conclusion

The distributed nature of blockchain makes it a promising technology to support the decentralization of the electricity grid and its transformation into a smart grid. As it is managed by the users themselves, without the intervention of a central agent, it enables various processes to be operated in a completely distributed and secure way.

The approach developed in this chapter illustrates the feasibility of using a local blockchain to optimize energy exchanges in a local energy community, by means of a distributed optimization algorithm. The combination of Python clients interacting with blockchain nodes for local optimization and a smart contract deployed on all network nodes constitutes a fully distributed environment for solving the optimization problem.

The ability to operate complex processes without a central agent is the main advantage of blockchain. Indeed, with this distributed approach, users themselves reach a consensus, without sharing their personal information, such as their consumption preferences.

In the present study, for example, actors' consumption or production profiles, once sent to the smart contract for the global optimization stage, were not accessible to other users. Indeed, in the smart contract, reading these profiles was not authorized, in order to guarantee actors' privacy. In this respect, blockchain is a relevant tool for increasing the acceptability of energy management solutions since it ensures user confidentiality.

The results of tests carried out with two types of consensuses (PoA and PoW) show that for a private blockchain with a limited number of actors, the energy consumption of such a system is slightly higher, but of the same order of magnitude as a centralized solution. The findings even suggest that, with a certain number of actors, a private blockchain operating with PoA can be more energy-efficient than a centralized solution. This illustrates the potential that can be derived from blockchain technology as a support for distributed algorithms, whose complexity remains constant with the number of users. In addition, this technology can offer considerable advantages for sustainable development. Consumers can choose renewable energy and have greater control and transparency over their energy purchases. Transactions are also more secure and faster, with lower costs. Blockchain technology also enables businesses to cut costs by automating some of their tasks and simplifying their processes.

However, the problem of scaling up is still holding back the development of this technology, and for the time being limits its use to simple smart contracts for a limited number of users. Even though they are evolving rapidly, current tools still lack the maturity and standardization required to deploy the technology on an industrial scale.

6.8. Acknowledgments

The work presented in this chapter was carried out in partnership with the Greenbirdie company as part of Junia's Smart Buildings as Nodes of Smart Grids chair, which was funded by MEL and nine companies: Accenta (which acquired Effipilot), EDF, Enedis, Energic, Greenbirdie, Projex, Stereograph, Vinci Construction and Vinci Énergie.

The smart grid demonstration sites received funding from the Hauts-de-France region via the Live TREE program.

The authors would like to thank Mattieu Stephant, who was the first at L2EP to explore the potential of blockchain technology for renewable energy communities, and Kahina Hassam, a teacher-researcher and computer scientist at Junia, for her contribution to the implementation of a blockchain.

6.9. References

[ADE 17] ADEME, L'effacement de consommation électrique en France, Report, 2017.

[ADE 19] ADEME, Autoconsommation photovoltaïque : comment produire de l'électricité et la consommer chez soi ?, available at: https://www.ademe.fr/sites/default/files/assets/documents/photovoltaique-autoconsommation-010839.pdf, 2019.

[AZA 19] Azarova V., Cohen J., Friedl C. et al., "Designing local renewable energy communities to increase social acceptance: Evidence from a choice experiment in Austria, Germany, Italy, and Switzerland", *Energy Policy*, vol. 132, pp. 1176–1183, 2019.

[BAS 17] Bashir I., *Mastering Blockchain*, Packt Publishing Ltd, Birmingham, 2017.

[BOU 14] Bouyaux P., "Économétrie et théorie des jeux", *Techniques de l'ingénieur*, 2014.

[BOY 11] Boyd S., Parikh N., Chu E. et al., "Distributed optimization and statistical learning via the alternating direction method of multipliers", *Foundations and Trends® in Machine Learning*, vol. 3, no. 1, pp. 1–122, 2011.

[BUR 18] Burkhardt D., Werling M., Lasi H., "Distributed Ledger", *2018 IEEE International Conference on Engineering, Technology and Innovation (ICE/ITMC)*, pp. 1–9, 2018.

[CHO 19] Chowdhury M.J.M., Ferdous M.S., Biswas K. et al., "A comparative analysis of distributed ledger technology platforms", *IEEE Access*, vol. 7, pp. 167–943, 2019.

[CHR 16] Christidis K., Devetsikiotis M., "Blockchains and smart contracts for the Internet of Things", *IEEE Access*, vol. 4, pp. 2292–2303, 2016.

[DEH 18] Dehghanpour K., Nehrir H., "Real-time multiobjective microgrid power management using distributed optimization in an agent-based bargaining framework", *IEEE Transactions on Smart Grid*, vol. 9, no. 6, pp. 6318–6327, 2018.

[ECU 17] E-CUBE, Autoconsommation collective et blockchain. Perspectives sur deux phénomènes émergents et liés, Report, 2017.

[ETH 23] Ethereum.org, Backend API libraries, available at: https://ethereum.org, 2023.

[EUR 17] European Commission, Clean energy for all Europeans package, available at: https://ec.europa.eu/energy/topics/energy-strategy/clean-energy-all-europeans_en, 2017.

[FRI 19] Frieden D., Tuerk A., Roberts J. et al.. "Overview of emerging regulatory frameworks on collective self-consumption and energy communities in Europe", *2019 16th International Conference on the European Energy Market (EEM)*, pp. 1–6, 2019.

[GIO 21] GIORGINI P., *The Contributory Revolution*, ISTE Ltd, London and John Wiley & Sons, New York, 2021.

[GON 18] GONZALEZ-LONGATT F., TORRES J.L.R., *Advanced Smart Grid Functionalities Based on Power Factory*, Springer, New York, 2018.

[GOO 80] GOODMAN J.C., "Notes and comments: A note on existence and uniqueness of equilibrium points for concave N-person games", *Econometrica*, vol. 48, no. 1, p. 251, 1980.

[HEC 20] HECK K., MENGELKAMP E., WEINHARDT C., "Blockchain-based local energy markets: Decentralized trading on single-board computers", *Energy Systems*, doi: 10.1007/s12667-020-00399-4, 2020.

[JAV 20] JAVED M.U., REHMAN M., JAVAID N. et al., "Blockchain-based secure data storage for distributed vehicular networks", *Applied Sciences*, vol. 10, no. 6, p. 2011, 2020.

[JIA 20] JIANG X., WU L., "A residential load scheduling based on cost efficiency and consumer's preference for demand response in smart grid", *Electric Power Systems Research*, vol. 186, p. 106410, 2020.

[KAN 17] KANG J., YU R., HUANG X. et al., "Enabling localized peer-to-peer electricity trading among plug-in hybrid electric vehicles using consortium Blockchains", *IEEE Transactions on Industrial Informatics*, vol. 13, no. 6, pp. 3154–3164, 2017.

[KUL 10] KULWORAWANICHPONG T., "Simplified Newton–Raphson power-flow solution method", *International Journal of Electrical Power & Energy Systems*, vol. 32, no. 6, pp. 551–558, 2010.

[LEE 20] LEE C.-Y., AHN J., "Stochastic modeling of the levelized cost of electricity for solar PV", *Energies*, vol. 13, no. 11, p. 3017, 2020.

[LÉG 17] LEGIFRANCE, Code de l'énergie, Article L315-2, 2017.

[LI 13] LI N., MARDEN J.R., "Designing games for distributed optimization", *IEEE Journal of Selected Topics in Signal Processing*, vol. 7, no. 2, pp. 230–242, 2013.

[LIC 20] LICHTENEGGER K., LEITNER A., MÄRZINGER T. et al., "Decentralized heating grid operation: A comparison of centralized and agent-based optimization, sustainable energy", *Grids and Networks*, vol. 21, p. 100300, 2020.

[LIU 17] LIU N., YU X., WANG C. et al., "Energy sharing management for microgrids with PV Prosumer: A stackelberg game approach", *IEEE Transactions on Industrial Informatics*, vol. 13, no. 3, pp. 1088–1098, 2017.

[LIU 19] LIU Y., LI Y., GOOI H.B. et al., "Distributed robust energy management of a multimicrogrid system in the real-time energy market", *IEEE Transactions on Sustainable Energy*, vol. 10, no. 1, pp. 396–406, 2019.

[LUT 15] LUTHANDER R., WIDÉN J., Nilsson D. et al., "Photovoltaic self-consumption in buildings: A review", *Applied Energy*, vol. 142, pp. 80–94, 2015.

[MA 14] MA K., HU G., SPANOS C.J., "Distributed energy consumption control via real-time pricing feedback in smart grid", *IEEE Transactions on Control Systems Technology*, vol. 22, no. 5, pp. 1907–1914, 2014.

[MAC 17] MACDONALD M., LIU-THORROLD L., JULIEN R., "The blockchain: A comparison of platforms and their uses beyond bitcoin", Working paper, 2017.

[MAR 10] MARLER R.T., ARORA J.S., "The weighted sum method for multi-objective optimization: New insights", *Structural and Multidisciplinary Optimization*, vol. 41, no. 6, pp. 853–862, 2010.

[MIN 19] MINISTÈRE DE LA TRANSITION ÉCOLOGIQUE, Arrêté du 14 octobre 2020 modifiant l'arrêté du 21 novembre 2019 fixant le critère de proximité géographique de l'autoconsommation collective étendue, Légifrance, available at: https://www.legifrance.gouv.fr/jorf/id/JORFTEXT000042434286, 2019.

[MOR 16] MORVAJ B., EVINS R., CARMELIET J., "Optimization framework for distributed energy systems with integrated electrical grid constraints", *Applied Energy*, vol. 171, pp. 296–313, 2016.

[MOR 19a] MORET F., PINSON P., "Energy collectives: A community and fairness based approach to future electricity markets", *IEEE Transactions on Power Systems*, vol. 34, no. 5, pp. 3994–4004, 2019.

[MOR 19b] MORSTYN T., MCCULLOCH M.D., "Multiclass energy management for peer-to-peer energy trading driven by prosumer preferences", *IEEE Transactions on Power Systems*, vol. 34, no. 5, pp. 4005–4014, 2019.

[MÜN 17] MÜNSING E., MATHER J., MOURA S., "Blockchains for decentralized optimization of energy resources in microgrid networks", *2017 IEEE Conference on Control Technology and Applications (CCTA)*, pp. 2164–2171, 2017.

[NAK 08] NAKAMOTO S., Bitcoin: A peer-to-peer electronic cash system, Technical report, 2008.

[NOC 06] NOCEDAL J., WRIGHT S., *Numerical Optimization*, Springer Science & Business Media, New York, 2006.

[PIL 19] PILZ M., AL-FAGIH L., "Recent advances in local energy trading in the smart grid based on game-theoretic approaches", *IEEE Transactions on Smart Grid*, vol. 10, no. 2, pp. 1363–1371, 2019.

[ROB 19] ROBYNS B., DAVIGNY A., BARRY H. et al., *Electrical Energy Storage for Buildings in Smart Grids*, ISTE Ltd, London and John Wiley & Sons, New York, 2019.

[ROB 23] ROBYNS B., LENGLET C., BARRY H. et al., *Smart Users for Energy and Societal Transition*, ISTE Ltd, London and John Wiley & Sons, New York, 2023.

[ROS 65] ROSEN J., "Existence and uniqueness of equilibrium points for concave n-person games", *Econometrica*, vol. 33, no. 3, pp. 520–534, 1965.

[SCH 19] SCHMIDT O., MELCHIOR S., HAWKES A. et al., "Projecting the future levelized cost of electricity storage technologies", *Joule*, vol. 3, no. 1, pp. 81–100, 2019.

[SED 20] SEDLMEIR J., BUHL H.U., FRIDGEN G. et al., "The energy consumption of blockchain technology: Beyond myth", *Business & Information Systems Engineering*, vol. 62, no. 6, pp. 599–608, 2020.

[SOL 20] SOLIDITY, Contracts — Solidity 0.7.3 documentation, available at: https://solidity.readthedocs.io/en/v0.7.3/contracts.html, 2020.

[SOR 19] SORIN E., BOBO L., PINSON P., "Consensus-based approach to peer-to-peer electricity markets with product differentiation", *IEEE Transactions on Power Systems*, vol. 34, no. 2, pp. 994–1004, 2019.

[SOU 19] SOUSA T., SOARES T., PINSON P. et al., "Peer-to-peer and community based markets: A comprehensive review", *Renewable and Sustainable Energy Reviews*, vol. 104, pp. 367–378, 2019.

[STE 20] STEPHANT M., ABBES D., HASSAM-OUARI K. et al., Increasing photovoltaic self-consumption with game theory and blockchain, Report, EAI Endorsed Transactions on Energy Web, 2020.

[STE 21a] STEPHANT M., ABBES D., HASSAM-OUARI K. et al., "Distributed optimization of energy profiles to improve photovoltaic self-consumption on a local energy community", *Simulation Modelling Practice and Theory*, vol. 108, p. 102242, 2021.

[STE 21b] STEPHANT M., Optimisation de l'autoconsommation dans une communauté énergétique locale via une blockchain, PhD thesis, ENSAM, Paris, 2021.

[THE 18] THEA E., "Les fondamentaux de la blockchain", *Revue de l'électricité et de l'électronique*, vol. 2018, no. 2, 2018.

[TUS 21] TUSHAR W., YUEN C., SAHA T.K. et al., "Peer-to-peer energy systems for connected communities: A review of recent advances and emerging challenges", *Applied Energy*, vol. 282, p. 116131, 2021.

[WOO 19] WOOD G., Ethereum yellow paper: A secure decentralised generalised transacton ledger byzantium version, available at: https://ethereum.github.io/yellowpaper/paper.pdf, 2019.

[WOU 15] WOUTERS C., FRAGA E.S., JAMES A.M., "An energy integrated, multi-microgrid, MILP (mixed integer linear programming) approach for residential distributed energy system planning – A South Australian case-study", *Energy*, vol. 85, pp. 30–44, 2015.

[YAN 10] YANG B., JOHANSSON M., "Distributed optimization and games: A tutorial overview", in BEMPORAD A., HEEMELS M., JOHANSSON M. (eds), *Networked Control Systems*, Springer, London, pp. 109–148, 2010.

[YAN 19] YANG J., HE S., XU Y. et al., "A trusted routing scheme using blockchain and reinforcement learning for wireless sensor networks", *Sensors*, vol. 19, no. 4, p. 970, 2019.

[ZHE 18] ZHENG Y., SONG Y., HILL D.J. et al., "Multiagent system based microgrid energy management via asynchronous consensus ADMM", *IEEE Transactions on Energy Conversion*, vol. 33, no. 2, pp. 886–888, 2018.

[ZHO 20] ZHOU Q., HUANG H., ZHENG Z. et al., "Solutions to scalability of Blockchain: A survey", *IEEE Access*, vol. 8, pp. 440–455, 2020.

7

Sustainable and Desirable Living Thanks to Smart Buildings

7.1. Introduction

Global warming is causing sea levels to rise around the globe, destroying and endangering many buildings along the coast. Global warming is however also responsible for droughts, disturbing the foundations of buildings on clay-rich soils that are shrinking as they dry out. In Lille in northern France, this is compounded by a drop in the water table. For the wooden pilings supporting buildings in water-bearing zones, this is leading to oxygenation and rot, and in the worst-case scenarios to collapse. Insurance companies are seeing the number and scale of claims soar. An Australian risk analysis company, XDI, which specializes in advising decision-makers, major corporations and banks, has drawn up a global ranking of the 100 regions where buildings will be most vulnerable to global warming in 2050 [COU 23]. Figure 7.1 shows these 100 regions on a map of the planet. Asia, North America and South America are particularly vulnerable. In Europe, the Netherlands and Lombardy in Italy are the most exposed; the most vulnerable region in France is Hauts-de-France.

The buildings in which we live, work and shop, as well as those used for leisure and sport, are the main emitters of CO_2. In France, in 2016, residential and tertiary buildings accounted for 45% of total final energy consumption. The French PACTE law passed in 2019 aims to reduce building consumption by 40% by 2030 and 60% by 2050. In 2019, the residential and tertiary sectors accounted for just over 20% of greenhouse gas emissions, with a downward trend thanks to energy efficiency and renovation efforts [GAS 22]. The technologies used to construct and operate these buildings play an important role in the scale of these carbon emissions, but the way in which they are lived in and used also contributes to these emissions. These buildings are equipped with devices that enable them to better identify and control energy input and consumption; they can become energy producers and store energy, thanks to the new materials used to construct these buildings, and also thanks to new energy and digital technologies, all the while ensuring the comfort of their occupants. Buildings are becoming smart.

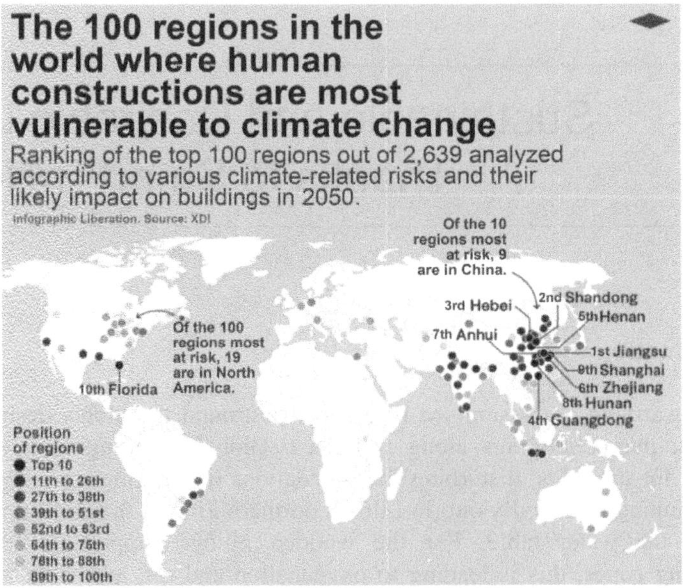

Figure 7.1. *The 100 regions of the world where human constructions are most vulnerable to climate change [COU 23]*

However, technology alone is not enough to optimize a building's operation and reduce its carbon footprint. The building's users also play an important role, in terms of their activities and their comfort needs, which vary from one person to the next, to the point where we might wonder whether, for smart buildings to achieve their low-carbon footprint objective, they should not be used and inhabited by smart users adapted to, or integrated into, the building's intelligence? In Chapter 5 of [ROB 23], sociologist Hervé Barry develops a socio-technical analysis of buildings becoming smart, based on feedback from two smart building demonstration sites at the Université Catholique de Lille.

In Chapter 1 (see section 1.8), the smart building concept was introduced after mentioning some futurist perspectives from the building sector towards 2050, positioning these perspectives in relation to the ADEME 2050 scenario. The low-tech approach was also introduced, as it is particularly relevant to the digital buildings of the future. These concepts and approach are explored further throughout this chapter.

The concepts of smart building and smart grid will be increasingly actively interrelated in the future, with buildings potentially becoming active nodes (producer, consumer, storage, flexible, integrating electric mobility) of future smart grids. This has been the subject of a research chair entitled Smart Buildings as Nodes of Smart Grids (SBnodesSG) supported by the Junia Engineering Graduate School in Lille in conjunction with nine companies from the private sector, between 2018 and 2021.

Smart buildings make increasing use of digital technology, making them a rather high-tech object, even if low-tech approaches also have their place within these buildings. Low-tech approaches are at the heart of ADEME's S1 and S2 scenarios (Table 1.1), while a high-tech smart building approach is more likely to be found in the S3 scenario (Table 1.2). Smart buildings can also be approached from a more socio-technical angle, which is not incompatible with ADEME's S2 or even S1 scenarios (Chapter 5 of [ROB 23]).

The evolution of buildings will have to integrate the environmental framework that is an integral part of the habitat in which we live; the aim is to preserve and even contribute to the development of biodiversity, and more generally to bring nature back into the city [ROB 23]. With this in mind, buildings undergoing renovation or construction must respond to UN Sustainable Development Goal 11 "Sustainable Cities and Communities" by becoming part of a sustainable urbanization, with a reduced or even better positive environmental impact, and contributing to the resilience of territories in the face of climate change.

In section 7.2 of this chapter, smart buildings are defined by illustrating the services they can offer thanks to the vast amount of data they can collect and manage, particularly in the case of tertiary buildings. These digital data, managed, for example, by a BIM digital twin, can offer new services to users and to the energy networks to which the building is connected, and help to integrate the building more effectively into its environment.

Section 7.3 develops several examples of the use of data in building management. These involve the dynamic energy optimization of a building using weather data and potentially data provided by BIM, taking into account room occupancy and user comfort. The interest of managing IAQ for ensuring user comfort is then discussed. Finally, the contribution of blockchain technology to smart buildings is illustrated with two examples.

The deployment of digital technology offers many new opportunities for the development, management and enhancement of buildings. However, the environmental impact of digital technology must be drastically reduced in the future. To illustrate this move towards more environmentally virtuous buildings, section 7.4 presents and compares two examples of tertiary buildings – one renovated, the other new – at the Université Catholique de Lille. Users have roles to play within a smart building, which contribute to its environmental performance while ensuring their comfort. This dimension is illustrated by the results obtained from a socio-technical approach to managing a building that was not initially designed as “smart”. Low-energy and low-material-use buildings and low-tech approaches are also important ways of reducing a building’s climatic impact. The case of residential buildings is discussed to illustrate the value of these approaches.

7.2. Smart building

7.2.1. *Definition of a smart building*

The smart building is often defined as a connected, communicating building. In section 1.8.3 (see Chapter 1), it is stated that a building’s level of intelligence can vary from process automation, centralized process control and fine-tuned equipment regulation to information exchange.

However, the climate crisis and growing urbanization mean that more is expected of buildings that become smart. The smart building represents a complex ecosystem that must combine a number of qualities: resilience, inclusiveness, collaboration with users, mutability (reversible, flexible, scalable), frugality (energy efficiency and low carbon footprint), service, sensitivity (adaptation to occupants and users), responsiveness and dynamism, and desirability [PEU 21]. It must be at the service of sustainable urbanization.

Smart buildings make increasing use of digital technology, and therefore process a growing amount of data. The trend is to turn them into high-tech objects.

However, the low-tech approach makes a lot of sense in the building sector, even if it is still not very widespread. Low-tech obviously concerns energy renovation, with a set of simple techniques and know-how that take into account what is already there, mobilize local, accessible and renewable resources of materials and energy, and call on transferable know-how for low-cost implementation, maintenance and use, accessible to all. The few key aspects of a low-tech approach are the definition of needs, user involvement, short circuits, reuse, etc., with constant questioning of energy, carbon, resources and biodiversity, while dealing with the constraints of the program, operation, financing, etc.; in short, this is techno-discernment [ADE 22b].

7.2.2. Services provided by a smart building

7.2.2.1. Sensors and data

By multiplying the number of sensors in a building, a large amount of data can be collected via a wired or wireless communication network. By processing these data, it is possible to improve the building's energy performance and reduce its carbon footprint, for example, by making it a self-producer and self-consumer of renewable energy, usually photovoltaic, and by involving users (which can reduce energy consumption in a commercial building by up to 10% through the adaptation of their behavior). This data processing enables the building's operating performance to be improved via the building management infrastructure (BMS) and services provided to the building's various users.

To illustrate the transformation of a tertiary building into a smart building, the example of the École des hautes études d'ingénieur (HEI), now part of the Junia school in Lille, is developed in the remainder of this section. Figure 7.1 shows a digital image (BIM) of the historic building dating from 1885 and its extensions over the years. The users are students, teaching, research, administrative and technical staff, visitors and building operators.

The transformation of the HEI's buildings into smart buildings[1] enabled the centralization of all building information (energy consumption, room temperatures, CO_2 concentration in classrooms, etc.), as well as the control and operation of equipment such as heating substations, air handling units, corridor lighting, heating valves and ventilation registers for all rooms in the buildings.

As the HEI buildings were renovated in 2015, the smart building transformation was not intended to replace previously installed equipment, but to improve its use, with the aim of reducing energy consumption while guaranteeing user comfort.

In order to optimize the heating periods of the various spaces, presence sensors were integrated in all the rooms of the buildings, so that temperature settings can be adapted according to actual occupancy, as well as connected window handles that switch off the heating when users wish to open them. An interconnection was set up with business software (e.g. enterprise resource planning (ERP), ticketing to manage task scheduling, etc.) and Wi-Fi terminals.

The monitoring of buildings and their environment was complemented by additional air quality and meteorological functionalities, with the installation of a

1 Details of the HEI "smart building" were provided by Grégory Vangreveninge, technical manager for demonstration sites at Junia.

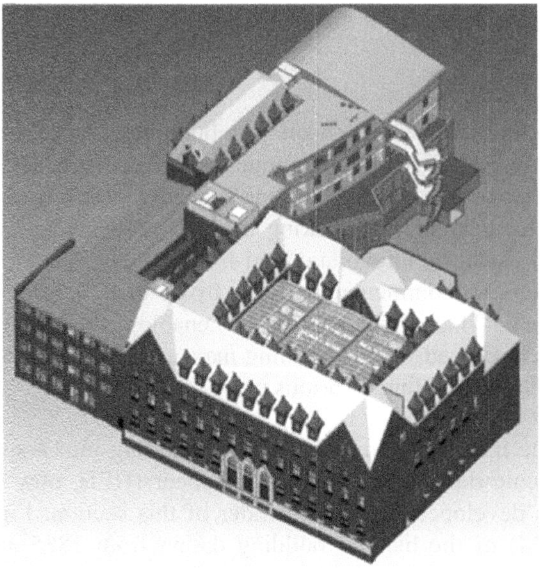

Figure 7.2. *Junia HEI building transformed into a smart building (photo credit: Grégory Vangreveninge)*

weather station and indoor and outdoor air quality sensors. Eight pollutants are monitored simultaneously (CO_2, CO, VOC, NO_2/O_3, PM1, PM2.5, PM10)[2], as well as indoor and outdoor humidity levels, and outdoor noise. These data can be viewed on screen, as shown in Figure 7.3, with the indication of threshold exceedance in orange for PM2.5 and PM10 pollutants.

Thanks to this new equipment, new user comfort control algorithms have been developed, as regulatory recommendations and obligations, for example, with regard to temperature set points, were not adapted to the use of certain spaces. Regulations require commercial heating systems to be sized to provide a temperature of 19°C +/−2°C in spaces. Temperatures can vary according to a range of individual parameters (gender, build, state of health, whether work is static or not, etc.), as well as meteorological and situational factors (air humidity, wind direction and speed, exposure to sunlight, etc.). The combination of these meteorological and situational parameters enables us to apply temperature settings that are better adapted to the actual use of our spaces.

---

2 CO_2 = carbon dioxide, CO = carbon monoxide, VOC = volatile organic compounds, NO_2 = nitrogen dioxide, PM1 = particulate matter less than 1 μm, PM2.5 = particulate matter less than 2.5 μm, PM10 = particulate matter less than 10 μm.

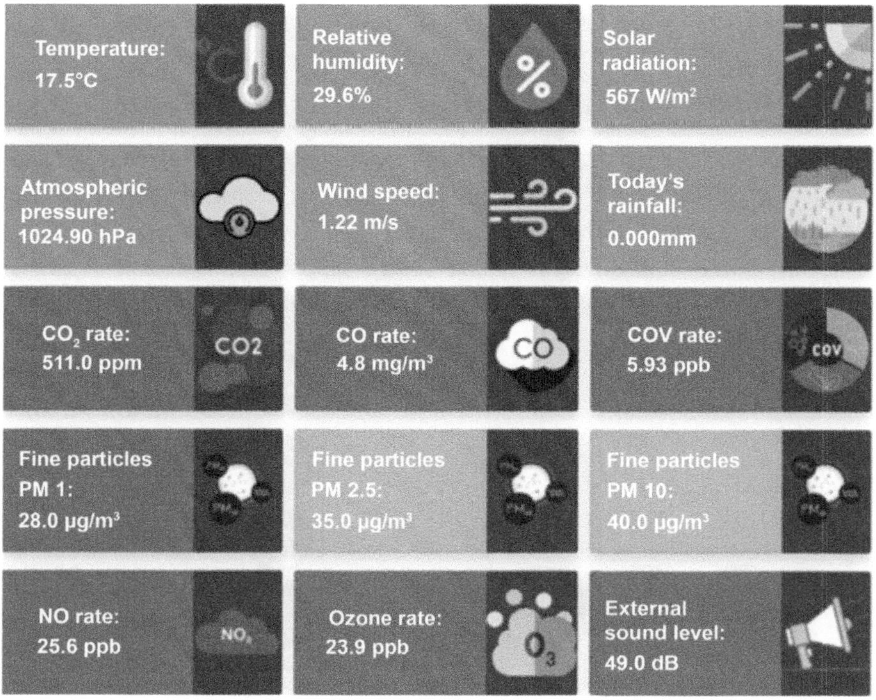

Figure 7.3. *Real-time collection of weather, outdoor air quality and noise levels at the HEI building in Junia, which is being built in Junia (photo credit: Grégory Vangreveninge)*

People counters at all building entrances and exits enable us to determine the number of people using the building. Thanks to these counters, the presence detectors in each room and the data transmitted by the Wi-Fi terminals located in the building, the concentrations of people in the spaces can be estimated while respecting the anonymity of the occupants. These data enable energy management algorithms to be optimized.

The performance of a smart building, in terms of service to users, operators and managers, essentially depends on the building's interconnection with its entire internal and external ecosystem. The building is thus able to collect and exploit all the data at its disposal to provide new services to its users. Figure 7.4 shows the interaction between the various devices that transform the HEI building into a smart building.

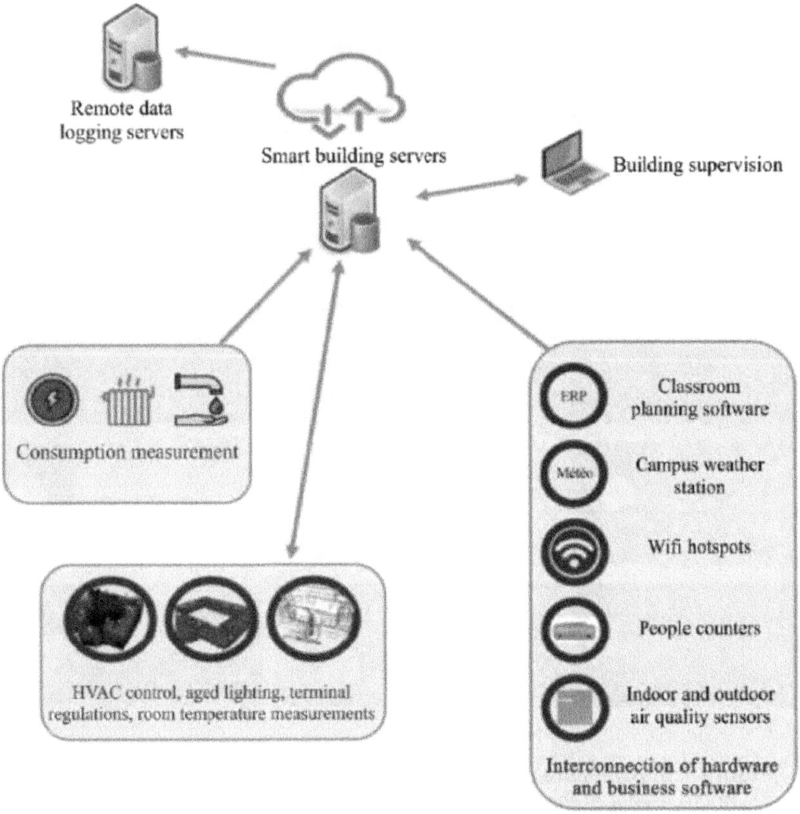

Figure 7.4. *Interaction between different types of equipment to transform HEI building into a smart building (source: Grégory Vangreveninge)*

7.2.2.2. *Building information modeling*

BIM provides a digital mock-up of a building, a kind of digital twin, which promotes collaboration between the various building stakeholders throughout the building's life cycle, from programming to design, construction, operation and deconstruction [PEU 22]. This digital twin is one of the levers in the fight against climate change. It enables the building's life cycle analysis (LCA) method to be integrated, optimizing the choice of materials, maintenance and even deconstruction, in order to limit the use of material resources and the building's overall carbon impact. During operation and maintenance, the digital twin can adjust to actual behavior and anticipate failures thanks to integrated sensors.

Figure 7.2 illustrates the BIM model of HEI's historic building. Figure 7.5 illustrates the BIM model of the interior of this building.

Figure 7.5. *BIM model of the interior of the historic HEI building in Junia (photo credit: Grégory Vangreveninge)*

The BIM model provides access not only to static building information but also to dynamic information that is updated in real time. For example, in the HEI model shown in Figure 7.6, it is possible to visualize temperature differences by changing the color of the rooms or to see whether or not they are occupied, thanks to the icon above them.

Figure 7.6. *Dynamic visualization of room-temperature variations on the BIM model of the historic HEI building in Junia (Photo credit: Grégory Vangreveninge)*

It is also possible to see directly on the model shown in Figure 7.7 whether any windows are open, according to the color of the room: the room is green when the windows are closed, or red if at least one window is open. In this case, ventilation and heating are switched off for this room to avoid wasting energy.

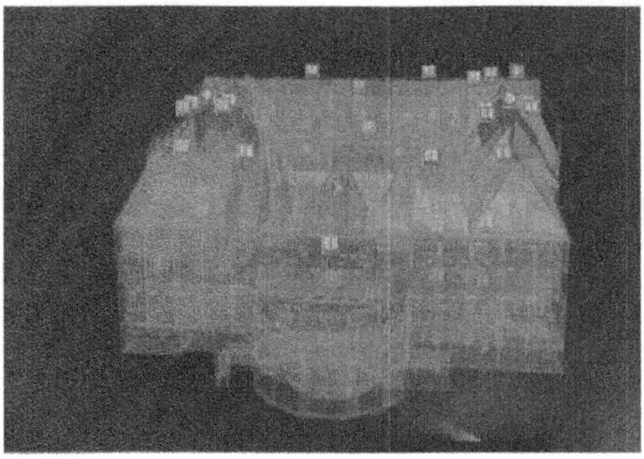

Figure 7.7. *Dynamic visualization of room occupancy and open windows on the BIM model of the historic HEI building in Junia (Photo credit: Grégory Vangreveninge)*

7.2.2.3. Customer services

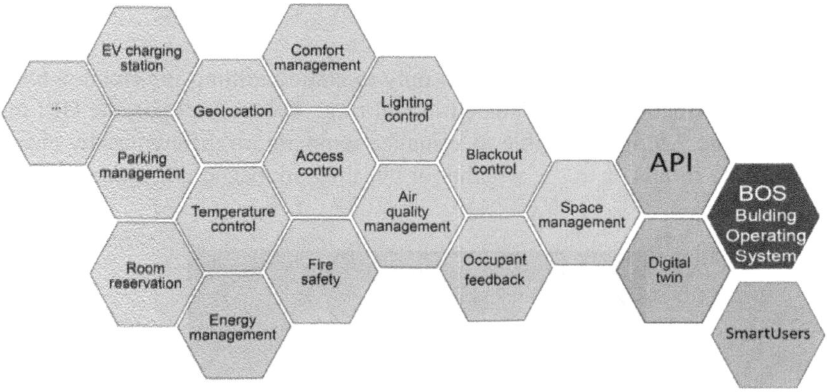

Figure 7.8. *A summary of the services that smart buildings can provide to users (source: Grégory Vangreveninge)*

Figure 7.8 summarizes the range of services that a smart building can provide for its users. To manage the digital twin of the building shared by several players, the building operating system (BOS) provides the building's digital foundation and represents the logistical data processing chain: extraction, transport, structuring, merging, distribution, securing and transformation. It also makes it possible to manage all data sources by using a single programming interface or API [PEU 22].

A mobile application enables building users to access a variety of services and news according to their user profile: employee, student or visitor, in the example of the HEI smart building, with the assumption that building users can become smart users.

Figure 7.9. *Ready to Service (R2S) label, dedicated to connected, supportive and humane buildings (sources: Certivéa and Grégory Vangreveninge)*

Improving building performance and user comfort, with varying degrees of user involvement, is the overall objective of smart buildings. To recognize the achievement of these objectives, a Ready to Service (R2S) label is introduced in France for 2022. This label is based on six major themes for a connected, supportive and humane building, linked together in Figure 7.9: network architecture, connectivity, equipment and interfaces, digital security, responsible management and services.

However, there may be conflicting interests in associating performance, particularly energy performance, with user comfort. In [ROB 23], sociologist Hervé Barry develops a socio-technical critique of the ability of smart buildings to involve users, based on feedback from two tertiary smart building demonstration sites of the Université Catholique de Lille, including the HEI building. Here is his conclusion:

> Performance in terms of use is certainly more difficult to achieve than performance in terms of energy, because the human factor does not

correspond to any scientific law. To achieve this, the sociology of energy favours promoting the involvement of users as far upstream as possible, in the decision-making process for building philosophy, and then in the design of socio-technical devices. So far, this has been done without much success, because what is being called for takes time, costs money and does not claim to guarantee an effective and virtuous response. Exploring the dynamics of use and operating methods in two buildings [demonstration sites at the Université Catholique de Lille] reveals, first of all, some encouraging signs for the future and a strong observation. Firstly, there is a growing *contradiction between calls for greater responsibility and commitment on the part of energy users, the growing awareness of and commitment to the environment among the general public, and the continued distancing of [users] from the working environment.* In the more or less long term, we can only hope for a paradigm shift in favor of user involvement.

Secondly, the timid but real development of socio-technical engineering, embodied by the profession of facility manager, demonstrates a new concern for dialogue with users. Time will tell whether this organizational evolution will be conducive to a more open consideration of the user's point of view. Our strongest observation concerns smart users. The advent of smart buildings, and the change in the socio-technical context that this implies, has given new impetus to the issue of taking users into account. The description of the dynamics of use and the two case studies show that occupants are as ready as ever to play with a more or less automated socio-technical environment, for their own comfort and to reduce their mental workload. Beyond this observation, it should be emphasized that the exploratory work carried out has enabled us to advance our understanding of what a "smart user" is or should be. In contrast to the occupant acting for their own comfort and according to their own free will, the "smart user" would be an "enlightened" actor, i.e. informed of the usage patterns to be respected and the energy management issues at stake, and also "cooperative". As free as ever, they would know when to inform technical maintenance [of the building] of their practices. The powerful and seductive technophile imaginations associated with "smart buildings" tend to accentuate the representation of an occupant to be neutralized. Sociological analysis applied to "smart buildings" leads us to believe that quite the opposite should be the case [ROB 23].

7.2.2.4. *Energy network services*

The concepts of smart building and smart grid will become increasingly interrelated in the future, with buildings becoming active nodes (producer, consumer, storage, flexible, integrating electric mobility) of future smart grids. An extension of the French R2S label called R2S-4Grids defines the services to be implemented in terms of energy management, such as energy control services with self-production and self-consumption of renewable energy, and energy flexibility aimed at modulating energy production and consumption according to the energy market and the constraints of the energy networks to which the building is connected, in particular electricity. This approach reduces technical constraints and generates financial gains. One of the challenges, of course, is to rely on a responsible, low-energy digital system. This last point involves a search for a compromise between the temptation to retrieve a large amount of data and the assessment of the right amount of data needed to optimize performance (energy, comfort, etc.) and the gains sought (financial, carbon, etc.).

The flexibility that a tertiary building island can provide in the interests of its users was put into practice in Chapter 6, using as a case study the historic block of the Université Catholique de Lille, including the HEI building. This same block is used to test flexibility as a service to the electricity grid in real-life conditions; the demonstration site is presented in Chapter 8.

By extending these flexibility potentialities to several buildings and building islands connected to different points on the power grid, the capacity for increased interaction between buildings and energy networks increases, to the shared benefit of both. This gives rise to the concept of smart buildings as nodes of smart grids, described in section 1.9 of Chapter 1.

7.2.2.5. *The smart building in its environment*

Beyond the interactions between buildings and energy networks, buildings are generally part of a wider ecosystem within a city. To illustrate this dimension, Figure 7.10 lists the different axes of the Junia Engineering Graduate School campus demonstration site within the Université Catholique de Lille. The aim of a zero-carbon campus is to increase energy efficiency and the use of renewable energies, to develop alternative forms of mobility that are less carbon-intensive and less invasive for the city, and to bring nature back into the city, both on campus and within the buildings. The campus also aims to be innovative (innovation being obviously very important for an engineering school) through the integration of digital technology (which must remain "responsible"), the development of urban agriculture (Chapter 3 of [ROB 23]) and the circular economy, as well as by taking into account well-being, inclusion, the social and solidarity economy (SSE), etc.

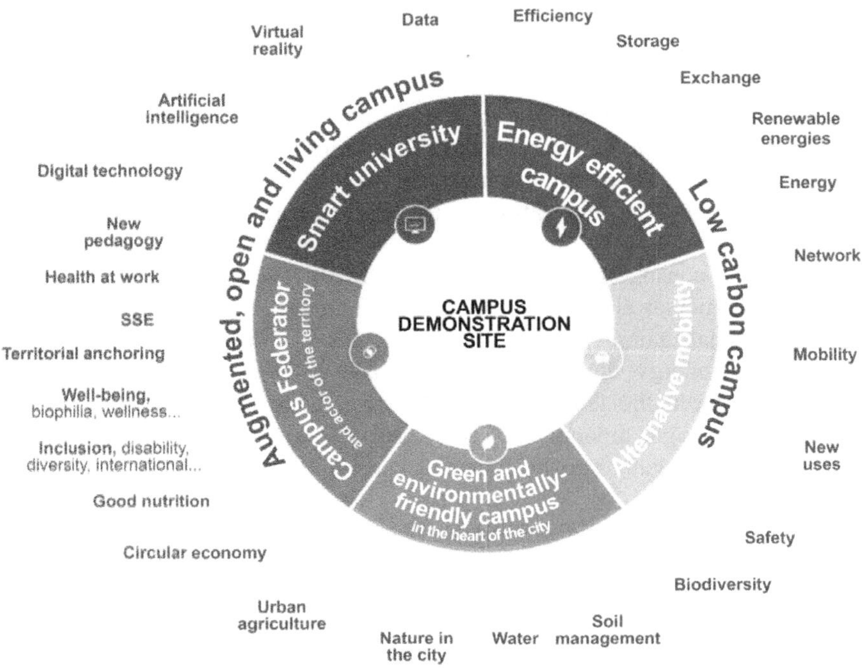

Figure 7.10. *Targets for Junia's demonstration campus in Lille (Source: Myrina Romero, Junia)*

To illustrate the return of nature to the city, particularly within buildings, Figure 7.11 shows the creation of a green agora at the entrance to the Institut Supérieur de l'Electronique et du Numérique (ISEN) building in Junia. Such a green interior helps to absorb noise and improve acoustic comfort. It will also have an impact on thermal comfort by reducing temperature contrasts and reinforcing the insulation effect, helping to combat extreme heat thanks to the evapotranspiration process. It also lowers the energy consumption of buildings and walls. On average, we spend 80–90% of our time in enclosed spaces, and indoor air often tends to be far more polluted than outdoor air. Paints, glues, varnishes and solvents used in homes and buildings release toxic volatile organic compounds (VOCs). Choosing plants capable of filtering and dissolving these VOCs will help to clean up and purify indoor air [ROB 23].

Working or living in a green environment brings people closer together. In particular, "small-scale green spaces" are proving to have a positive effect on social cohesion. Neighborhoods with more green spaces are less subject to aggression, violence and vandalism.

Figure 7.11. *Green agora at the entrance to the Institut supérieur de l'électronique et de l'électronique et du numérique (ISEN) building in Junia (Photo credit: Benoît Robyns)*

7.3. Data processing and building management

7.3.1. *Introduction*

The wealth of data acquired through the digital transition of the building industry, combined with the development of digital twins thanks to BIM, complements an older method known as dynamic thermal simulation (DTS) in the building industry. DTS appeared in the 1970s and 1980s, in parallel with the application of the first thermal regulations. It can be used to describe the temporal evolution of a building's thermal state, and in particular to assess the thermal phenomena of systems that are difficult to apprehend globally over a full year's cycle: the impact of free energy gains, the effects of masking or shading, etc.

Building occupants often notice a more or less significant difference between predicted energy performance, during the building design phase, and actual energy performance, during the operating phase. This difference is known as the energy performance gap (EPGAP) [ABD 19]. This EPGAP can be due to several factors throughout a building's life cycle [KAL 17]:

– during the design phase: poor dimensioning and unsuitable simulation tools;

– during the construction phase: poor-quality equipment and materials;

– during the verification phase: lack of verification of systems and work execution;

– during the operating phase: occupant behavior, weather and system control.

The rise of BIM as a collaborative working tool helps to avoid errors in sizing and execution. Furthermore, BIM models are rich in information, providing a faithful representation of the building and thus improving the accuracy of energy consumption estimates.

On the other hand, the widespread use of building instrumentation provides real-time monitoring of the building's interior conditions and energy consumption, and therefore more information on its actual behavior. These smart buildings often use networks of intelligent sensors, known as connected objects, which are easy to install. This is known as the Internet of Things (IoT) [ABD 19].

This section presents an initial case study illustrating the dynamic energy optimization of a building using weather data and, potentially, data provided by a BIM model, taking into account room occupancy and user comfort. The BIM process has proved its worth in the construction phase, since it saves considerable time and money. However, BIM models are rarely used in the operating phase. The methodology we have developed will help to make the most of these models in the operating phase, and to take advantage of their wealth of information to manage the buildings.

Section 7.3.3 presents the correlation between outdoor air quality (OAQ) and indoor air quality (IAQ) data, and the benefits of managing IAQ to ensure user comfort.

Since so much data can be processed in a smart building, blockchain technology is of interest within smart buildings, illustrated by two examples at the end of this section.

The work presented in this section was carried out as part of Junia's Smart Buildings as Nodes of Smart Grids chair (SBnodesSG chair between 2018 and 2021, supported by the MEL) in collaboration with nine companies. This chair has benefited from the support of a building demonstration site developed as part of the Université Catholique de Lille's Live TREE program [ROB 23].

7.3.2. *Dynamic energy optimization for buildings*

There are various ways of predicting building behavior: DTS, statistical methods, learning algorithms and hybrid approaches.

DTS software describes classical thermal exchanges (convection, conduction, radiation, change of state). Recent developments have also given them capabilities to[3]:

– shape complex energy solutions using a global approach (taking into account different energy sources, carbon footprint, gray energy, etc.);

– evaluate winter and summer comfort, with a fine time interval (down to a few minutes);

– take a global approach to lighting (taking into account natural light and modulation of artificial lighting, solar protection, etc.);

– simulate the impact of wind on facades, natural ventilation phenomena and the benefits of a heat sink;

– estimate actual energy consumption based on information on the building envelope, inertia, energy systems (whatever the energy used, and taking into account free external or internal energy inputs), occupant behavior and local climate.

Evaluating building performance, both in the design phase and during operation, is essential to improving efficiency. Due to the complexity of the problem, numerous studies have been carried out in different fields (physics, mathematics, computer science, etc.), leading to the distinction of three main methods [ABD 21]:

– White box: studies physical phenomena and estimates them at a given moment and in a given space, using dynamic thermal simulation software (e.g. TRNSYS, EnergyPlus, etc.). This method requires a good knowledge of the building (materials, geometry, HVAC description, control strategies, occupancy, location, etc.). It can provide very detailed results, but requires a lot of computing time, usually calibration, and is impacted by uncertainties at various levels (model, parameters, etc.) [FOU 13].

– Black box: model based on mathematical and statistical methods (machine learning, Markov chain, multiple linear regression, etc.). Training time depends on the method used, and computation time is generally low [FOU 13].

– Grey box: approach combining the two previous models, requiring less training data and less knowledge of physical phenomena.

The aim of the study carried out by [ABD 21], as part of Junia's Smart Buildings as Nodes of Smart Grids chair with Vinci Construction and Vinci Énergie (Cegelec), is to implement a methodology for thermal control of the indoor heating systems in the rooms of a tertiary building that optimizes comfort and energy consumption in real time, taking advantage of the large amount of data measured in modern

3 See: https://cegibat.grdf.fr/dossier-techniques/simulation-thermique-dynamique.

buildings that are becoming smart buildings. This methodology can be applied to any type of building, provided it is instrumented, usable all year round, whatever the season, and self-adjusting by improving its accuracy over time. The implemented method illustrated in this section is multiple linear regression (MLR), part of the black box family of models for predicting indoor temperature.

Linear regression models seek to establish a linear relationship between two variables: the input, known as the "explanatory variable", and the output, known as the "explained variable". More generally, these models can have several explanatory variables (or predictors) and are often referred to as MLR models. MLR models are generally written in the following form:

$$y_i = \alpha_0 + \alpha_1 . x_{i1} + \alpha_2 . x_{i2} + \cdots + \alpha_p . x_{ip} \quad [7.1]$$

where i varies between 1 and n.

The integer n represents the population size, often the number of observations, and p is the number of predictors. The p values of α are the model parameters and α_0 is the model bias. From the recorded data, these parameters can be estimated. The least squares method is one of the most widely used methods for estimating model parameters.

The advantage of MLR is its ease of implementation. Once the parameters have been calculated, no further parameterization is required. It is appreciated for its simplicity and ease of programming. MLR coefficients are often used to express the relationship between the explained variable and the explanatory variables. However, MLR does not deal with the nonlinearity of the parameters under study.

MLR is therefore a regression model with at least two predictors [ABD 21]. MLR models are commonly defined by equation [7.2], or in matrix form, by equation [7.3]:

$$\hat{y}_i = \alpha_0 + \alpha_1 . x_{i1} + \alpha_2 . x_{i2} + \cdots + \alpha_p . x_{ip} + b + \boldsymbol{\varepsilon} \quad [7.2]$$

$$Y_i = \beta . X + b + \boldsymbol{\varepsilon} \quad [7.3]$$

In equations [7.2] and [7.3], $\hat{y}_i$ is the predicted variable, x_i are the p predictors, α are the p+1 model parameters, b is the intercept and ε is called the error term or perturbation. In matrix form, Y_i is the vector of the variable to be predicted, X is the matrix grouping all predictors and β is the matrix grouping the regression parameters.

The least squares method for calculating the regression parameters is used. In the case of a simple linear regression (X: predictor and Y: variable to be explained), the least squares method consists of representing the data on a plane (x, y) by a cloud of n points. This is known as the scatter plot. The aim is to find a line that best fits the point cloud. Of all the possible lines, the one that minimizes the sum of squared deviations between observed values y_i and predicted values $\hat{y}_i$ is selected. These deviations are called residuals. The aim of this method is to identify the values of a and b by minimizing the sum of squares of the residuals (equation [7.4]):

$$E = \sum_{i=0}^{n} \varepsilon_i{}^2 = \sum_{i=0}^{n} (y_i - (a.x_i + b))^2 \quad [7.4]$$

Generally speaking, the calculated values of a and b are denoted $\hat{a}$ and $\hat{b}$ and are obtained by equations [7.5] and [7.6], respectively. In these equations, $\bar{x}$ and $\bar{y}$ are, respectively, the averages of x_i and y_i:

$$\hat{a} = \frac{\sum_{i=1}^{n}(x_i - \bar{x})(y_i - \bar{y})}{\sum_{i=1}^{n}(x_i - \bar{x})^2} \quad [7.5]$$

$$\hat{b} = \bar{y} - \hat{a}\bar{x} \quad [7.6]$$

In the case of multiple linear regression, estimating the parameters α_1, … , α_p, which we will note $\hat{\alpha}_0$,…, $\hat{\alpha}_p$, amounts to minimizing the quantity presented in equation [7.7]:

$$\sum_{i=0}^{n} (y_i - (\widehat{\alpha_0} + \widehat{\alpha_1}.x_{i,1} + \cdots + \widehat{\alpha_p}.x_{i,p}))^2 \quad [7.7]$$

This minimization problem can be solved by transforming it into a matrix form.

Several metrics can be used to evaluate the performance of different prediction models [ABD 21]:

– Adjusted R^2 (adjusted coefficient of determination): the coefficient of determination R^2 provides an estimate of the quality of the model's prediction, by measuring the resemblance between observed and predicted data. The advantage of R^2_{aj} over R^2 is that it takes into account the number of predictors. The R^2_{aj} is calculated according to equation [7.8], where n and k are, respectively, the number of realizations (sample size) and the number of predictors:

$$R^2{}_{aj} = 1 - \frac{(1 - R^2)(n-1)}{n-k-1} \quad [7.8]$$

– Mean absolute percentage error (MAPE): average of absolute deviations from observed values. It is a percentage and therefore a practical indicator for comparison. The normalized MAPE is defined by equation [3.6] in Chapter 3.

– Root mean square error (RMSE): the mean square root of the MSE, which is the arithmetic mean of the squared differences between model predictions and observations. The normalized RMSE is defined by equation [3.5] in Chapter 3.

In the rest of this section, a method to help control heating systems is presented that uses multiple linear regression to predict the indoor temperature of a specific zone to anticipate possible thermal and energy comfort anomalies, such as over-consumption of energy [ABD 21].

Based on this methodology, a platform is proposed in [ABD 20, ABD 21], shown in Figure 7.12. It is composed of three levels:

– A data level: first, data measured in the building, such as indoor parameters (e.g. indoor temperature) and outdoor parameters (e.g. weather data), supplemented by weather forecasts and the initial heating strategy over a predefined horizon, are transmitted to the central platform. Then, after processing and at each iteration, the optimized heating strategy is received and transmitted to the building management system (BMS) for application, and to the BIM 3D digital model or twin for display.

– A main level: at each iteration, measured and predicted data are received from the data preprocessing level and used to train and update the MLR temperature prediction model. The indoor temperature is predicted every hour for a time interval depending on the predefined prediction horizon (e.g. 10 hours in this study). At each time interval (e.g. 1 hour), the resulting temperature prediction is used to predict the next hour. Each iteration consists of n predictions (for example, n = 10 for a 10-hour horizon). At each iteration, these predictions are updated, reducing uncertainty as the predicted data are replaced by measured data. Then, using the prediction model, each generated heating strategy is evaluated. For each heating strategy, n predicted temperatures are obtained, constituting a solution. Each solution is scored according to criteria defined in the third level.

– A configuration level: depending on the method chosen for evaluating the heating strategy, a prediction horizon is defined, allowing for reasonable calculation times. The criteria for rating heating strategies are defined by the user (e.g. keeping indoor temperatures between 20 and 24°C, minimizing energy consumption, encouraging energy consumption during low-price periods, etc.).

Firstly, MLR models are developed to diagnose individual rooms based on past data (sensors, user feedback, etc.). Secondly, future consumption is predicted from the trained model, taking into account future data (weather and room usage planning). This model is used to detect anomalies (temperatures too high or too low, for example) and to correct them in advance. The study uses MATLAB and TRNSYS simulation software.

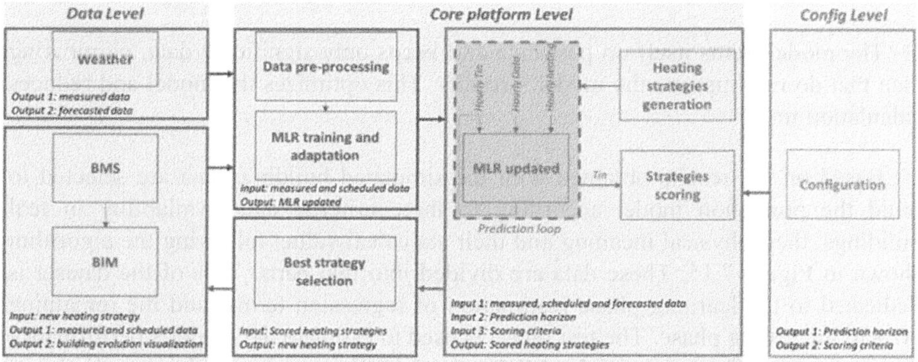

Figure 7.12. *Operating principle of the thermal management platform of a commercial building [ABD 20, ABD 21]*

Simulations were carried out on a fictitious building, shown in Figure 7.13, to validate hypotheses (choice of parameters, for example), as well as to study the model's ability to react and anticipate [ABD 20, ABD 21].

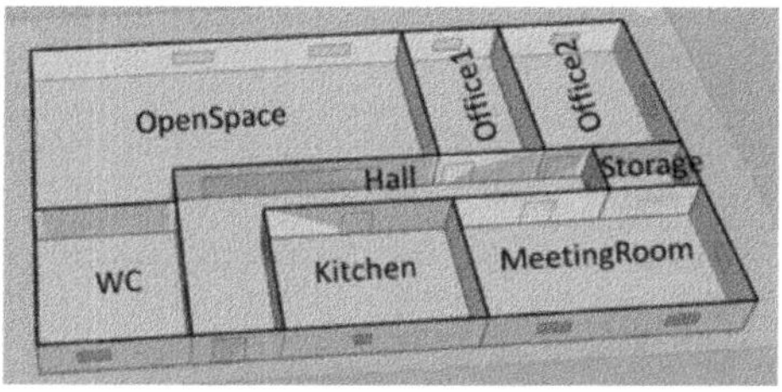

Figure 7.13. *Fictitious building simulation [ABD 20, ABD 21]*

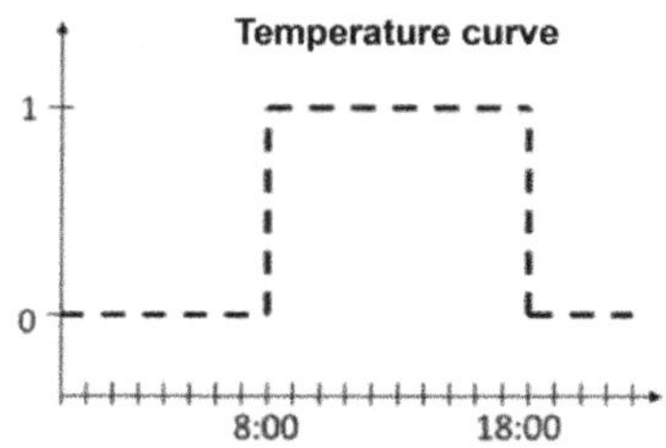

Figure 7.14. *Initial temperature curve. 0: low temperature; 1: high temperature [ABD 20, ABD 21]*

The basic heating scenario for the building, shown in Figure 7.14, assumes heating between 8 a.m. and 6 p.m. with a setpoint temperature ranging from 20 to 24°C, and between 6 p.m. and 8 a.m. with a setpoint temperature of 16°C.

The model trains itself on past data and keeps only significant data, eliminating data that do not improve the model's results. This optimizes the model and reduces calculation time.

Based on the results obtained with the simulated building, data are selected to build the prediction model according to three criteria: data availability in real buildings, their physical meaning and their statistical value, following the algorithm shown in Figure 7.15. These data are divided into two parts: 70% of the dataset is dedicated to the learning phase (estimation of regression terms) and the remaining 30% to the testing phase. The test phase is used to estimate the model's performance with new data. The terms of the MLR model [7.3] are calculated using the least-squares method [CIU 19].

The input data selected (illustrated in Figure 7.16) are heating power demand (*Pheat*), outdoor temperature (*Tout*), outdoor relative humidity (*RHout*), global horizontal radiation (*Rad*), sky opacity (*OpaSky*), wind speed (*WindS*), wind direction (*WindDir*) and output indoor temperature (Tin).

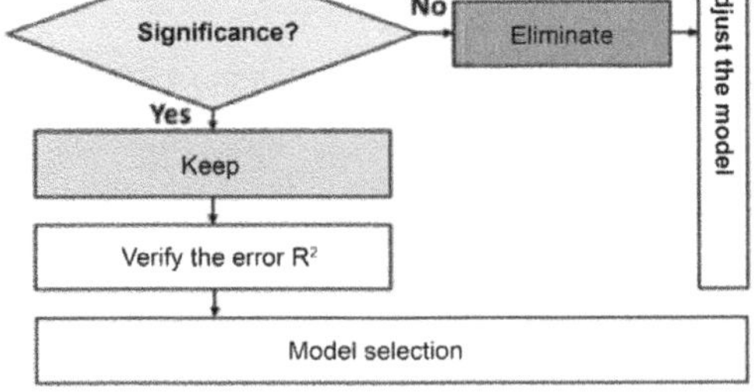

Figure 7.15. *Prediction model construction process [ABD 20, ABD 21]*

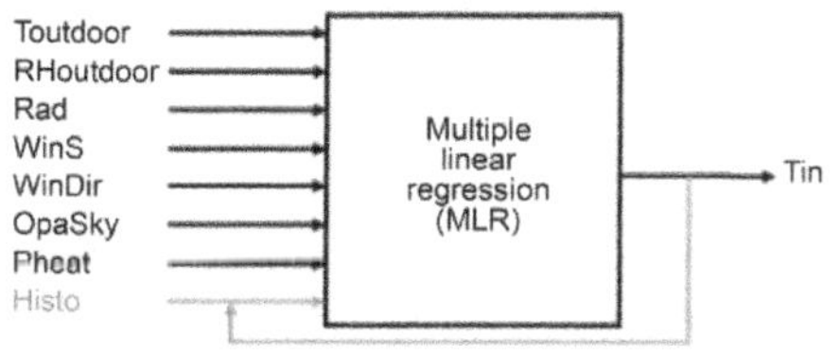

Figure 7.16. *Architecture of the prediction model [ABD 20, ABD 21]*

Model accuracy is greatly improved if data at T-3 h and T-6 h are added to the initial model data. Figure 7.17 compares observed and predicted temperature using three MLR model variants. Model 1 (red curve) corresponds to a model obtained without considering previous data, Model 2 (blue curve) corresponds to Model 1 with data at T-3 h and Model 3 (green curve) corresponds to Model 1 with data at T-3 h and T-6 h. It can be observed that Model 3 is much more accurate, being the closest to the actual curve observed in black, and that its MAPE (less than 3%) and RMSE (worth 1.06) are the smallest [ABD 20].

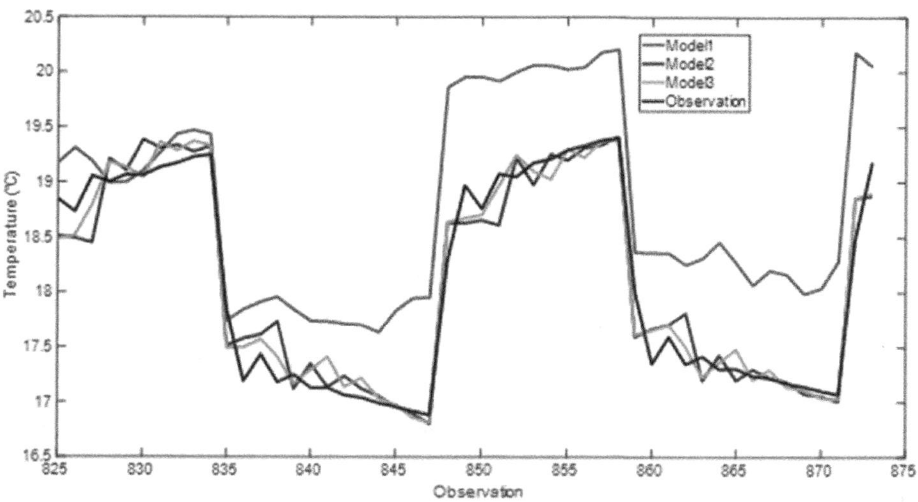

Figure 7.17. *Comparison between observed and predicted temperature using the three MLR models [ABD 20, ABD 21]. For a color version of this figure, see www.iste.co.uk/robyns/smartgrids.zip*

Based on this trained and validated model, the indoor temperature is optimized by changing the heating setpoints, thereby improving temperature comfort in the zones under study.

At each time interval (e.g. 1 hour), the chosen heating strategy is updated, and the setpoints are sent in real time to the BMS and the BIM model of the building to be displayed.

Every hour, data are collected from the simulated building, and the MLR model is trained according to the approach described above. At each iteration (1 iteration = 1 hour of sampling), the platform attempts to improve the heating strategy by predicting the indoor temperature with different values of *Pheat* (heating on: *Pheat* = 1 and heating off: *Pheat* = 0, as illustrated in Figure 7.14), with a time horizon of 10 hours that limits the number of strategies to be tested to 1,024 (2^{10}).

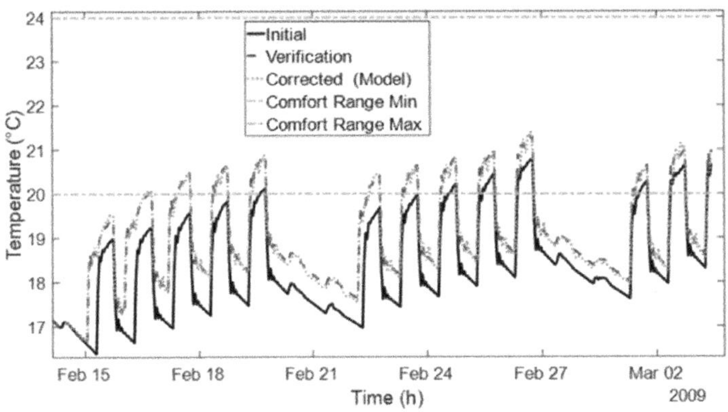

Figure 7.18. *Platform-determined temperature in green vs. original temperature in black [ABD 20, ABD 21]. For a color version of this figure, see www.iste.co.uk/robyns/smartgrids.zip*

Figure 7.18 shows the results obtained (with a sampling time of 1 hour) between 23:00 on February 14 and 23:00 on March 3, from a typical simulation year. The MLR model is determined with data from 00:00 on January 1 to 22:00 on February 11. As shown by the green temperature curve in Figure 7.18, the platform attempted to improve thermal comfort by minimizing the energy consumed, anticipating heating to achieve an indoor temperature within the predefined comfort range between 20 and 24°C during occupancy (8:00 a.m. to 6:00 p.m., blue lines) and above 16°C during unoccupied periods. Genetic algorithms are used to perform the optimization [ABD 21]. Focusing on the first day processed (between February 15 and 16), the platform activated heating 7 hours earlier than initially programmed (curve in black). Comfort was improved, but the objective was not reached on the first day (green curve below the blue line in Figure 7.18), because the 10-hour prediction horizon was not sufficient. In fact, the building did not have enough time to react and activate heating early enough. For the following days, the objective was achieved, as the green curve in Figure 7.18 becomes greater than the green line marking the low limit of the comfort temperature. February 20 and 21, as well as February 27 and 28, correspond to two weekends.

To analyze the platform's ability to improve the thermal comfort of the test building (of the Algeco type described in [ABD 21], a modular building the size of a container), according to user-defined criteria, an indicator of discomfort (*Ind*) is proposed, calculated according to equations [7.9], [7.10] and [7.11]:

$$\text{If } Tin_i > Tin_{Max}\ Ind_i = Tin_{Max} - Tin_i \quad [7.9]$$

$$\text{If } Tin_i < Tin_{Min}\ Ind_i = Tin_i - Tin_{Min} \quad [7.10]$$

$$\text{If } Tin_{Min} < Tin_i < Tin_{Max}\ Ind_i = 0 \quad [7.11]$$

This indicator is equal to 0 when comfort is respected and is equal to the difference between the measured indoor temperature and the nearest comfort threshold (e.g. if Tin = 19°C, then the discomfort indicator = 19–20 = 1°C, and if Tin = 25°C, then the discomfort indicator = 24–25 = −1°C), when comfort is not respected. This indicator can be used to identify discomfort situations and measure the deviation from the comfort threshold. Following a 21-day test under real-life conditions in May, Figure 7.19 shows the evolution of this indicator (in red), outdoor temperature (in blue) and occupancy hours (shaded slots). At first glance, we can see two peaks around May 9 and May 11, with maximum values of −6.5°C and −6.34°C, respectively. The first is explained by an unexpectedly sharp rise in outside temperature (25.63°C) at the end of the day. This is certainly due to the building's low inertia and does not call into question the platform's ability to control the building's heating in advance. To avoid this discomfort, a cooling system would have been required. The second peak in discomfort was caused by a power cut, accompanied by a computer bug that prevented the heating strategy from being updated and the indoor temperature from being predicted for the day of May 10.

The platform proposed by [ABD 21] optimizes the building heating control according to outdoor conditions (weather forecast) to satisfy two criteria: the thermal comfort criterion (between 20°C and 24°C) and the heating energy consumption criterion. Testing of the platform in real-life conditions over 21 days (from May 3 to May 24, 2021) ensured thermal comfort for 67% of the relevant occupancy hours, or with a deviation of less than 0.5°C, it remained acceptable for 76% of the time. However, the remaining 33% or 24% do not necessarily call into question the platform's ability to anticipate thermal comfort, since on the one hand, the building tested proved to be a difficult case to deal with due to its extremely low thermal inertia, and on the other hand, during the study period, outside temperatures in excess of 20°C were recorded, which corresponds to summer conditions. It is therefore impossible to guarantee thermal comfort in this type of building without recourse to a cooling system.

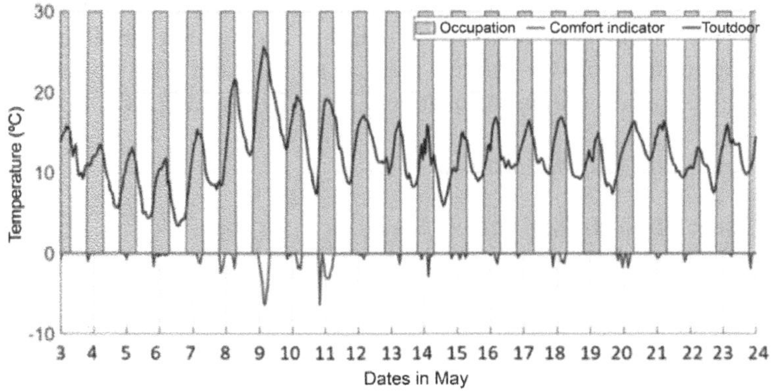

Figure 7.19. *Evolution of the indoor temperature measured in blue and the discomfort indicator in red during the test period [ABD 21]. For a color version of this figure, see www.iste.co.uk/robyns/smartgrids.zip*

The author of [ABD 21] also implemented an artificial intelligence tool to develop a model for predicting the indoor temperature of a building. Using ANNs, in [ABD 22], a nonlinear model was obtained that overcomes the linearity constraint of MLR. However, for the application considered, [ABD 21] found that the two models, RNA and MLR, performed equivalently in terms of prediction errors, but that the RNA model required 3.6 times more computation time than the MLR model, which is why the latter was chosen.

This section illustrated the potential of a smart building to better control the temperature of rooms in a building, by controlling energy consumption while ensuring the thermal comfort of building users. For the latter objective, a discomfort indicator can be used to take better account of users' needs and feelings.

7.3.3. *Indoor and outdoor air quality in a building*

More and more, attention is being paid to the quality of the air we breathe, whether interior (IAQ) or exterior (OAQ). In cities, air quality is a growing problem, with pollution induced mainly by the circulation of petrol and diesel vehicles. This degraded air quality can have an impact on the air inside buildings. This impact can be exacerbated by high occupancy levels in certain rooms of the building during certain hours of the day (meeting rooms, classrooms, etc.), and ventilation that is ill-suited to these occupancy levels and indoor air quality. The impact on the comfort of building users can become significant in the event of high concentrations of undesirable particles.

A user survey highlighted the symptoms perceived by users, as illustrated in Figure 7.20. These include nasal congestion, difficulty concentrating, headaches and fatigue. Analysis of the measurement results revealed high CO_2 concentrations of up to 2,250 ppm, due to ventilation that was insufficient and unsuited to the building's needs (for a maximum recommended value of 1,500 ppm). As for relative humidity, measurements revealed that it was below 50%. As a result, hygrothermal comfort was impacted, and occupants experienced sensations of discomfort and dryness. It should be noted that comfort was not studied, and ventilation control was based on temperature and/or CO_2 measurements, while concentrations of the main air pollutants, such as CO, benzene, formaldehyde and PM2.5, were low without exceeding recommended values. However, although the concentrations of certain pollutants measured in the room were not high, their effects should not be underestimated, as they may cause discomfort for the occupant and affect their health [QAB 20b].

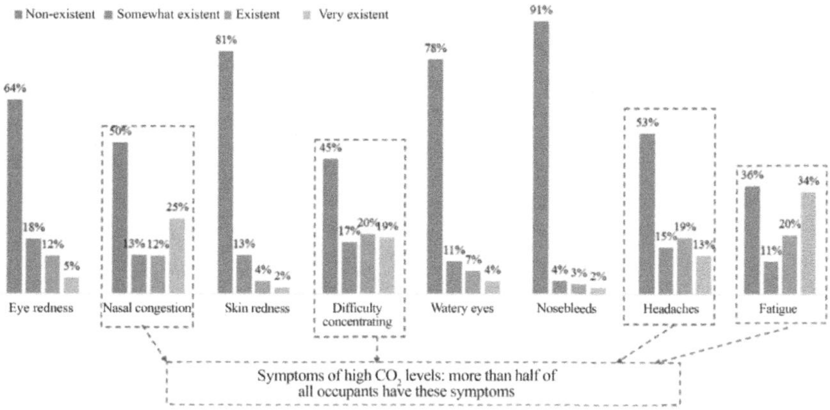

Figure 7.20. *Distribution of responses according to symptoms perceived by users [QAB 20b]. For a color version of this figure, see www.iste.co.uk/robyns/smartgrids.zip*

Sensors for measuring the various pollutants and parameters affecting interior and exterior air quality were installed around and inside the HEI building. Figure 7.21 shows PM2.5 and PM10 pollutants measured outside HEI over 25 days. The figure shows that the monitoring thresholds represented by the orange line were regularly exceeded outdoors for fine particulates PM2.5 and PM10.

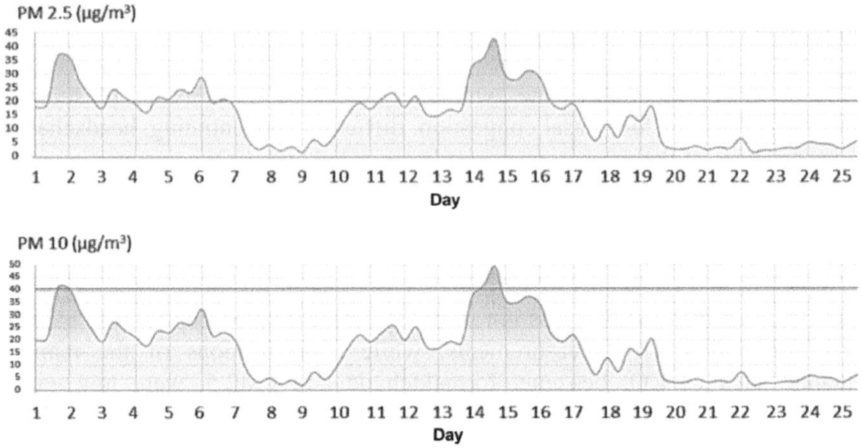

Figure 7.21. *Pollutants PM2.5 and PM10 measured outside HEI (source: Grégory Vangreveninge)*

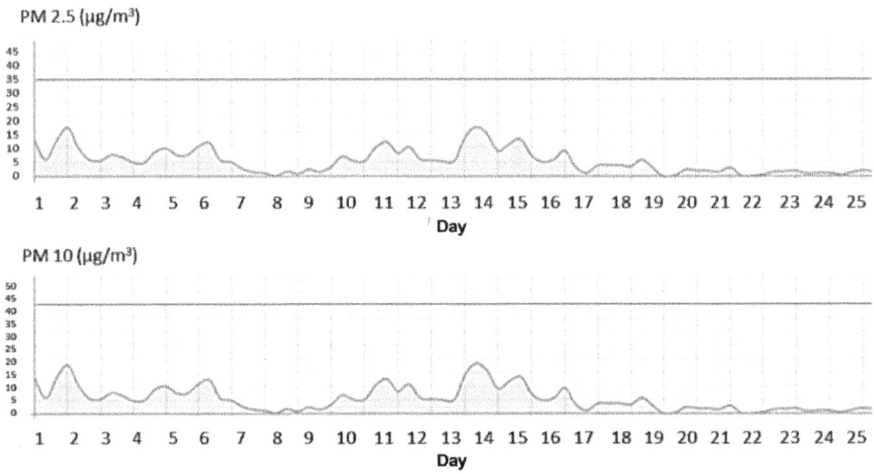

Figure 7.22. *Pollutants PM2.5 and PM10 measured inside HEI (source: Grégory Vangreveninge)*

Figure 7.22 shows PM2.5 and PM10 pollutants measured inside HEI over 25 days. Comparison of Figures 7.19 and 7.20 shows a direct link between the quantity of fine particles inside and outside the building. This shows the effectiveness of air handling unit filters, as well as their weaknesses, as we can see that they are not sufficient to block all these particles. Outdoor air quality can therefore be an aggravating factor in indoor air quality.

These IAQ measurements show the importance of taking them into account in the intelligent management of a smart building. The principle of this management is shown in Figure 7.23. It is structured in four phases: building instrumentation, data storage, data processing and, finally, decision-making, with the generation of setpoints for the air ventilation system (temperature, humidity, flow rate, etc.). This system can be complemented by a device for real-time data visualization. The authors of [QAB 20a, QAB 20b] developed a study of intelligent ventilation for improved indoor air quality in smart buildings.

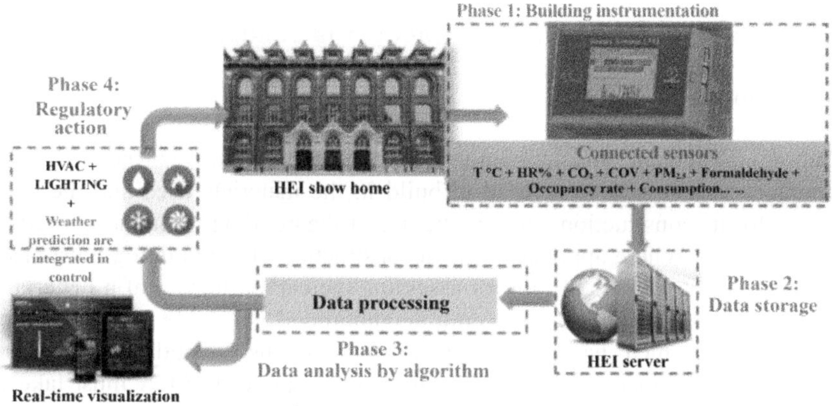

Figure 7.23. *Indoor air quality management system of a tertiary building [QAB 20b]*

7.3.4. *Blockchain and buildings*

Given that so much data need to be processed, blockchain technology can be experimented with for smart buildings. This technology was introduced in Chapter 6, in section 6.5; it operates on the basis of a registry consisting of a large, decentralized database, enabling ultra-secure storage and transmission of information between users who are members of the registry, without the need for an intermediary. In Chapter 6, this technology was implemented between several consumer buildings and some energy producers, organized as an energy community, in an energy exchange management system to maximize the self-consumption of photovoltaic energy, while taking into account the actor profiles inspired by Chapter 4. This community integrates storage and recharging of electric vehicles considered as actors in their own right.

Blockchain technology is also beginning to be associated with BIM. Reference [PEU 22] gives the example of the Magma platform, which enables digital real estate assets to be created from Non-Fungible Tokens (NFTs)[4], guaranteeing their uniqueness. In this platform, each stakeholder (owner, manager, service provider, tenant) contributes according to the access rights granted to them. All validated data thus become immutable, and the building's digital twin contains contractual documentation, thanks to the smart contract, and architectural documentation via the BIM digital model. On the basis of the pre-determined economic value of the NFT, each contributor of the data receives a fair remuneration, encouraging them to upload secure data to the platform thanks to the blockchain. The building model thus becomes a showcase for the real estate asset, documentation safe and task manager.

7.4. Environmental and climate impact of the building

7.4.1. *Introduction*

The environmental and climatic impact of a building is induced by its construction, via the site work used to build it, the materials used and the energy consumed for its construction, and via the use of the building throughout its life, its maintenance, etc. Where applicable, the deconstruction of the building must also be taken into account. The main tool for assessing the carbon impact of a material, an item of equipment, a building or even a neighborhood is the life cycle assessment (LCA) method, which has become mandatory for new buildings, and which compares the performance of different building components. LCA must take into account the waste and pollution generated by the building during its construction.

As illustrated in the introduction to this chapter, global warming will increasingly affect buildings, particularly in different regions of the world (Figure 7.1). Projections on this climatic evolution will have to be taken into account for future building constructions, whether to ensure the resistance of the building over time, or to ensure the comfort of users as temperature peaks become higher and higher, by reducing energy consumption and using renewable energies as much as possible.

To illustrate this trend towards more environmentally friendly buildings, two examples of tertiary buildings – one renovated, the other new – at the Université Catholique de Lille, are presented and compared in the remainder of this section.

---

4 NFTs represent the digital version of things that exist in the real world. They are unique assets that cannot be replaced and can be used as a certificate of authenticity.

The user has a role to play in a building, contributing to its environmental performance while ensuring its comfort. This dimension is illustrated here by the results obtained from a socio-technical approach to managing a building that was not initially designed as "smart".

Sufficient building and low-tech approaches are also important ways of reducing the climatic impact of buildings. The case of residential buildings in particular is discussed to illustrate the value of these approaches.

7.4.2. *Renovating instead of building new*

The first renovated building, called the Rizomm, has become a smart building demonstration, illustrated in Figure 7.24. The features, described in [ROB 23], can be summarized as follows:

– external insulation and replacement of external joinery;

– terracotta sunshade, blinds and external roller shutters to limit direct solar gain in summer and enhance user comfort;

– ventilation system with energy recovery;

– high-efficiency LED lighting and automatic management of lighting by presence detection and natural light;

– photovoltaic roof and local production plant: 1,200 m^2 of photovoltaic panels enable Rizomm to produce and consume its own energy.

The technologies deployed are designed to ensure that occupants retain some control over the management of their own comfort:

– individual control units for fine-tuning room temperature and brightness within pre-programmed tolerances;

– energy coaching application on smartphone.

A detailed socio-technical analysis based on the first years of feedback is presented in Chapter 5 of [ROB 23].

The second building, shown in Figure 7.25, is the new building of the School of Political Science (École de sciences politiques (ESPOL)) at the Université Catholique de Lille, which is in fact a reconstruction.

Figure 7.24. *Smart building demonstration at the Université Catholique de Lille, called the Rizomm (source: Media Lab UCL)*

This is the first of the university's buildings to have undergone a full LCA during the design phase. Its construction generated around half the GHG emissions of a conventional building, thanks to the use of bio-sourced or low-carbon materials (mixed wood/concrete, wood-fiber cladding), the reuse of part of the old building's structure, etc. Spaces have been optimized in terms of use to limit the m^2 required, as each additional m^2 generates CO_2. In terms of energy, the building benefits from high-performance insulation; it uses Lille's heating network (60% carbon-free), and is equipped with 200 m^2 of self-consuming photovoltaic panels, etc.

Figure 7.25. *New building for the School of Political Science (ESPOL) at the Université Catholique de Lille (source: Thibaud Foucray, Wonk architectes)*

The LCA for the ESPOL building is 998 $kgCO_2e/m^2$ per year over 50 years. This is below the recent French standard of 1,365 $kgCO_2e/m^2$. Construction products and equipment make up 85% of this balance, energy consumption 9%, building site operations 5% and water consumption and discharge 1%.

Figure 7.26 compares the LCA-derived GHG emissions of the ESPOL and Rizomm buildings. Emissions from the Rizomm building are almost 36% lower than those from the ESPOL building (640 $kgCO_2e/m^2$ instead of 998 $kgCO_2e/m^2$). As shown in Figure 7.26, the main difference is due to the foundations, infrastructure and superstructure. There are also savings in the construction of access roads, the installation of water, electricity and telecommunication supply networks and the roofing and waterproofing of the building.

In general, renovating a building rather than rebuilding it reduces its GHG emissions by around 30%, which obviously encourages us to plan for future building renovation rather than destroying them to rebuild new ones.

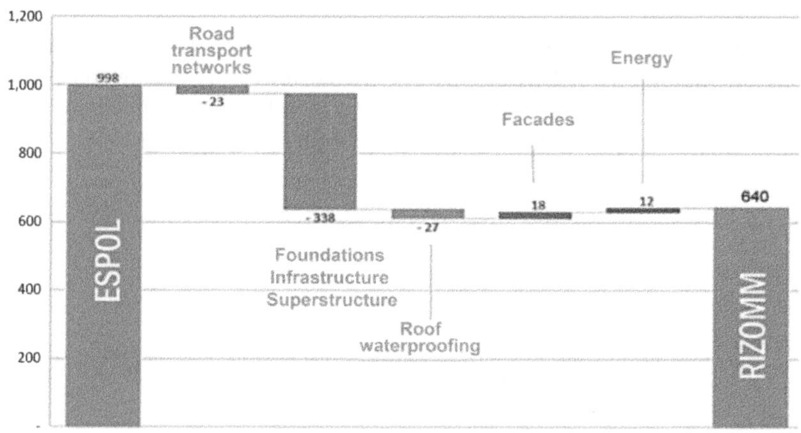

Figure 7.26. *Difference in GHG emissions between a new building (ESPOL) and a renovated building (Rizomm) in $kgCO_2e/m^2$ (source: Grégoire Destombes)*

7.4.3. *Socio-technical management of a building*

In order to illustrate the impact of user behavior on the reduction of energy consumption within a building, and therefore of its GHG emissions, the Junia ISA building is used as an example in this section. Figure 7.27 shows the ISA building, which was built in the early 2000s and is not considered a smart building.

Figure 7.27. *The Junia ISA building (source: Benoît Robyns)*

Some equipment has been replaced by more energy-efficient materials, such as conventional lighting with LEDs. However, it is above all the awareness-raising campaigns aimed at staff and students that have significantly reduced consumption of consumables: energy, water and other non-mobility consumables.

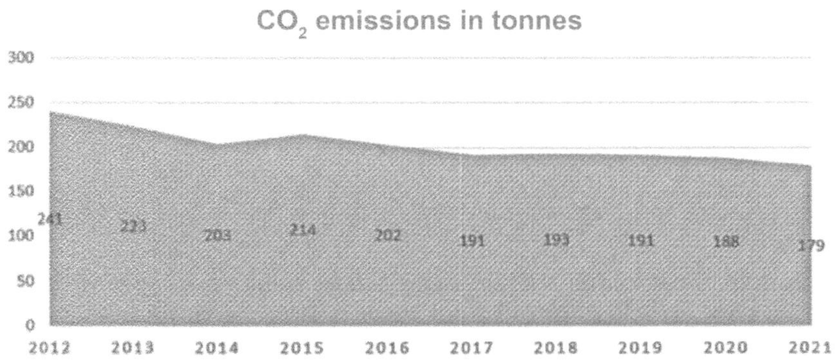

Figure 7.28. *CO_2 emissions from energy, water and other consumables for the Junia ISA building between 2012 and 2021 (source: Junia)*

Figure 7.28 shows a 21% drop in CO_2 emissions between 2012 and 2019, a good half of which was achieved through changes in user behaviors, with the remainder the result of more efficient equipment. This result illustrates the importance of the user behaviors on energy and environmental performance, with potential gains in performance at around 10%. The decrease in consumption shown in Figure 7.28, which continues in 2020 and 2021, is partly due to a drop in building use, a consequence of the Covid-19 pandemic. Despite this steady decline in consumption, vigilance must be maintained, as the trend towards the growth of new electrical uses (digital, with an increase in videoconferencing and distance learning, electric mobility by means of scooters, bicycles and electric cars) is likely to induce a rebound effect with increased energy consumption that must continue to be controlled, even if the latter is decarbonized (of renewable or nuclear origin).

7.4.4. *Sufficiency in residential buildings*

Dominique Bourg, philosopher and honorary professor at the University of Lausanne, defines sufficiency as individual and collective self-limitation, and the best way to reduce inequalities. In contrast to renunciation, or even punishment, sufficiency enables us not only to rediscover our roots and meaning but also to reach a spiritual ideal that nothing can spoil [PEU 22].

In France, in 2020, the building sector accounted for 44% of energy consumption and nearly 19% of greenhouse gas emissions. The environmental consequences of construction and housing are numerous: soil artificialization, consumption of non-renewable raw materials, emissions of polluting gases, energy consumption, etc.

Also in 2020, according to INSEE, housing was the leading item of contracted expenditure for French households, accounting for almost 27% of final consumption. This weight on the budget is particularly significant for the most precarious households: for the 20% least well-off, housing represents almost a third of income [ADE 22a].

Despite the benefits that connected solutions can bring to building management (resource savings, increased comfort and well-being for residents, access to services, etc.), the issue of the environmental impact of digital technology must be taken into account in future choices. While the benefits of these solutions are fairly well identified, assessing their costs and environmental impact remains complex. Intelligent buildings should offer the right mix and subtle balance of high-tech and low-tech approaches to enhance residents' quality of life while preserving their living environment. It is a question of moving from a technocentric approach to one that puts people back at the heart of projects, in order to respond first and foremost to the needs of residents. The challenge is to reconcile collective intelligence, embodied by the behaviors of all stakeholders, with artificial intelligence, which puts digital technology at the service of a sustainable and desirable building project [PEU 22].

In particular, low-tech offers the housing sector an opportunity to reflect on the necessity of the act of building and how to go about it, in order to propose environmental solutions that allow maximum energy and financial autonomy. Here are a few examples of low-tech approaches in the residential sector, identified in [ADE 22a].

As already illustrated in section 7.4.2, building should be avoided wherever possible, which is the first lever for reducing a building's environmental impact. Promoting shared housing solutions helps to optimize occupancy and create social ties by sharing common spaces (kitchen, guest room, laundry room, garden, etc.). To optimize the use of existing, potentially high-quality buildings, Philippe Bihouix [BIH 21] suggests designing solutions for sharing these buildings. This type of thinking is emerging in new construction and development projects that emphasize chronotopy[5].

Thanks to their smaller surface area, lightweight housing can help limit the artificialization of land and real estate pressure on an area. The reduced surface area of housing, as well as the choice of construction methods, also reduces the need for building materials and, when used, water and energy. As a result, lightweight housing tends to cost less. Therefore, promoting lightweight housing also helps to improve the affordability of housing. Indeed, industry actors note that residents of lightweight housing may be motivated by difficulties in accessing home ownership in conventional housing, whether in terms of access to land or loans. Finally, light housing is generally associated with issues of autonomy. As they can sometimes be self-built, or because their small size makes it easier for them to be self-sufficient in water or electricity, many inhabitants claim the freedom that their light habitat affords them. In France, light habitats are classified according to three regulatory categories: light leisure dwellings (cabins, tiny houses, etc.), mobile homes and finally caravans characterized by permission to transport them elsewhere [ADE 22a].

When construction is necessary, low-tech techniques can minimize the environmental impact of building and using habitats. Certain bio-sourced or geo-sourced building materials, marginal at present, offer numerous advantages over conventional materials, particularly in environmental terms; wood and straw, as well as rammed earth, can be used as alternatives to conventional concrete. There are also low-tech solutions for recycling waste materials into building materials, such as textile waste. The materials to be used in low-tech construction are highly dependent

5 In urban planning, chronotopy refers to the temporal planning of spaces, in order to create a "city that is malleable, flexible, supple and adaptable in its spaces and times" [GWI 13, authors' translation]. The aim is to pool the use of buildings and enable multiple uses of a single space.

on the context and the abundance of materials available locally. Low-tech insulation techniques include cork, hemp, cellulose wadding and wood fiber. Solutions can also be found to improve energy performance without replacing the entire insulating layer, such as hot-air collectors, sunshades or green roofs.

However, low-tech experts warn against the rebound effect that can result from different forms of renovation: better insulation should not be used as a justification for raising temperatures inside buildings just because heating is now cheaper.

Furthermore, more than one in four households in France suffers from energy poverty, that is, having an energy bill that accounts for at least 10% of household income. There are no regulations governing heating temperatures. However, ADEME [ADE 22a] recommends heating to 19°C in living rooms and 16°C in bedrooms. ADEME recommends that air-conditioning should not be used before 26°C in the home. In fact, by raising the set temperature from 22°C to 26°C, electricity consumption is halved.

In conclusion, intelligence in tomorrow's buildings will not be driven solely by the deployment of digital technology. Intelligence is also to be found in ingenious bioclimatic design, with the judicious use of eco-sourced materials and energy to save resources and ensure frugal construction and renovation. Intelligence will lie in optimized use, following the principles of the economy of functionality, favoring the use of a good rather than its ownership, and the concept of chronotopy already mentioned. The smart building will also have to be conceived as an integrated living organism, adapted to its environment and capable of providing ecosystem services, such as welcoming biodiversity, ensuring good air quality, purifying and storing rainwater or converting sunlight into reusable energy. This diverse conception of a building's intelligence should encourage the transformation of lifestyles towards more sustainable and resilient consumption behaviors and production strategies [PEU 22].

7.5. Acknowledgments

Junia's Smart Buildings as Nodes of Smart Grids chair has received funding from the MEL and nine companies: Accenta (which acquired Effipilot), EDF, Enedis, Energic, Greenbirdie, Projex, Stereograph, Vinci Construction and Vinci Énergie.

The smart building demonstration sites have received funding from the Hauts-de-France region via the Live TREE program.

7.6. References

[ABD 19] ABDELLATIF M., CHAMOIN J., DEFER D., "Du BIM au BEM en phase d'exploitation", *AJCE*, vol. 37, no. 1, 2019.

[ABD 20] ABDELLATIF M., CHAMOIN J., NIANGA J.-M. et al., "A thermal control methodology based on a multiple linear regression predictive model for indoor heating", *5th International Conference on Smart and Sustainable Technologies*, Split, IEEE SpliTech2020, 2020.

[ABD 21] ABDELLATIF M., Optimisation énergétique dynamique d'un bâtiment grâce à l'intelligence artificielle exploitant les données d'une maquette BIM, tenant compte du comportement et du confort de l'usager, PhD thesis, Artois University, Béthune, Lyon, 2021.

[ABD 22] ABDELLATIF M., CHAMOIN J., NIANGA J.-M. et al., "A thermal control methodology based on a machine learning forecasting model for indoor heating", *Energy and Buildings*, vol. 255, 2022.

[ADE 22a] ADEME, État des lieux et perspectives des low-tech, Report, 2022.

[ADE 22b] ADEME, "Low-tech : une nouvelle façon d'innover", *Ademe*, no. 154, 2022.

[BIH 21] BIHOUX P., *L'âge des low tech. Vers une civilisation techniquement souhaitable*, Le Seuil, Paris, 2021.

[CIU 19] CIULLA G., D'AMICO A., "Building energy performance forecasting: A multiple linear regression approach", *Applied Energy*, vol. 253, 2019.

[COU 23] COULAUD A., SCHAUB C., "La région Hauts-de-France est la plus menacée par le changement climatique dans l'Hexagone", *Libération*, available at: https://www.liberation.fr/environnement/la-region-hauts-de-france-est-la-plus-menacee-par-le-changement-climatique-dans-lhexagone-20230220_WS4PERISEFDBBLYZNHH2NRFPI4/, 2023.

[FOU 13] FOUCQUIER A., ROBERT S., SUARD F. et al., "State of the art in building modelling and energy performance prediction", *Renewable and Sustainable Energy Reviews*, vol. 23, pp. 272–288, 2013.

[GAS 22] GASPARD A., VISIER J.-C., "Le bâtiment à l'horizon 2050", *Futuribles*, no. 449, pp. 25–41, 2022.

[GWI 13] GWIAZDZINSKI L., "Urbanisme des temps. Premières chorégraphies de la métropole hypermoderne", *L'Observatoire*, no. 43, pp. 3–8, 2013.

[KAL 17] KALLAB L., CHBEIR R., BOURREAU P. et al., "HIT2GAP: Towards a better building energy management", *Energy Procedia*, vol. 122, pp. 895–900, 2017.

[PEU 12] PEUVOT A., "Le smart building : pour un bâtiment durable et désirable", *Les dossiers techniques de la construction*, no. 88, 2012.

[PEU 21] PEUVOT A., Le smart building : vers un bâtiment durable et désirable, Report, Le guide des services généraux et achats hors production, 2021.

[QAB 20a] QABBAL L., YOUNSI Z., NAJI H., "Indoor air quality investigation in a ventilated demonstration site building *via* a smart sensor", *International Journal of Ventilation*, vol. 21, no. 2, pp. 89–104, 2020.

[QAB 20b] QABBAL L., Étude de la ventilation intelligente au service d'une meilleure qualité de l'air intérieur dans les smart buildings, PhD thesis, La Rochelle University, La Rochelle, 2020.

[ROB 23] ROBYNS B., LENGLET C., BARRY H. et al., *Smart Users for Energy and Societal Transition*, ISTE Ltd, London and John Wiley & Sons, New York, 2023.

8

Demonstration Sites

8.1. Introduction: full-scale implementation

Implementing the energy and societal transition will require full-scale experimentation with a wide range of technological, economic and societal solutions, while accepting trial, error and adjustment. It is a question of validating the viability of technologies and their effectiveness and impact on the environment, social acceptance (which is by no means automatic), finding new economic models, devising new legal frameworks, etc. The success of the project also depends on effective persuasion. Persuasion plays a role in the degree of resistance to change.

Science and technology are generally developed in laboratories where testing conditions are under control. This generally leads to the development of prototypes for innovations. The complexity and speed of the transition means that emerging solutions need to be tested earlier and more systematically in real-life conditions, where not everything is under control, particularly climatic conditions and human behavior. For example, in the case of smart buildings, technologies will have to be tested and, ideally, designed with the users of a building in normal operation.

Confronting the reality on the ground, that is, the real world, requires us to engage with a wide range of disciplines: electrical and communication techniques for smart grids, for example, as well as sociology, to assess the acceptability and involvement of stakeholders, economics, because new viable models need to be found, law, ethical issues, etc. Interdisciplinarity, and even transdisciplinarity, will be essential to the success of the transition.

This chapter presents an approach to the development of demonstration sites to showcase new solutions aimed at the better control the consumption and renewable production of electrical energy, via the self-production and self-consumption of energy, electrical mobility, electrical energy storage and the flexibility of electrical loads in buildings. The approach is both technological and sociological. As part of the assessment of the potential for electrical load shedding in the demonstration site, the possibilities for load shedding are provided by a socio-technical context. This philosophy sets out the respective responsibilities of technical and human factors in the proper energy operation of a building. Next, a socio-technical approach is taken to the potential for energy savings in the context of commercial buildings, by looking at the savings potential of hot water tanks and electric vehicle charging stations.

The development of highly interdisciplinary demonstration sites represents a significant contribution to SDG 17 regarding "Partnerships for Achieving the Goals" of sustainable development, and to the development of responses to climate and environmental challenges.

8.2. Technology Readiness Level

Technology Readiness Level (TRL) is a method developed by NASA in the 1970s to assess the degree of maturity of a product under development [SAD 89]. This method is based on a graduated scale of nine steps, 1 corresponding to "Basic principles observed and reported" and 9 to "Actual system 'flight-proven' by successful mission operations".

In 2014, the European Association of Research and Technology Organizations (EARTO) published a policy guidance document on the TRL scale for research and innovation [EAR 14]. Table 8.1 shows the TRL levels as used by the European Union in the Horizon 2020 R&D program.

TRL Level	Description
TRL 1	Basic principles observed
TRL 2	Formulation of a technological concept
TRL 3	Experimental proof of concept
TRL 4	Technological validity in the laboratory
TRL 5	Validated technology in the relevant environment
TRL 6	Technological demonstration in the relevant environment
TRL 7	Demonstration of a prototype system in an operational environment
TRL 8	Completed and qualified system
TRL 9	Real system tested in an operational environment

Table 8.1. *Technology Readiness Levels according to the European Union*

In the context of smart grids and smart buildings, manufacturers and other actors in these sectors are setting up demonstration sites to test and validate, "in an operational environment", the feasibility and viability of innovative solutions (TRLs 6 and 7). The next stage is qualification by industry before mass-marketing the resulting solution. The demonstration sites developed by the Université Catholique de Lille correspond to TRLs 5 and 6.

8.3. Development of a smart grid demonstration site

Section 6.3 presents the electrical network of the Université Catholique de Lille's historical block demonstration site, which serves as a case study for the application of the energy management methods developed in this chapter. The Université Catholique de Lille, through the Live TREE program [ROB 23], has embarked on an energy renovation of the campus, promoting energy sufficiency, energy efficiency and the installation of renewable energy sources, including the encouragement of energy-saving behavior and a significant increase in the self-production and self-consumption of electrical energy.

The electricity network in question, shown in Figure 6.3 in Chapter 6, has the following infrastructure:

– two photovoltaic power plants with outputs of 185 kWp and 28 kWp;

– a 250 kWh Li-ion battery, with a charging capacity of 40 kW, a discharging capacity of 80 kW and an efficiency of 97%;

– six 22 kW charging stations for electric vehicles;

– four buildings containing different types of loads (air conditioning, lighting, heating, etc.);

– a connection point to the HVA (15 kV) distribution network.

An information system dedicated to the development and experimentation of the Live TREE program has been set up to provide extensive data storage and processing facilities, while ensuring data confidentiality and protecting the privacy of all those involved in the experiments. This information system for the historical island is shown in Figure 8.1.

Photovoltaic production is forecasted over the medium term (15 minutes to 6 hours) using satellite weather data (see section 3.3.1 of Chapter 3), and over the short term (2 to 15 minutes) using a camera monitoring the sky (Figure 3.5, see section 3.3.2 of Chapter 3). Long-term forecasts (over several days) of building production and consumption are made using artificial intelligence tools such as neural networks, developed in Chapter 3. These forecasts are important for optimizing self-consumption, controlling storage, electric vehicle charging and interactions with the public distribution network. These forecasts need to be supplemented by weather and room occupancy predictions (the historic block can accommodate 5,000, students, staff and visitors at peak times), which influence consumption, as well as predictions of building users' usage and behavior.

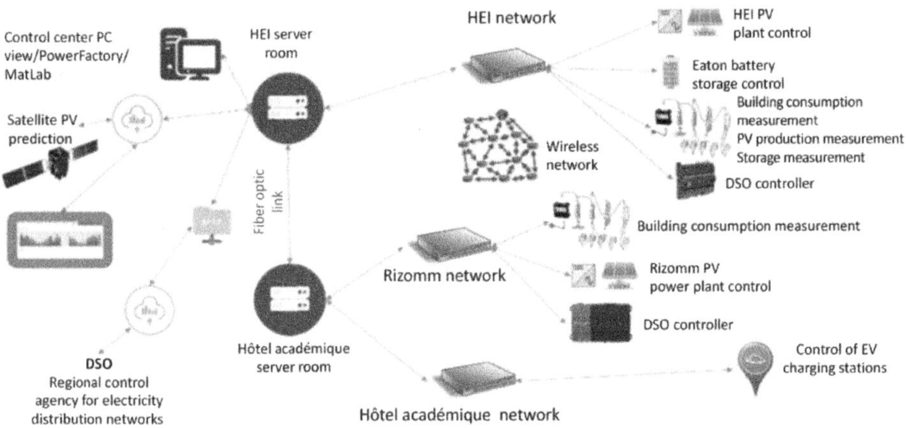

Figure 8.1. *Information system developed as part of the Live TREE program initially extended to the historic block (source: Grégory Vangreveninge)*

An experimental wireless communication network with relay sensors is deployed in parallel with the conventional wired network, with the aim of comparing different modes of communication and ensuring redundancy of this communication, which is crucial in the control of all systems that become smart.

An energy control center (shown top left in Figure 8.1), as well as a smart buildings control center, has been developed to collect, manage and store data retrieved from electrical equipment and buildings, to process consumption and production forecasts and to control photovoltaic sources, storage and controllable loads such as electric vehicle loads. Figure 8.2 shows part of the control center equipped with a SCADA system. Figure 8.3 shows the first block of demonstration buildings (the historic block of Junia HEI and faculty buildings) equipped with photovoltaic power plants, storage systems and electric vehicle charging stations, as seen from the SCADA system.

Figure 8.2. *Part of the control center equipped with a SCADA system*

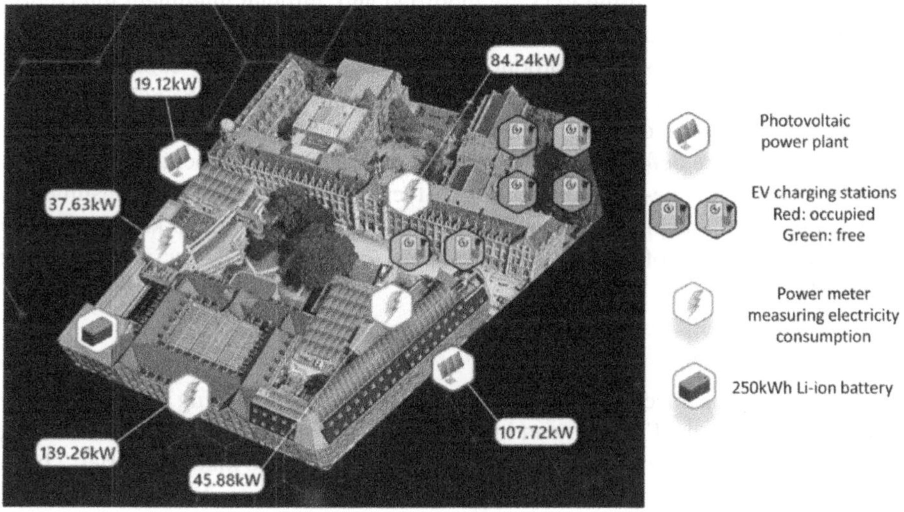

Figure 8.3. *Block of demonstration buildings as seen through the SCADA system*

8.3.1. *Demonstration projects*

8.3.1.1. *So Mel So Connected*

The development of new forms of self-production and self-consumption of energy, particularly photovoltaic energy, contributes to the necessary energy transition by promoting the use of local renewable energy. However, the large-scale implementation of self-consumption raises new issues concerning the modified uses of the public distribution network, the need for accurate predictions of energy production and consumption, the deployment of efficient communication networks and the identification of new business models. In order to test these new modes of consumption in an optimized way, it is necessary to move on from theoretical studies on a reduced scale in laboratories, to tests under real conditions (climate and real uses) by means of demonstration sites. One of the original features of the work carried out in the So Mel So Connected[1] project, in addition to being confronted with real-life conditions, is that it is interdisciplinary, bringing together electricians, computer scientists and economists.

Junia and its demonstration sites have made it possible to test and analyze individual self-consumption in a tertiary building using building-mounted photovoltaics, in intelligent interaction with the public distribution system operator (DSO), and to develop a business model for this type of self-consumption that takes into account the services rendered to the power distribution systems (PDSs) and to society. In fact, the uses of the PDS are modified and, through intelligent interaction with the system operator, it is possible to modulate the photovoltaic production injected into the PDS (surplus production not consumed by self-consumption) in prediction and real time, in order to avoid potential constraints on the public network (overvoltage, power overload). This type of test and evaluation has never been carried out in France before. It requires interdisciplinary developments in interaction with the DSO, the fine-tuned predictions of energy production and consumption (see Chapter 3), the deployment of efficient communications networks (Figure 8.1) and the identification of new business models (see Chapter 3, section 3.4.4.3 of [ROB 23]).

8.3.1.2. *ebalanceplus*

The H2020 ebalanceplus[2] project, entitled "Energy Balancing and Resilience Solutions to Unlock the Flexibility and Increase Market Options for Distribution Grid", brings together 15 corporate and academic partners in nine countries. It is coordinated by Cemosa in Spain. The Université Catholique de Lille (Junia and Institut Catholique) is one of the project's four demonstration sites (based on the electrical and communications networks shown in Figure 6.3 and Figure 8.1, respectively). The ebalanceplus project aims to:

– increase the energy flexibility of distribution networks by modulating load consumption, local generation and storage;

– forecast available flexibility;

1 This project is supported by the MEL and financed by ADEME as part of France's future investment projects. Junia and the Université Catholique de Lille worked on it in partnership with Enedis, EDF, Intent and the Lille Economie-Management Laboratory (LEM). See: https://www.smartgrids-cre.fr/projets/so-mel-so-connected.

2 See: https://www.ebalanceplus.eu/.

– increase distribution network resilience;

– design and test new models for power system services (voltage and frequency regulation, etc. [ROB 15, ROB 19]) to promote new markets based on energy flexibility.

The system consists of units that implement algorithms to forecast and manage available flexibility in order to encourage demand response programs, and to increase the capacity of the distribution network with a view to avoiding congestion and optimizing its use. The project is testing various flexibility solutions: electrical storage, V2G systems (vehicle-to-grid, or reversible electric vehicle charging [ROB 23]), electric hot water storage and systems connected to the IoT.

8.4. An all-in-one demonstration site

8.4.1. *Introduction*

Innovative property renovations combined with research and development projects involving a variety of partners, and benefiting from various sources of funding, have led to the creation of smart grid and smart building demonstration sites. As mentioned in the introduction, these developments draw on expertise from a wide range of disciplines, including electrical engineering, construction, IT, sociology and economics. These demonstration sites enable us to test innovative supervision technologies and processes in real-life conditions, with the aim of developing these solutions on a large scale.

This section illustrates the control of self-consumption photovoltaic power plants in intelligent interaction with the public DSO, the control of electric vehicle charging and storage potential, the control of loads within a building such as hot water tanks and heating systems, the contribution of an electrical storage system in a self-producing and self-consuming building and the associated communication networks and computer programming interfaces or APIs.

8.4.2. *Controlling photovoltaic production*

As part of the So Mel So Connected project, tests were carried out on shaving the photovoltaic (PV) electricity production injected into the distribution network, by controlling the inverters of the PV plant following receipt of a setpoint from the public DSO, initially several hours before the setpoint was applied.

In a second phase, the experimental campaign aimed to test self-regulation devices for self-generating sites and dynamic capping devices for power injected into the PDS, via an inverter control device, in order to optimize connection costs for this type of installation.

The aim of the demonstration experiment was to set up a system for the smart management of the power generated, in line with consumption requirements, and possibly using a storage vector. This system should make it possible to limit power calls and energy re-injection into the grid.

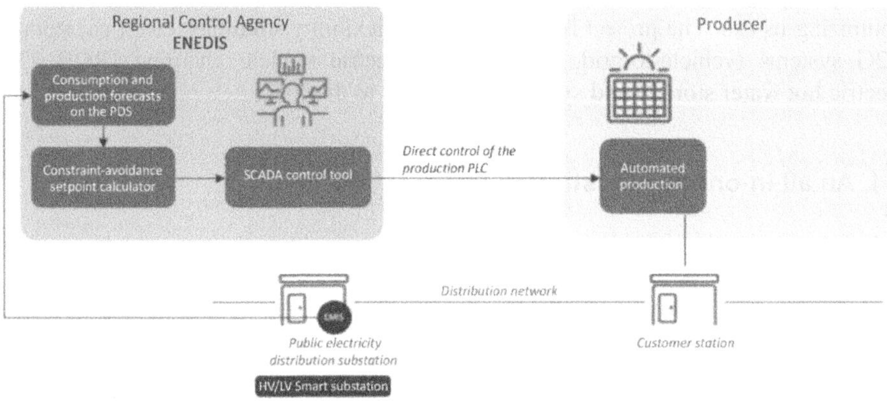

Figure 8.4. *Simplified control diagram for a photovoltaic power plant from the regional control center of the public distribution system (source: Enedis)*

In the So Mel So Connected project, given that nearly 98% of the self-generated PV energy is self-consumed [ROB 23], the objective of dynamic curtailing on PV power reinjected into the distribution grid is met by a storage solution implemented with a maximum discharge power of 80 kW, and with more precise knowledge and control of the load, we aim for 100 kW of reinjection into the grid, which would correspond to nearly 20% of the consumption peak of the demonstration site.

The DSO's Regional Control Agency has a tool for anticipating constraints on the PDS using consumption and generation prediction tools, as illustrated in Figure 8.4. HV/LV smart substations always provide detailed knowledge of the HV network, enabling prediction models to be improved.

Instructions are drawn up to anticipate constraints: they are then sent to the producer via a secure data exchange server for processing in real time or on D-1 (or day-ahead), without human intervention.

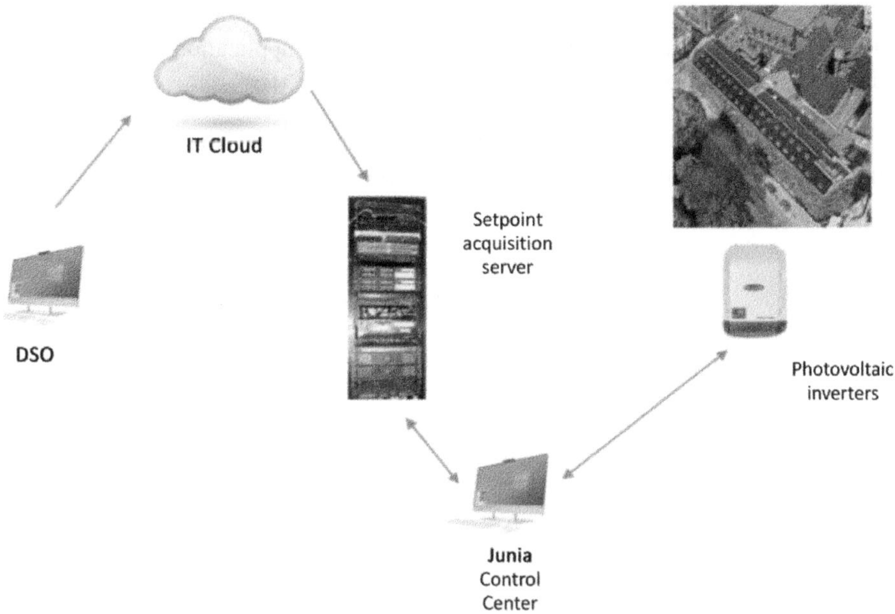

Figure 8.5. *Path showing the photovoltaic production curtailing setpoint from the regional control agency of the public distribution system to the inverters controlling the Rizomm photovoltaic power plant (source: Grégory Vangreveninge)*

The operation implemented between the DSO and Junia for communication and setpoint application is as follows (illustrated in Figure 8.5):

– the DSO sends a clipping instruction file to Junia's file transfer protocol (FTP) server;

– the acquisition server receives the file and integrates its elements into a database;

– the control center interrogates the database every minute to check whether it has been updated. As soon as a setpoint is received, the control center recovers the power to be applied to the inverters. It then sends the setpoint to be applied at the date and time specified in the file;

– once the setpoint has been sent to the inverters, the clipping power is applied within 3 minutes;

– the DSO visualizes the feedback from the application of the clipping setpoint using the injection counter setup.

For IT security reasons, the operator station receives a generic e-mail from Junia's control system, transmitting information on the application of the setpoints and the associated measurement.

Figure 8.6 shows an example of dynamic capping of Rizomm's photovoltaic output following the application of a setpoint received from the DSO.

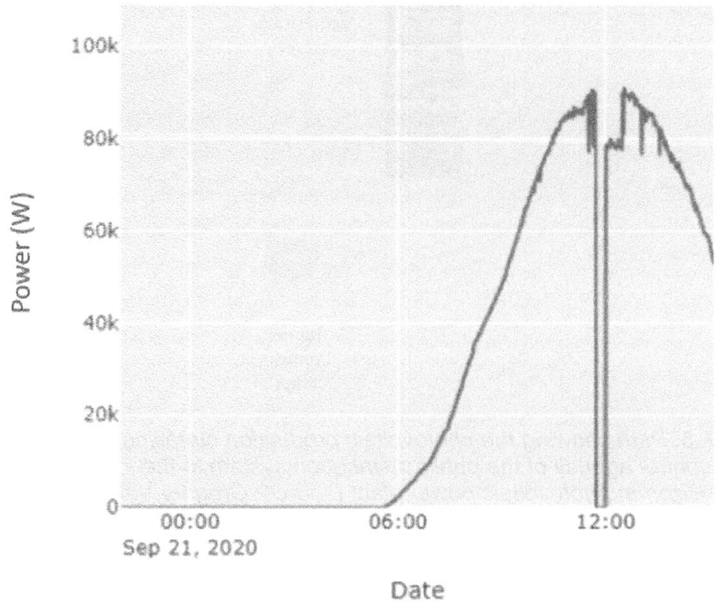

Figure 8.6. *Example of dynamic curtailing of photovoltaic production from Rizomm following the application of a setpoint received by the DSO (source: Grégory Vangreveninge)*

8.4.3. *Integration and control of electric vehicle charging*

In 2019, France's transport sector remained the largest emitter of greenhouse gases, responsible for 31% of national emissions. Slightly more than half (51%) of transport's CO_2 emissions come from private cars[3]. The development of electric vehicles is an important contribution to the energy transition. One of the major advantages of electric vehicles is their potentially low CO_2 emissions. However, the

3 See: https://www.statistiques.developpement-durable.gouv.fr/chiffres-cles-du-transport-edition-2021#:~:text=En%202019%2C%20le%20transport%20reste%20le%20premier%20secteur,des%20transports%20sont%20%C3%A9mises%20par%20les%20voitures%20particuli%C3%A8res.

CO_2 emissions of this type of vehicle depend on the CO_2 content of the electrical energy used to power them. Ideally, the charging of these vehicles should be based on renewable energy, to minimize all environmental impacts, not just greenhouse gases. In [ROB 16], a strategy for managing the charging of electric vehicles in an area was presented, with the following objectives:

– limit the environmental impact, induced by greenhouse gas emissions, by avoiding charging electric vehicles during peak hours associated with highly "carbon-intensive" electricity production sources and by favoring electric vehicle charging during favorable periods when renewable energies are injected into the grid;

– limit the economic impact by avoiding charging electric vehicles during peak hours, to minimize the costs of electricity transmission and distribution network reinforcement.

As part of the Université Catholique de Lille's Live TREE program [ROB 23], the "Low Carbon Mobility" project, in collaboration with TotalEnergies, aims to develop new recharging services for electric vehicles using non-carbon energy, while enabling users to save on the cost of recharging by targeting the hours when electricity prices are lowest.

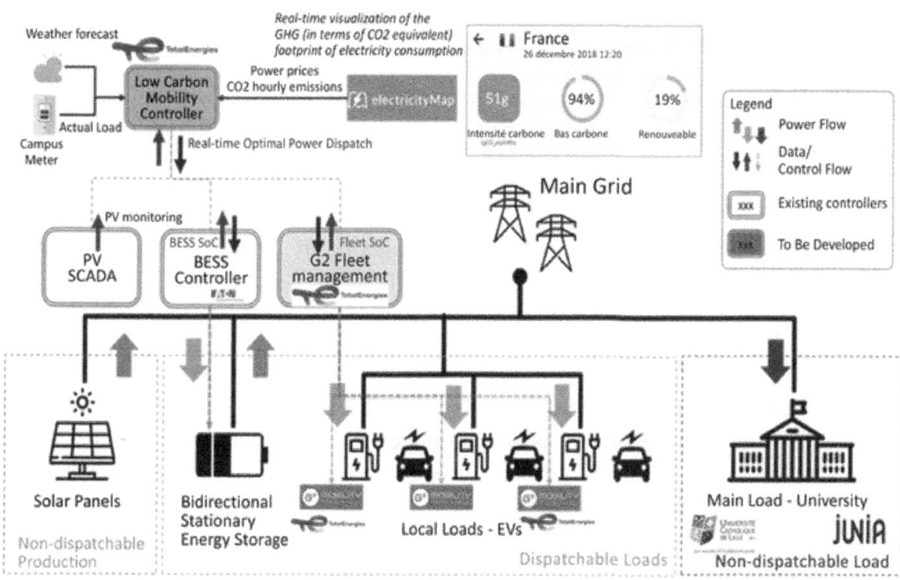

Figure 8.7. *Schematic view of the electrical network of the Université Catholique de Lille demonstration block, as implemented to control the recharging of electric vehicles (source: TotalEnergies)*

On a first part of the demonstration site, the university has decentralized generation (PV) and "centralized" stationary electrochemical storage and electric vehicle charging stations (Chapter 6, Figure 6.3). Figure 8.7 shows a schematic illustration of the demonstration site power grid as implemented to control electric vehicle charging.

The aim of this project is to offer users a low-carbon recharging service via a mobile application. This service requires the development of a "Low Carbon Mobility" control system, capable of interfacing with the various measuring devices on the campus to control centralized storage devices and electric vehicle charging in real time. In real time, this system controls the various devices in an optimal way, maximizing the campus' self-consumption rate and taking into account user consumption conditions, electricity market conditions and CO_2 emission rates. The proposal for this new mobility service is based on a data management platform and an optimization system designed to minimize the TCO_2 of recharging at any given moment (TCO_2 = cost of electricity including the cost of the carbon footprint). This optimization system ("TCO_2 optimization") also relies on prediction solutions:

– solar production, presented in Chapter 3;

– consumption of the electric vehicle fleet (amount of energy required for recharging, and prediction use of recharging stations);

– consumption of the buildings, presented in Chapter 3;

– electricity prices;

– CO_2 emissions.

Three charging modes are available to users:

– boost mode: immediate recharge at maximum power for 30 minutes, corresponding to x% recharge;

– medium mode (default): recharging at normal power for 2 hours, corresponding to y% recharge;

– "low-carbon" type: modulation of electric vehicle charging to reduce its carbon content, favoring charging from locally produced PV energy and/or from the public grid when the latter is low in carbon. The time of departure (end of recharge) is requested from the user. The recharging rate in terms of TCO_2 is given to the user at the end of recharging.

Figures 8.8 and 8.9 show two graphs illustrating medium (default) charging with no time limit. This is a conventional charge starting as soon as the vehicle is connected to the terminal, recorded on the charging terminals of the Live TREE demonstration site. Figure 8.8 shows the constant charge of an electric Fiat 500

(42 kWh battery, 320 km range)[4] with a maximum power of 11 kW. The blue surface shows the value of the charging current, while the black curve represents the accumulated energy. The charge of almost 30 kWh took 3 hours and 5 minutes. Figure 8.9 shows a non-constant charge of a Renault Zoé (52 kWh battery, 390 km range). The charge is also 30 kWh, with a maximum charge power of 20 kW and an average power of almost 11 kW. Charging took 2 hours and 45 minutes. These charging curves show that charging depends not only on the electrical characteristics of the charging station but also on the charging characteristics of the various electric vehicles.

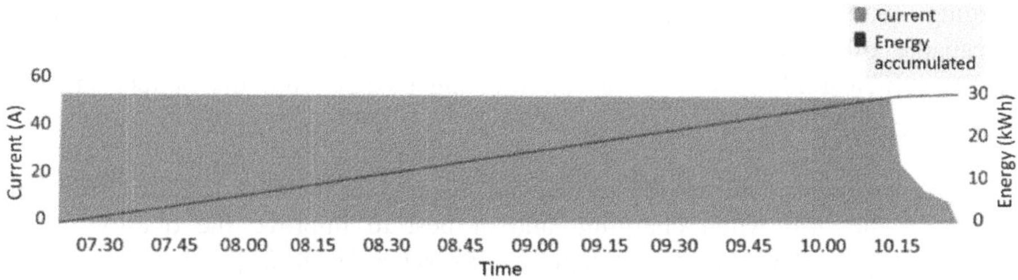

Figure 8.8. *Load curves for an electric Fiat 500 measured on the Université Catholique de Lille demonstration site. Current in blue and stored energy in black (source: Grégory Vangreveninge)*

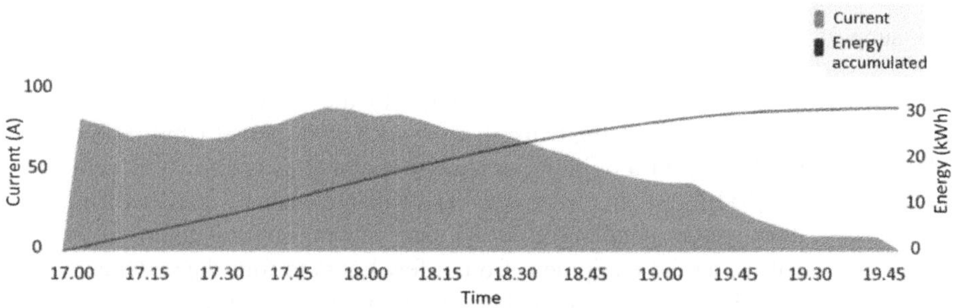

Figure 8.9. *Charging curves of a Renault Zoé measured on the Université Catholique de Lille demonstration site. Current in blue and stored energy in black (source: Grégory Vangreveninge)*

The on-board batteries of an electric vehicle could be used to store energy, as a private vehicle spends most of its time parked (according to EDF, 50% of vehicles

4 See: www.automobile-propre.com.

are parked permanently at home, and 69% of working people park for an average of 6 hours a day in a reserved space). The grid could draw electricity from the batteries to meet high demand (during peak consumption in the early evening, for example) or to compensate for a one-off lack of production (typically when the weather does not allow renewable energies to be used). This is known as V2G: the electric vehicle feeds the grid according to the needs of the power system (bidirectional model) and offers it a flexibility service.

The major difference for V2G, compared with passive charging (no adjustment possible) and dynamic charging (unidirectional on–off controllable or scalable charging [ROB 16]), is the reversibility of charging. The charger must be capable of drawing energy from the electrical grid to recharge the battery (grid-to-vehicle (G2V)), as well as of injecting energy from the battery into the electrical grid (V2G).

Electric vehicles can provide services to the grid, at the very least to limit their impact on the grid when charging, and at best to improve the quality of the electricity supply. There are four categories of services [ROB 23]:

– improving electricity supply;

– participation in network operations (system services such as voltage and frequency control);

– direct benefits to users through lower energy costs;

– support for renewable energies.

With V2G, the charger feeds energy back into the electrical grid. The energy can also be fed back into a small installation such as a house (this is known as vehicle-to-home (V2H)), or a building or service sector (this is known as vehicle-to-building (V2B)) [ROB 23]. In this case, the vehicle(s) can be used as a back-up power supply for a given installation. In the case of a house, this emergency power supply can supply essential equipment from a few hours to four days, depending on the initial state of the battery. This technology is particularly interesting in regions subject to critical situations (earthquakes, climatic hazards, etc.) that can isolate homes for several days, as it can provide emergency power to buildings. It can also enable buildings to be disconnected from the power grid during critical peak periods, thereby contributing to its reliability. As part of the "Low Carbon Mobility" project, V2B charging stations will be tested on the university's demonstration site.

8.4.4. *Controlling electrical loads in buildings*

8.4.4.1. *Hot water tanks*

Controlling consumption is nothing new. As early as the 1970s, a system was introduced in France to control domestic hot water tanks in order to limit electrical peaks in the morning and evening. To achieve this goal, two systems were implemented: a tariff incentive with the introduction of different rates for peak and off-peak hours, and the possibility of remotely controlling the start-up of hot water tanks during periods of low consumption. Figure 8.10 illustrates the positive impact of these measures on the power demand profile of hot water tanks over the years. It should be noted that these solutions were made possible thanks to the thermal storage capacity of these domestic hot water (DHW) cylinders over a 24-hour horizon [ROB 23].

The load profile and energy requirements of hot water tanks can be determined using a single-element thermal model inspired by [SHA 12], and depend mainly on the users' hot water demand, which is determined statistically. Incoming cold water enters the hot water tank at temperature T_{in} = 15°C and is heated by an electric resistance and outgoing hot water is delivered at temperature T_{out} = 60°C, as illustrated in Figure 8.11. The water temperature is considered uniform throughout the hot water tank [BEN 21].

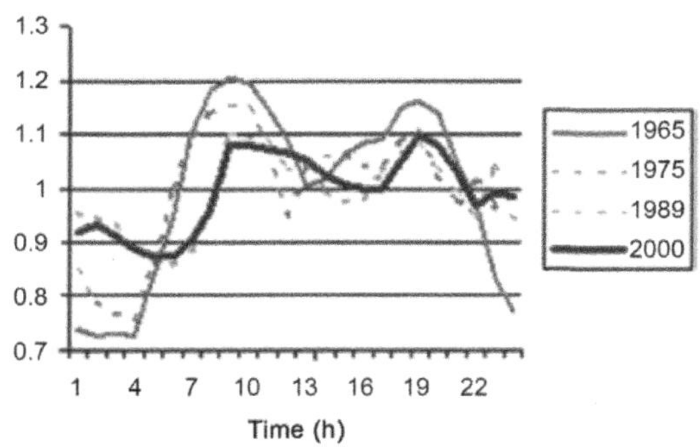

Figure 8.10. *Influence on power demand by electric hot water tanks on the (normalized) power profile for metropolitan France [DOB 13]*

The hot water tank is activated or deactivated by a thermostat, which means that the electrical power supplied is equal to its nominal value or zero. The only parameter that varies over time is hot water demand. Hot water use was determined on an hourly basis from a study [ADE 12] carried out by ADEME in France. An average three-person household consumes around 150±50 liters of water at 40°C per day, which corresponds to 83±28 liters at 60°C. In [BEN 21], a 24 h hot water consumption model for a group of 3,107 hot water tanks (in a residential area) was carried out; it is shown in Figure 8.12.

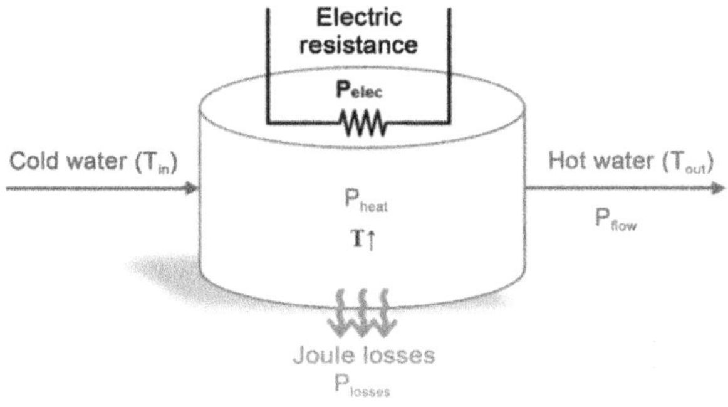

Figure 8.11. *Schematic diagram of a hot water tank [BEN 21]*

Figure 8.13 shows the range of power profiles for a 3,107 hot water tank unit. The same physical tank parameters are considered for all hot water tanks, namely a capacity of 150 liters and a thermal conductance of 1.5 W/°C. The hot water tanks are activated during off-peak hours only (from 11:30 p.m. to 7:30 a.m.) and controlled by their individual thermostats according to water temperature trends.

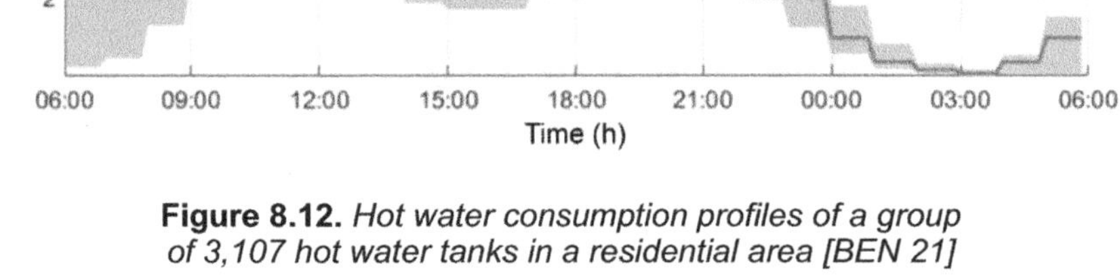

Figure 8.12. *Hot water consumption profiles of a group of 3,107 hot water tanks in a residential area [BEN 21]*

With the development of self-generation and self-consumption of energy, in particular photovoltaic energy, it is becoming interesting to modify the load profile of a DHW cylinder by considering a midday load when PV production is at its maximum and consumption is low (because it is off-peak). The modulation of this load can also be adapted to the constraints of the electrical distribution system.

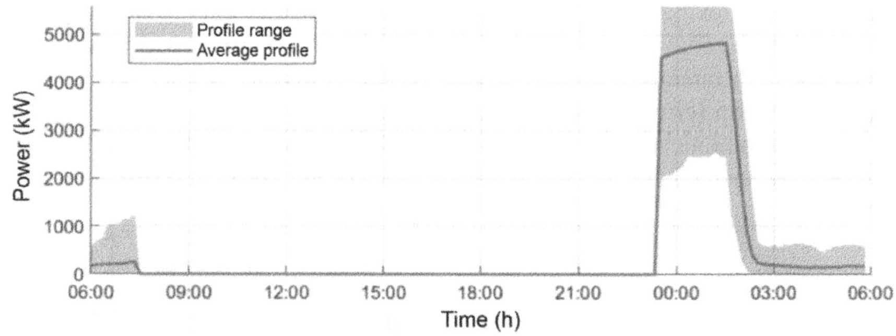

Figure 8.13. *Off-peak load profiles for a 3,107 hot water tank unit operating at 1.8 kW [BEN 21]*

The demonstration site for the Université Catholique de Lille's historic block of three tertiary buildings, illustrated in Figure 8.3, includes 30 electric storage tanks of varying volumes and power ratings, as shown in Table 8.2. They represent a total installed electrical power of 46.7 kW, to heat a total water volume of around 1 m^3. Their main use is building cleaning, with the exception of two locker rooms, a few laboratories (equipment cleaning) and a staff canteen.

	Unit volume (L)	Unit power (kW)	Number	Total volume by DHW cylinder category (L)	Total power by DHW cylinder category (kW)
HEI	15	2	1	15	2
	50	1.2	5	250	6
	HEI total		***6***	***265***	***8***
Hôtel Académique	15	1.2	5	75	6
	15	2	2	30	4
	100	1.2	2	200	2.4
	100	1.5	1	100	1.5
	150	1.2	1	150	1.2
	Hôtel Académique total		***11***	**555**	***15.1***
Rizomm	15	1.6	5	75	8
	15	2	7	105	14
	50	1.6	1	50	1.6
	Rizomm total		***13***	***230***	***23.6***
	Total for the three buildings		***30***	***1,050***	***46.7***

Table 8.2. *Domestic hot water tanks for the demonstration site*

To provide flexibility to the distribution network, the DHW cylinders (Figure 8.14a) have to be made controllable. Control units are developed and produced for both operational and research purposes (Figure 8.14b). They include electrical protection (circuit breaker) and opening (relay) elements, as well as wired and wireless communication elements. By receiving commands from the flexibility algorithm, they open and close the electrical circuit of the associated storage unit.

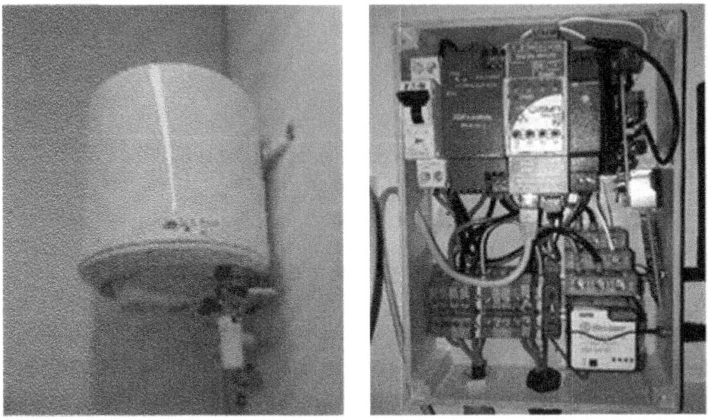

a) b)

Figure 8.14. *a) DHW cylinder; b) control box for a DHW cylinder*

In section 8.5.3.1, the potential of DHW cylinders in a commercial building is analyzed from a socio-technical point of view. The focus is on hot water tanks used by cleaning staff. Hot water tanks in washrooms also offer potential for flexibility.

8.4.4.2. *Air handling unit*

Heating and ventilation in the HEI building are provided by coupled air-handling and hot-water circulation systems, supplied by the district heating network. Pumps circulate hot water in different areas of the buildings, feeding underfloor heating networks and heating coils in the air handling circuits shown in Figure 8.15.

For example, Figure 8.15 shows that pump no. 1 supplies hot water to the heating circuits of 25 rooms in the HEI building, 21 of which are also supplied with warm air via air handling unit (AHU) no. 2.

Figures 8.16 and 8.17 show, respectively, the pumps of a substation supplying the heating circuits of the HEI building with hot water and two AHUs.

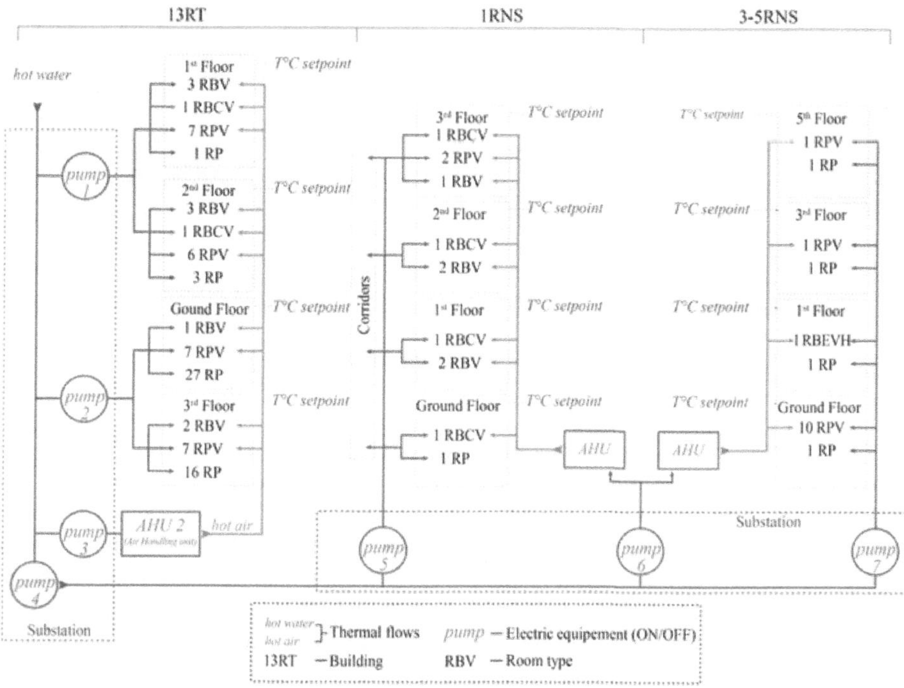

Figure 8.15. *Diagram of the heating and ventilation system in the HEI building (source: Benoit Durillon). For a color version of this figure, see www.iste.co.uk/robyns/smartgrids.zip*

Figure 8.16. *Sub-station housing the pumps supplying the heating circuits with hot water*

Figure 8.17. *Air handling units (AHUs)*

Figure 8.18 shows a cross-section of an AHU. It shows four air flows: fresh air (from outside), returned air (from rooms), supply air (to rooms) and exhaust air (to outside). The heat exchanger, located at the center of the system, exchanges heat between the returned air from the rooms and the fresh air. In the situation shown in Figure 8.18, the fresh air, initially at 6.9°C, is reheated to 15.8°C. This recovery improves the AHU's efficiency. Then, before being blown into the rooms, the air coming from the exchanger undergoes a temperature increase, using a coil supplied with hot water by one of the substations (Figure 8.16), with the aid of a three-way valve (3WV), the opening of which is adjustable (69.5%). The air blown into the ducts finally reaches a temperature of 23.7°C. Figure 8.19 shows the curve used to set the supply air temperature setpoint, taking into account the outside temperature measurement, and then the laws used to activate the heat recovery process on the one hand, and free cooling on the other, for AHUs equipped with this feature.

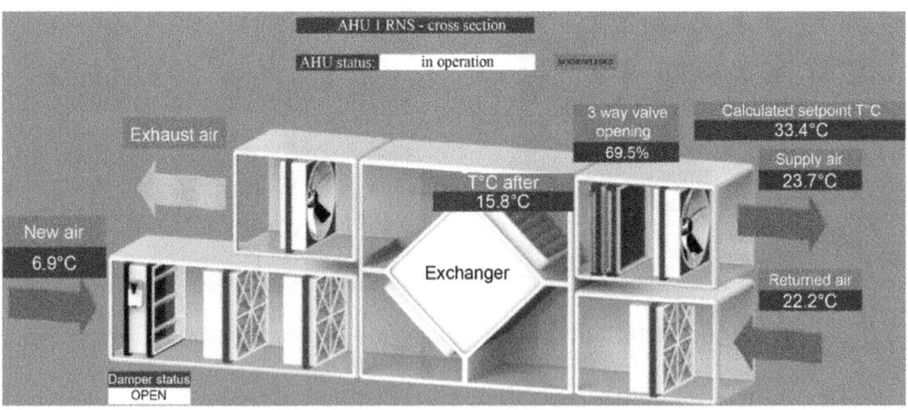

Figure 8.18. *Cross-section of an air handling unit (AHU)*

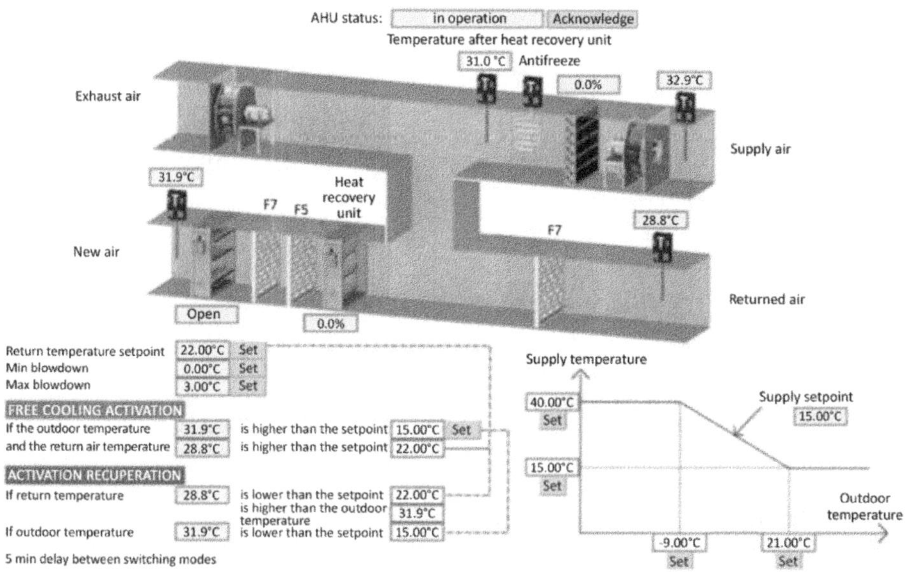

Figure 8.19. *The operation principles of an air handling unit*

In addition to underfloor heating, four types of heating system are used in the rooms of the HEI building, illustrated in Figure 8.20. The four variants are as follows:

– RP type: radiant panel supplied with hot water, equipped with valve control (Figure 8.20a);

– RBV type: air circulation from the AHU, whose temperature can be controlled locally, by means of heating and cooling coils, supplied, respectively, by a hot and freshwater circuit. Air circulation depends on the speed of the AHU fans, which is controlled according to the pressure inside the ventilation ducts, itself a function of the opening amplitude of the associated dampers (Figure 8.20b);

– RBVC type: similar to the RBV type, without cooling coil (Figure 8.20c);

– RPV type: combination of the RP type and air circulation from the AHU (Figure 8.20d).

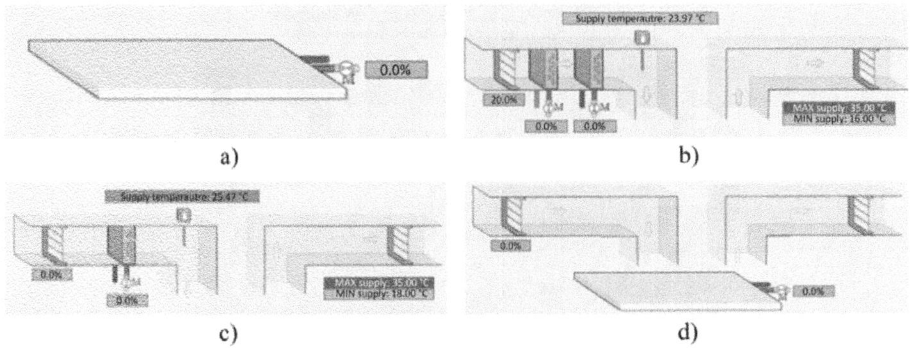

Figure 8.20. *Different heating modes for rooms in the HEI building: a) RP; b) RBV; c) RBVC; d) RPV*

This description shows that there is a link between room air treatment and heating, and between rooms supplied by the same AHU and the same pump. Flexibility will therefore be sought by grouping several rooms together.

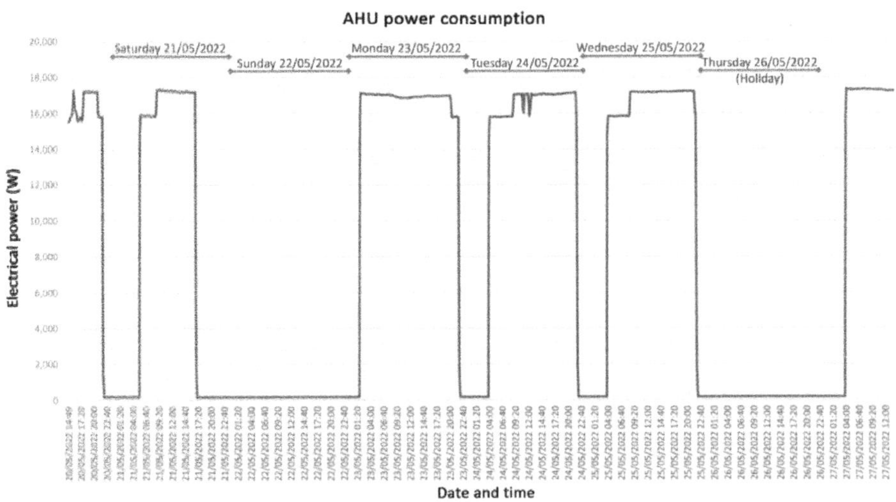

Figure 8.21. *Air handling unit power consumption measurement in "on/off" mode*

Tests carried out on this equipment showed a nominal electrical power for the pumps of around 3 kW, and 17 kW for an AHU. A significant difference in power consumption can therefore be achieved by controlling the start-up and shutdown of each AHU. Figure 8.21 shows the power consumption of an AHU controlled in "on/off" mode. On Sunday May 22, 2022, as well as on Thursday May 26, 2022, a public holiday, there was no electrical consumption. At this time of year, as the AHU is used for room ventilation only, the absence of occupants on these two days limits the need for forced ventilation.

8.4.5. *Electrical energy storage control*

Buildings supplied directly by the grid do not a priori need electrical energy storage, except in the case of certain sensitive buildings equipped with emergency power supplies to ensure the safety of buildings open to the public (essentially for lighting purposes) and health needs in hospitals or to meet the economic challenges of maintaining business activity in certain companies (data servers, sensitive industries, etc.). Renewable energies, which can be produced locally using solar panels for example, create new needs. Indeed, the variability of this production, the uncertainties of its prediction and the prospect of consuming it locally, combined with a reduction in transmission losses, are encouraging the development of local electricity storage [ROB 23].

The storage systems which will increasingly equip buildings in the future will meet the needs of these applications and ensure the compensation of variable renewable energies, but will also offer the possibility of providing services to other actors in the power system, such as consumption flexibility and participation in system services via a contribution to the voltage or frequency regulation of the distribution network [ROB 15]. Even if their price is falling, as storage systems remain costly, this represents a means of enhancing their financial value, provided that the constraint of controlling their ageing is taken into account [ROB 23].

In [ROB 23], several examples of the contribution of storage in tertiary and residential buildings were developed, highlighting both economic and environmental gains by reducing CO_2 emissions associated with electricity consumption by maximizing the use of mainly photovoltaic renewable sources.

In [AOU 22], a study of the contribution of the storage system installed in the historical campus demonstration site, illustrated in Figure 8.3, was carried out. This 250 kWh storage system consists of two storages that can be controlled manually or automatically from the SCADA control center. Figure 8.22 shows the storage control interface. An energy management strategy has been developed with a number of objectives in mind: economic gain by reducing electricity bills by optimizing peak and off-peak tariffs and environmental gain by reducing CO_2 emissions, while maximizing self-consumption. The supervision strategy exploits the predictions of photovoltaic production and building consumption presented in Chapter 3.

Further to these predictions is added a prediction of the potential CO_2 emissions of the electricity coming from the public electricity grid. The CO_2 content of the electricity kWh corresponds to the CO_2 emissions generated by the production of this kWh of electricity, taking into account the life cycle analysis of the photovoltaic power plant, that is, the energy required to build, maintain and dismantle it at the end of its life [BRU 14]. The prediction system uses data on CO_2 emissions of electricity from the public transmission grid (provided, for example, by a grid operator such as RTE), as well as photovoltaic predictions, with learning based on data from the last two weeks. Figure 8.23 compares the prediction and actual values of CO_2 emissions in grams per kWh during a week in August 2021.

Figure 8.24 shows the production of the main photovoltaic plant (of the Rizomm building) in orange, and the consumption of two buildings on the block, the Rizomm and the Hôtel Académique in blue. From July 29 to August 1, 2019, the storage system is not in operation, and from August 2, storage is managed to maximize self-consumption and thus prevent production from exceeding consumption, which corresponds well to the result shown in this figure.

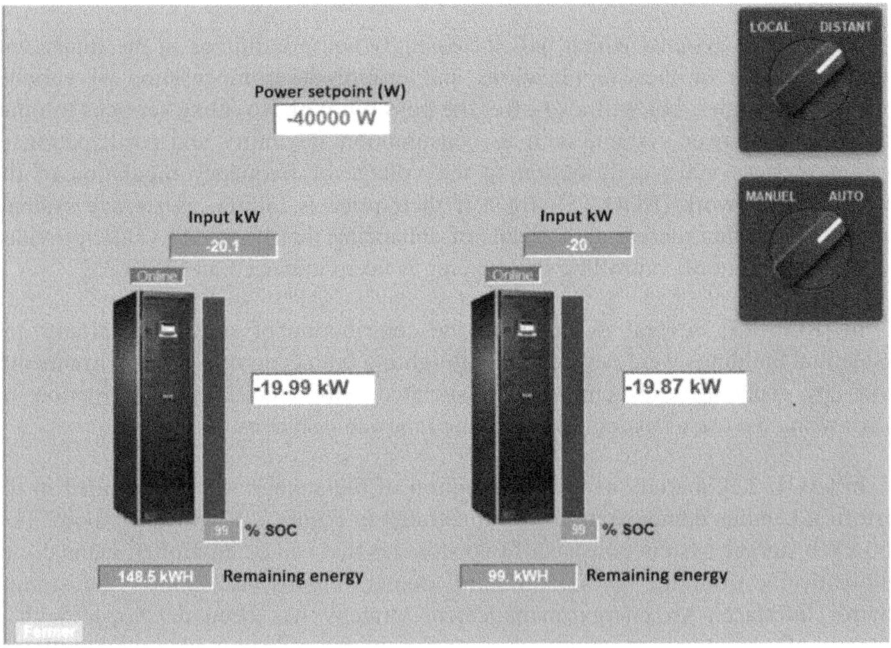

Figure 8.22. *Manual storage control interface for the Université Catholique de Lille's historic block (source: Grégory Vangreveninge)*

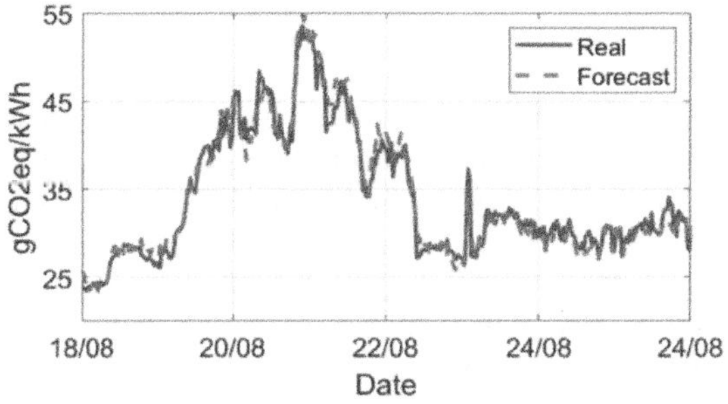

Figure 8.23. *Forecast in grams of CO_2 equivalent per kWh for one week in August 2021 [AOUI 22]. For a color version of this figure, see www.iste.co.uk/robyns/smartgrids.zip*

Finally, without the presence of storage, 4,481 kWh of photovoltaic production must be reduced because it exceeds the consumption of the island, representing 3.19% of total annual production. The control method developed in [AOU 22] reduces the reduction to 20 kWh, or 0.014% of annual production.

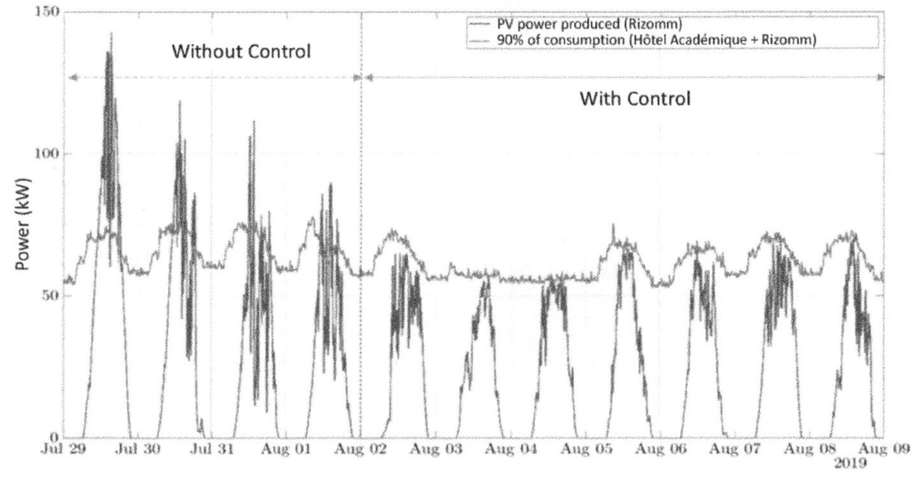

Figure 8.24. *Production by the main photovoltaic plant (on the Rizomm building) in orange, and the consumption of two buildings on the block, the Rizomm and the Hôtel Académique in blue, without and with the contribution of energy storage (source: Augusto Edgardo Nieto Castro). For a color version of this figure, see www.iste.co.uk/robyns/smartgrids.zip*

8.4.6. *Communication networks*

Releasing flexibility by acting on electricity-consuming equipment therefore requires the circulation of "incoming" information (setpoints, reference values) and "outgoing" information (measurements, status feedback, etc.), between the equipment and the servers where the databases and flexibility optimization algorithms are located. Two developments therefore need to be considered: the communication network that will enable data to circulate, and the API that will manage these data flows (Figure 8.25).

If the equipment is connected to a BMS, communication is already in place, and can be adapted if necessary. Otherwise, it needs to be installed. For this purpose, two types of communication network – wired and wireless – were considered. The aim was to experiment and compare these solutions. The sensors developed were based on a Raspberry PI combined with a Zolertia module, resulting in a completely open solution that can be adapted to suit requirements (Figure 8.26). The Zolertia module enables bidirectional inter-sensor communication using the ZigBee protocol. A deployment plan was drawn up for the buildings, taking into account the short range of this type of solution and the configuration of the site, particularly with the thickness of certain walls, which can reach several tens of centimeters in older buildings. This diagram is shown in Figure 8.27.

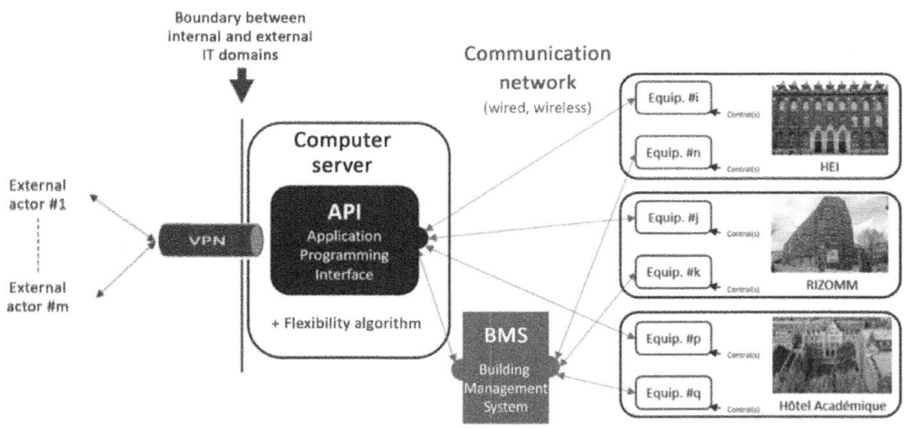

Figure 8.25. *API and communication network*

Figure 8.26. *Sensors based on Raspberry PI and Zolertia module*

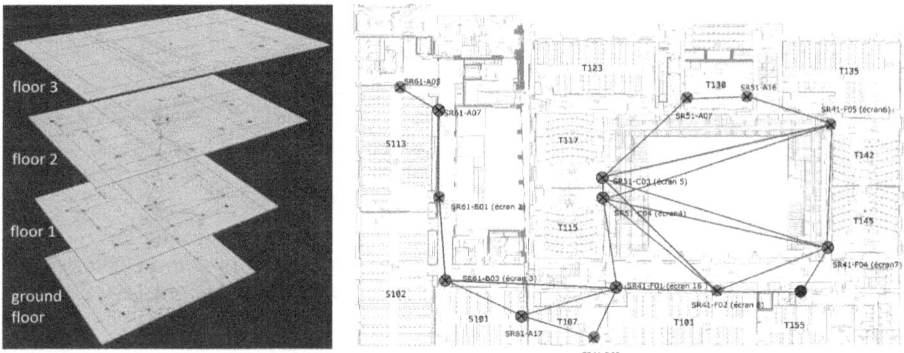

Figure 8.27. *Schematic diagram of IoT sensor deployment in the HEI building*

8.4.7. *IT developments*

A specific information system was developed to support the deployment of the demonstration sites. In this system, shown in Figure 8.1, various components need to communicate with each other. To achieve this, APIs or computer programming interfaces were developed. APIs enable applications to communicate with each other and exchange services or data. In the case of demonstration sites, for example, APIs need to be developed to enable communication between a user of an electric vehicle charging station, the charging station and the control center that manages the energy. APIs are also needed between the loads to be controlled, such as the hot water tanks in a building, and the control center.

The demonstration API was developed in Python. It integrates all the variables required to control the equipment (read variables, such as measurements, status feedback) and write variables (setpoints, start/stop commands). It lies at the interface between the optimization algorithm, the databases, the BMS and the equipment. It responds to incoming and outgoing write and read requests.

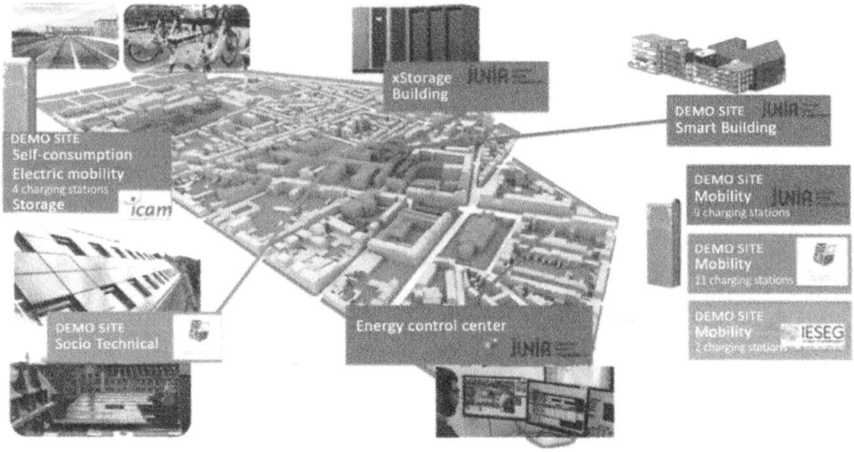

Figure 8.28. *Smart grid and smart building demonstration sites at the Université Catholique de Lille in Lille's Vauban district (source: Grégory Vangreveninge)*

8.4.8. *Perspectives*

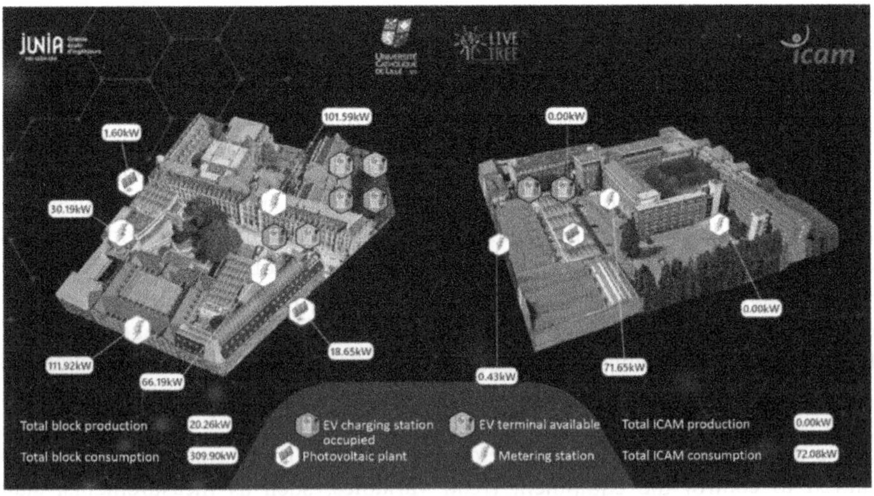

Figure 8.29. *Two demonstration islands 500 m from the control center (source: Grégory Vangreveninge)*

The first smart grid and smart building demonstration site to be built on the Université Catholique de Lille's historic block is being extended to other blocks in

Lille's Vauban district. This extension is illustrated in Figure 8.28, with the historic block at the center of the image. A second demonstration island on the Institut Catholique des Arts et Métiers (ICAM) site, to the left of Figure 8.28, is linked to the control center shown through the SCADA system illustrated in Figure 8.29.

In section 8.5, we take a socio-technical approach to the potential for electrical load shedding in the context of commercial buildings, dealing with the shedding of hot water tanks and the charging of electric vehicles.

8.5. The contribution of occupants of a service sector site to electricity savings

The search for potential for electrical load shedding within organizations has focused primarily on large consumer installations. Cases of load shedding practices in socio-technical contexts such as industrial equipment or food storage platforms have been identified [ROB 23]. In both cases, load shedding is made possible by the presence of energy "cushioning" resources that eliminate or limit any disruption to the workflow: for example, generator sets in industry and thermal inertia time in cold stores.

Socio-technology in a building refers to all the devices, actions and tricks devised to obtain, improve, support, transport, make last, control the cost, etc., of energy inputs, ideally with the least possible mental and physical effort. The inputs required vary according to the context of use (industry, services, housing, transport, etc.). Taking an interest in socio-technical energy engineering means understanding the functional interactions between an energy-related installation and the way it is used, as well as examining the resulting effects (efficiency, costs, consumption, carbon emissions, etc.).

Box 8.1. *Definition of socio-technical energy [ZEL 15]*

This was followed by a phase of more diffuse energy-saving measures. Work was carried out in homes, as well as in workplaces with no large electrical installations [ADE 17]. In this section, we propose to retrace the steps taken to assess the potential for electrical load shedding in a tertiary installation. The installations in question are of two types: free-access charging stations for electric vehicles, and hot water tank cumulus heaters used mainly by cleaning staff[5].

The pedagogical exploration of potential energy-saving measures in our study site involves two preliminary stages. Firstly, we look at developments in the

5 Another potential savings area still to be explored is the adjustment of ventilation intensity for heating and air quality.

philosophy behind building design, which sets out the respective responsibilities of technical and human factors in ensuring a building's energy performance. Next, we take a theoretical look at the potential for load shedding in the commercial sector.

8.5.1. *Evolution of the sources of energy consumption reduction*

8.5.1.1. *Evolving the construction model*

The socio-technology of energy is an anthropological reality dating back to the dawn of time. Human beings have always relied on energy for light, heating, food, protection, to transform matter, for transport and to move around. Over time, these uses have been shaped by two dynamics:

– the combination of energy sources (human, animal, hydraulic, wind, solar, etc.) and their technical tools (rope, plough, mill, etc.);

– progress in terms of efficiency, reliability, safety and ease of use.

The combined effect of these dynamics has been a steady lightening of the physical and mental demands placed on users to switch on and adjust their energy-consuming equipment. This evolution can be clearly seen in the domestic and working worlds, where since the beginning of the 20th century we have moved from the kerosene lamp to the electric switch for lighting, and from the fireplace (which must be fed and maintained) to easily adjustable and programmable central heating. This dynamic of socio-technical progress, closely associated with the development of electricity, is at the root of the decline in the involvement of energy users – let us think of them as "building occupants" – in the operation of their energy equipment.

However, there remains a limit to the virtuous effects of these movements, particularly in terms of thermal comfort. For this priority criterion of well-being in a building depends more on technical qualities (insulation and ventilation) than on heating equipment, so when they feel the need, occupants instinctively seek to solve their comfort problem. From a sociological point of view, these necessary initiatives are akin to "constructions" (covering up, changing rooms, installing auxiliary heating, etc.). They can be found wherever discomfort (thermal, lighting, ventilation, noise, smell) is experienced.

It is a well-known fact that the emergence of energy concerns in buildings, first for cost reasons (in France since 1973), then for environmental reasons (internationally since 1995), has triggered a steady reduction in the energy consumption set by successive thermal regulations. The constant need for regulatory compliance has led to a constant evolution in the design principles, materials and appearance of new and renovated buildings. Most of them are built on the same model: a thermally efficient, airtight box, ventilated for indoor air quality and equipped with a high-efficiency, right-sized heating system [BES 15].

8.5.1.2. *The increasing and indispensable involvement of occupants*

One might have thought that the concern to control consumption would be accompanied by a lesser involvement of occupants in the socio-technical operation of buildings. In fact, the opposite is true, except in the few buildings designed according to an "entirely technical" logic, since the reduction in heating requirements made possible by better-insulated buildings brings with it the risk of overheating in summer and thermal discomfort in winter, the control of which falls to the occupants in the socio-technical script. Occupants must be vigilant: to limit overheating with blackout blinds in summer, to limit heat loss through prolonged ventilation in winter and to regularly maintain their ventilation systems to control humidity levels and air quality.

The increased involvement of occupants is provided for in the socio-technical script of new or renovated high-performance buildings (which are, in fact, more comfortable), and does not totally obviate the need to "build in" comfort. There is therefore more of an accumulation than a modification of occupant interventions in high-performance buildings. This change in context is illustrated in Figure 8.30, with the inevitable rise in mental workload in three stages.

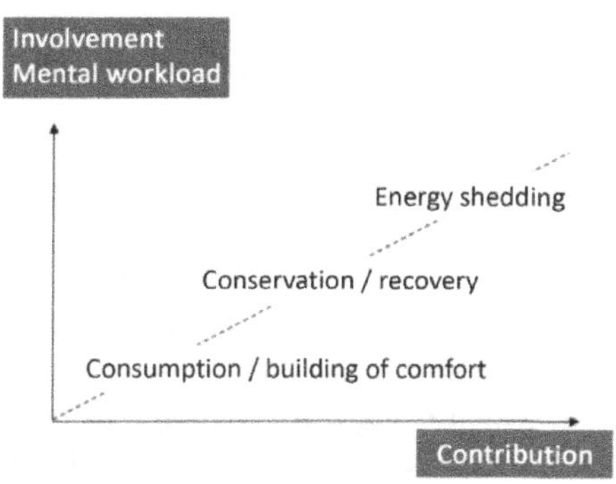

Figure 8.30. *Increasing the occupants' mental workload*

The first stage is specific to all buildings, where occupants are mobilized to activate their energy systems (heating and lighting in particular), as well as to "build" in the face of thermal deficiencies. The second stage relates to high-performance buildings, with the management of heat or cool inputs, to be conserved or recovered according to the season. Although not yet widespread, energy-saving practices to facilitate the inclusion of renewable energies in the energy mix represent an additional mental burden for energy-using occupants, since they involve the indispensable planning of consumption times.

Analysis of the conditions for moderating energy consumption in a building becoming increasingly decarbonized highlights the need for greater cooperation on the part of occupants. A major question is whether this cooperation exists spontaneously. Certainly not, since we know that effective sufficiency is progressing, but remains uneven between individual consumers. Support measures such as "social engineering" are designed to facilitate their awareness and commitment, in the interests of better control of energy consumption.

8.5.1.3. *The challenges of load shedding for businesses*

Although it may be trite to say so, it cannot be repeated often enough that energy is, and will remain, indispensable simply because it makes everything possible. In the corporate context, this reality is more pressing than for households, and is reflected in operating costs that geological, economic and geopolitical contexts have ended up driving inexorably upwards.

Many companies have responded to the threat of rising costs. They have done so on their own initiative but have also been strongly encouraged by more stringent energy regulations, with the explicit aim of reducing environmental degradation. This reaction initially took the form of major energy efficiency actions on buildings, equipment and sufficiency, followed by stronger measures in terms of efficient management of energy flows (rationalization, optimization, local production, self-consumption, etc.).

The potential adoption of a "load shedding" measure is part of this drive to control energy consumption and reduce carbon emissions and is designed to be compatible with the development of business activity. It does, however, offer a number of specific advantages that make it even more attractive to businesses. The inclusion of off-peak time requires a thorough review of the way work is actually organized and carried out, which is generally a source of rationalization of methods and increased profitability. Load shedding is also an opportunity to initiate a less asymmetrical commercial relationship with energy suppliers, insofar as it facilitates the inclusion of renewable energies; this changes the conditions under which tariffs are negotiated. Last but not least, load shedding remains an original form of energy management, which can be promoted to all stakeholders.

8.5.2. *Shaving potential at tertiary sites*

The tertiary sector is home to three-quarters of the French population, spread across a very broad spectrum of activities. To make our demonstration easier, we have chosen to project ourselves into a context of office work, the most traditional and widespread in this sector (public administration, education, health and social care, IT and communications, banking and insurance, various consultancies, etc.).

8.5.2.1. *Activity-related constraints*

8.5.2.1.1. The business and management system

The "activity system" corresponds to the use of material, managerial and human resources mobilized to carry out the vocation of an organization (business, administration, association). While the vocation in question may be simply stated (to sell, manufacture, care for, accommodate, etc.), its realization may go through very different stages, the deployment of which must be understood to identify the potential for energy efficiency.

Components	Content	Impacts/Demand for Energy
Vocation	Welcoming the public? Food preservation? Sample preservation? Hours of operation?	Need heat? Need power? Continuous power? Security of supply? Need for luminosity? Safety constraints? Regulatory requirements?
Personnel	Numbers? Diversity? Posture? Physical activity?	Presence schedule? Stability? Permeability to instructions?
Working Conditions	Outdoor/indoor? Day/night? Mobile/static?	Heat and light? Working conditions? Regulatory requirements?
Regularity of Activities	Operating hours? Seasonality? Regularity?	Chronogestion? Programming? Anticipation of needs?

Table 8.3. *The components of an activity system*

Table 8.3 shows a typical list of the components of the activity system, with questions to ask in order to assess the scope for deletion. The first step in analyzing this system is to take a close look at what is needed to do the work. If we take a close look at each building and the equipment used (machines, IT, etc.), it is not so much the inventory that is important as the energy usage. For example, the presence of a freezer room for food products may require a permanent cold supply, summer and winter alike. Another example is a health care center, where good reception, waiting and examination conditions are essential. So, it is clear that each stage of a business process is associated with specific requirements in terms of energy intensity and permanence (over the course of the day, week, month or year). And it is precisely this energy requirement, its variability and predictability, that offers more or less potential for load shedding.

The focus here is on the technical aspects of production. But we must also consider the way the organization functions, that is, the actual workflow. In this respect, management is of crucial importance, since it largely determines work rhythms, and consequently energy requirement curves. Load shedding potential is not assessed according to what the organization of the work shows, but according to the load shedding opportunities that an organized operation is capable of releasing, taking into account the consumer teams involved.

8.5.2.1.2. Work equipment

Work equipment in the tertiary sector can be divided into two categories. Table 8.4 details them and describes their conditions of use, as well as the ease with which deletion can be envisaged.

<table>
<tr><th></th><th>Estimated electricity consumption (RTE 2016)</th><th>Control</th><th>Impact</th><th>Operating hours</th><th>Feasibility of recovery</th></tr>
<tr><td>Personal automation</td><td rowspan="2">36%</td><td>Decentralized</td><td>Process</td><td>Day</td><td>Difficult/To be built</td></tr>
<tr><td>Computer servers</td><td>Centralized</td><td>Process</td><td>24/24</td><td>Easy/Limited</td></tr>
<tr><td>Lighting</td><td>21%</td><td>Centralized</td><td>Working conditions</td><td>Day</td><td>Difficult/Limited</td></tr>
<tr><td>Air conditioning/ ventilation</td><td>20%</td><td>Centralized</td><td>Working conditions</td><td>24/24</td><td>Easy/If inertia</td></tr>
<tr><td>Heating</td><td>19%</td><td>Variable</td><td>Working conditions</td><td>24/24</td><td>Easy/If inertia</td></tr>
<tr><td>Distributors</td><td rowspan="2">4%</td><td>Centralized</td><td>Working conditions</td><td>24/24</td><td>Easy/To be built</td></tr>
<tr><td>Distributed consumption</td><td>Decentralized</td><td>Working conditions</td><td>Day</td><td>Difficult/To be built</td></tr>
</table>

Table 8.4. *Use and control of equipment in commercial buildings*

The most numerous are of an IT nature. They are intimately linked to the process, and without them, nothing can be done. Most are individual (computers) or semi-individual (printers), with an operating range based on presence at work. Control is described as "decentralized", since switch-on times depend on each individual's schedule, making it difficult to coordinate times and tasks in order to launch a load shedding period. Added to this difficult context is the impossibility of relying on decentralized energy reserves, due to their unreliability (laptops) and even more so to their absence (fixed computers, printers, telephones).

Deletion is more feasible for centrally controlled usage stations, but with the following important nuances:

– Modulating or switching off lighting is difficult for reasons of workplace comfort, safety and regulatory compliance.

– Reducing heating intensity is possible, provided the building offers good thermal inertia.

– Air flow reduction (air conditioning and ventilation) is possible in the absence of overheating and high occupancy density.

– Servers are the only obvious potential for energy reduction. Fewer in number than computer workstations, but consuming much more power, these can easily be connected to a decentralized electrical relay.

A review of the energy-consuming equipment present in most commercial buildings reveals limited and conditioned load-shedding capacities, concentrated on heating (if thermal inertia) and computer servers (if decentralized batteries).

8.5.2.2. *The social factors*

The presence and use of the equipment reviewed here reflect the necessities of work. The reality of these necessities is the result of a social context in the strict sense of the term, with, on the one hand, a managerial dimension that controls the essential functioning of the organization, and, on the other hand, the employees who do the work. A review of the main social factors favorable and unfavorable to load shedding highlights a certain form of continuity between the worlds of work and home. As the individuals involved are the same, this continuity has a certain logic to it.

8.5.2.2.1. Some social factors favorable to the acceptance of load shedding

– *A high level of environmental sensitivity*: this is essentially the result of the process of making connections between energy consumption and the negative effects induced in the domestic environment (including bills) and beyond (emissions, environmental impact, etc.). In this respect, the responsibility of the "worker" is engaged, as these reconciliations are not spontaneous. They require cognitive efforts (experience, interest, observation, knowledge, deduction) and information searches on what consumption produces.

– *A system of values and priorities that gives a place to the environment*: depending on the individual, the aforementioned environmental awareness has an impact on the hierarchy of action criteria. There is nothing automatic about the relationship. Rather, it is all a matter of conviction, to the benefit of the acceptability of load shedding.

8.5.2.2.2. Some social factors hindering the acceptance of load shedding

– *Energy and environmental concerns* (consumption, carbon footprint) are still far removed from the major performance indicators (profitability, cash-flow, production, productivity, etc.), despite the progress made by what are generally large companies. As a result, the influence of stated environmental ambitions is still too detached from the frame of reference governing the operation of organizations at the human level.

– *Time management*: time spent at work and the pace of production are major issues for employees, as time management is a lever for regulating the balance between effort and remuneration. As a result, anything that impacts working conditions and potentially attendance time is experienced as a constraint.

8.5.2.2.3. Some favorable factors, depending on the organization

– *The correspondence between the organization's stated commitment to the environment and its actual mobilization through the allocation of human and/or material resources*: this circumstance is favorable for everyone, that is, employees who are "sensitive" to the coherence between words and deeds, as well as "legitimist" employees who expect their organization to give directions, and who support them in all cases.

– *The quality of relations at work, particularly with management*: in a poor climate, the energy question becomes inaudible, as it is too disconnected from the traditional issues facing organizations.

A review of the main social factors shows that their real influence is the result of an unstable equilibrium. Since we are talking about the social, that is, the reality of work (a function, a role, relationships, supervision, management), it is all a question of intimacy: how work is organized, carried out and articulated with private life. The connection between the world of work and the world of intimacy is not obvious when it comes to load shedding, but it does exist. The bridge is that of the desire to control on the one hand, the content of one's activity, and on the other hand, the organization of one's time. A good illustration of this reality is provided by the operation of a car assembly line. The production context always conjures up an imaginary world of extreme constraints, whereas in reality, as soon as they can, operators carry out their tasks faster than the rate of progress of the line, in order to take small breaks (they are said to "move up the line").

As we can see, it is perilous to look for areas where work can be erased without getting as close as possible to the actual way in which work is carried out, not as management has planned it to be done, but as it actually takes place. This difference between the planned and the actual varies according to the entity's activity and jobs. However, it does highlight the need to take an interest in the actual execution of tasks. "Social" does not preclude anything; it merely conditions and profiles the actual potential for energy shedding.

8.5.3. *Exploring potential for load shedding in commercial sites*

Two items of electrical equipment on the study site were chosen for exploration: hot water tanks and electric vehicle charging stations. The educational value of these choices lies in their specific socio-technical issues.

The many hot-water DHW cylinders scattered around buildings heat continuously and are used during the day to clean spaces and surfaces (floors, office furniture, washbasins and washrooms). France has a culture of cleaning with hot water, which is not true everywhere. Charging stations for electric vehicles are freely available, with a usage logic more in keeping with the domestic sphere than with the conduct of business. Table 8.5 summarizes the usage and control patterns of the two families of equipment, hot water tanks and charging stations.

Given the decentralized control of usage, in both cases, the aim is to create the conditions for acceptance of a switch-off that could have repercussions on users' mental workload (the switch-off introduces a new parameter of use), and therefore on their conditions of comfort at work.

	Control	**Impact**	**Operating Hours**	**Recycling feasibility**
Hot water tanks	Decentralized	Process	24 h/24 h	Easy if inertia/To be built
Electric vehicle charging stations	Decentralized	Domestic finance	Day	Easy/To be built

Table 8.5. *How to use and control hot water tanks and charging stations*

The methodology followed a two-stage approach. Following a detailed inventory of the various installations, two panels of users were defined and interviewed using semi-directive questionnaires.

8.5.3.1. *Preliminary exploration: the use of domestic hot water tanks*

8.5.3.1.1. Overview of the socio-technical context

The Rizomm's 14 DHW cylinders are spread over six levels. The building accommodates a large number of visitors, so it comes as no surprise that the bulk of the draw is devoted to cleaning. This is carried out by a dedicated university department. The team is made up of seven people, all assigned to a given surface equipped with a specific DHW cylinder, usually 15 liters. The building has a lunch area (4th floor) and a shower on the first floor, accessible to mid-day athletes. It is supplied by the large 100-liter water tank. Table 8.6 lists the hot water tanks in the experimental building.

Hot Water Tank Capacity	Power	Number
15 liters	1,600 W	5
15 liters	2,000 W	7
50 liters	1,500 W	1
100 liters	1,200 W	1

Table 8.6. *Hot water tanks in the experimental building*

All DHW cylinders heat up continuously, with no possibility of standby or shutdown. In theory, this lack of control means that there is potential for load shedding, at least at night. It all depends on what happens during the day in terms of working hours, the diversity of needs and actual drawing practices.

8.5.3.1.2. Use of DHW cylinders

The cleaning staff can choose their starting time between 5 a.m. and 9 a.m., the bulk of the cleaning (floors and sanitary facilities) having to be done before the 8 a.m. arrival of staff and students.

They each have an assigned area (of the order of one floor) with a dedicated DHW cylinder. They organize their work on the basis of a set of specifications, without the intervention of management: at the very least, floors in circulation areas and work areas (floors and desks) must be washed once a week. Only the washrooms need to be completely cleaned every day. Figure 8.31 shows the cleaning equipment and a 15-liter water tank.

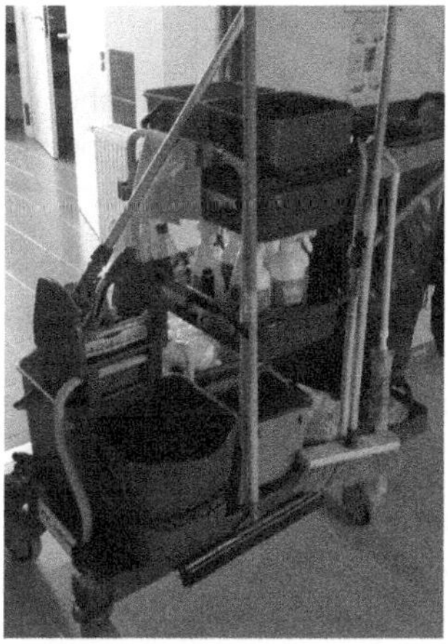 

Figure 8.31. *Cleaning equipment and 15-liter water tank*

In practice, corridor floors are washed almost every day. They account for the largest hot water draws of around 8 liters maximum for weight reasons, since the buckets have to be placed on a one-meter-high mobile cart. Smaller 15-liter cumulus heaters are therefore more than sufficient for the first draw. The smaller ones (around 5 liters) are intended for offices and washrooms. Over the course of the day, the number of drains varies according to the size of the area, the intensity of traffic and rainy days: for floors (8-liter bucket), there are generally four daily drains (two per half-day). After the first wash, surface maintenance is a question of assessing need and tactics, from two points of view. Firstly, staff need to find cleaning and drying opportunities according to the course schedule and the day's high-traffic events. Secondly, the cooling time in the bucket must be taken into account, as the water has to be hot enough to ensure its effectiveness.

This means that prints have to be made judiciously, when it is deemed necessary for the dirt and possible according to traffic and drying time. Taking these parameters into account on a day-to-day basis explains the variability in the size and time slots of the print runs. However, the habit of limiting the volume of sockets for weight reasons, combined with permanent heating, means that all cumulus heaters (large and small) always have sufficient hot water.

Under these socio-technical conditions, the supply of hot water is not a problem. This is evidenced by the fact that new recruits do not need to be told how to operate the cumulus. The red light on the front of the DHW cylinder is enough to reassure that hot water is always being produced. The precise heating capacities of the DHW cylinders are of no interest to anyone.

8.5.3.1.3. Lessons learned

The information gathered from the cleaning staff reveals a number of key aspects of the socio-technical operation of the hot water tanks. They point the way to exploring the potential for load shedding:

– Building cleaning is organized according to the principle of personal autonomy. Each person has their own area, equipment and DHW cylinder, and assesses cleaning needs and times. This arrangement makes cooperation unnecessary, although it does not rule out exchanges or services rendered between individuals.

– The organization explains why the supervisor is not very present when it comes to talking about how the work is done. They are only mentioned in connection with spaces to be recovered in the event of absence, but rarely more than that.

– So, there is a sort of pact between management and cleaning staff, regulated by a framework with two sources: on the one hand, the specifications that spell out the services required and the rules to be followed in the work, and on the other hand, the ever-present complaints of building occupants in the event of a one-off or recurring problem.

– This freedom to work in routine conditions is facilitated by the permanent heating of the water in the hot water tanks. It provides comfort and flexibility for staff, who have enough hot water at their disposal at all times, without fear, to do whatever they need to do, even if it is unexpected. In more sociological terms, this permanent heating is to be seen as an arrangement in the system of action (i.e. a social tension-reducing device) linking management, cleaning staff and building occupants. And like all conciliatory solutions, it is costly in financial terms and in terms of carbon emissions, due to the permanent operation of the heaters.

8.5.3.1.4. Feasibility and implementation conditions

The load shedding principle is to suspend heating times whenever possible. Based on an analysis of work organization and practices, we understand that it is compatible with the suspension of water heating at night, since this campus building is definitely unoccupied from 9 p.m. to 4 a.m. The necessary technical modification of the cumulus heaters would not pose any particular difficulty. On the other hand, the implementation of the daytime cut-off time is more problematic because, as we have seen, the socio-technical operation of the cleaning system relies in part on the permanent heating of the water. Considering how to get around this constraint calls for action in two main directions, in consultation with management staff.

When it comes to DHW cylinders, it is essential to have a better understanding of their technical features, to be able to assess the range of possible reductions. In particular, we need to measure for each DHW cylinder:

– the water temperature in the full tank;

– the temperature when filled after drawing 8 liters of water, then the temperature rise curve;

– the temperature drop curve when heating is switched off.

When it comes to the cleaning staff, we need to work together to better define the necessary conditions for working without adding to mental workload or fatigue. In particular, we need to:

– test the minimum temperature required for cleaning;

– better understand whether the water drawn is handled as it is, or after mixing with cold water, or after a voluntary cooling period;

– test the acceptability of modifying the socio-technical framework with the provision of an on–off control on cumulus heaters;

– test the acceptability of grouping two draws on the same DHW cylinder, based on measured heating times.

The aim of gathering these data is to model the potential for load shedding within today's socio-technical constraints. Then, we explore the possibilities of load shedding made possible, first by the technical performance of the DHW cylinders, then by those worked out in collaboration with cleaning staff.

8.5.3.2. *Secondary exploration: electric vehicle charging stations*

The experimental building is located on a university campus committed to a policy of carbon neutrality in its operations by 2050 [ROB 23]. One of the key actions is to optimize self-consumption of the electrical energy produced on site by photovoltaic panels installed on a growing number of buildings. The creation of many charging stations servicing a fleet of electric vehicles will also help to develop soft mobility and contribute to this self-consumption.

8.5.3.2.1. Overview of the socio-technical context

A total of six charging stations are available around the reference building. These are available free of charge to people who have registered as coming to work in an electric vehicle. The activity report for 2022 identifies 19 regular users with a total of 1,190 connections, that is, an average of 63 individual connections, or one every three working days. The power of the charging stations means they can be fully recharged over half a day. The average draw is 17.8 kWh. Figure 8.32 shows an electric vehicle charging station with two sockets.

Figure 8.32. *Bollard with two electric vehicle charging stations*

As the campus has no application for monitoring charging, this is done using the manufacturer's application installed on the vehicle owner's phone. This situation prevents optimal use of the charging stations, and everyone complains about "sucker" cars, plugged in all day despite being fully charged.

8.5.3.2.2. The use of charging stations

Usage was analyzed in two ways. First, the connection history for the year 2022 was used, followed by a focus on the number of connections and charging times. This enabled us to identify user profiles, and then to interview five of the most frequent users of electric charging.

Identification of four user profiles

Analysis of the connection data reveals four profiles of unequal importance, which are difficult to describe with any degree of precision, since vehicle autonomy, distance from home, weekly telecommuting and use of the electric vehicle outside work are not exhaustively known.

The profiles are structured around the "number of connections" and "charging volume" axes. They are classified according to their position below (– sign) or above (+ sign) the average number of annual connections (n = 62) and the average electrical draw per connection (18 kWh). These are described in scenario form for better understanding:

– *Group 1 (n = 6), "Connections –/Withdrawal +"*: this group includes people with different types of use. Some come from a household with only this small vehicle, living no more than 10 kilometers away. These are "short distance" drivers, motivated by a more or less stable balance between financial calculations and environmental awareness. Other people have a longer-range vehicle, which allows them to recharge less often but more consistently. Their rationality is mainly economic.

– *Group 2 (n = 5), "Connections –/Withdrawal –"*: this group is made up of people for whom the electric vehicle is explained more by a financial opportunity (purchase bonus, for example) than by the desire to reduce running costs and/or to protect the environment.

– *Group 3 (n = 4), "Connections +/Withdrawal –"*: this group is made up of owners of large plug-in hybrid cars. The electric capacity is not essential for use but contributes to the economic benefits and comfort in town. This type of motorization requires short, close-quarters charges.

– *Group 4 (n = 2), "Connections +/Withdrawal +"*: very much in the minority, they own a small, low-range vehicle, which they use every day to get to work. Their home is 30 to 40 kilometers away, so they generally have to recharge three times a week. Their vehicle moves very little outside of this use, and always over short distances. Their motivation is mainly economic.

Autonomy, a determining factor in the logic of use

Sociological interviews with five users provide a more detailed understanding of the dynamics of use of the study site's electrical charging stations, with a view to identifying the conditions for creating the capacity for electrical load shedding. A number of key findings emerge from this collection of qualitative data:

– *Connecting to a charging station is a daily problem for everyone.* Firstly, the number of charging stations is insufficient for the ever-increasing number of electric vehicles. The problem is further exacerbated by the lack of vigilance of those who leave their cars plugged in despite a full charge (the famous "suction cars"). Different strategies are then adopted: for example, arriving early to get a plug or handing over your keys to free up your space once the charge is complete. The use of charging stations is less a problem of charging than of availability.

– *Managing autonomy comes with a heavy mental burden*, except for rechargeable hybrid owners. The concern may be to maintain autonomy for the most anxious (with repeated small charges), or to restore it for the most "playful". This difference does not alter the feeling of an imperative necessity.

– *Economic interest largely dominates the reasons for purchase.* A detailed comparison of costs between motorizations was carried out. The fact that charging is free on campus is a decisive factor and is experienced a posteriori as compensation for the effort of anticipation required to manage autonomy. The certainty of being able to recharge at work is another buying factor for those who do not have a charging station at home. Those who mention an environmental concern are characterized by strong discipline in releasing their charging station as soon as charging is complete.

– *Recharging requires attention.* Charging progresses via a manufacturer's application installed on a phone. Variations in charging speed, or a bug at a charging station, are enough to keep users on their toes when it comes to choosing a location and actually charging. Attention is redoubled in cold weather, which is not very kind to batteries: charging is less effective, sometimes slower, and autonomy is reduced.

8.5.3.2.3. Lessons learned

Charging stations are currently made available to staff without any management tools. The uncertainty of being able to access a plug and the less intense uncertainty of charging time make them difficult to use. The fact that they are free of charge compensates for this, more or less consciously. This mitigated assessment upsets the imaginations behind the acquisition of an electric vehicle, and the sustainability of the choice of motorization is sometimes called into question. If the choice is maintained, it will rather be for a larger battery model to alleviate the imperiousness of recharging.

The supply of charging stations, which are not easily accessible except early in the morning, is faced with growing demand. In this context, the prospect of reduced charging times would mean a deterioration in an already low-quality service. The measure would seem paradoxical in the current supply situation.

8.5.3.2.4. Feasibility and implementation conditions

While load shedding is not considered appropriate at present, information on the conditions of service and use is also enabling us to identify a number of avenues for change, with a view to improving implementation conditions. These relate to charging station services, thanks to the development of an ad hoc application. Several items of information could be provided:

– the availability of charging stations to make it easier to find free spaces, if possible, with a reservation schedule closed on D-1 so that arrivals know when and where to plug in their vehicle;

– the quality of the charge, with an estimate of the time required so that the owner can anticipate and organize the movement of their vehicle;

– a full-charge alert.

Improving service conditions, with the help of a wide range of stakeholders, would be an essential step towards making load shedding more acceptable. To this could be added information to reaffirm what is really at stake. After all, load shedding is merely a technical lever designed to facilitate the integration of on-site renewable energy. It is important to place it within the more global project of decarbonizing the campus. The acceptability of load shedding also depends on the meaning it is given.

8.5.4. *Concluding remarks on both case studies*

The two cases explored on the study site demonstrate how difficult it is to release electrical load-shedding capacity without modifying the current socio-technical context. This is due to the close intertwining of energy supply (sources and intensity) and modes of use, in the service of pretexts for consumption that impose their own requirements (production, comfort, cleaning, logistics, IT, regulations, etc.). Planning for load shedding is therefore tantamount to seeking to overturn the logics of action, the socio-technical rationalities, favored by the actors at work or imposed on them. As a result, the implementation of effacement times calls for a realistic approach, that is, with the collaboration of the actors involved. They are the only ones who know why things are done as they are, and the likely effects of even the slightest change in energy supply.

The other main lesson to be learned from the case study concerns the "technical actors" [LAT 88], in this case the hot-water tanks and charging stations. To erase, it is not enough to reduce or suspend the flow of energy. It also requires a better understanding of their operation and performance. In this respect, we should not be surprised by the lack of interest in the quality of service provided to employees once the equipment has been made available to them. The reason for this is that technical departments remain mainly focused on "operation" as a criterion of performance in their mission [SUB 15], whereas the evolution of energy systems requires a more detailed interest in the capabilities of building devices, as well as those of technical objects used in the workplace.

From the point of view of employees at work, and of the departments in charge of the technical and energy-consuming environment, the introduction of load shedding times requires a veritable revolution in their socio-technical cultures.

8.6. Conclusion

The development of demonstration sites requires a highly interdisciplinary approach to reality, which is hardly possible in the laboratory. The demonstration sites presented in this chapter involve expertise in the fields of energy and power grids, buildings and information systems, in addition to the sociological expertise required to understand human functioning and potential flexibilities, compatible with the diversity of human modes of functioning and their possibilities for appropriating these potentialities. The solutions developed in response to the necessary energy and social transition will be those that integrate technological, human and social dimensions in their development. The demonstration site approach is a fine illustration of the "working together" approach advocated by MDG 17.

8.7. Acknowledgments

The development of the Live TREE demonstration sites has benefited from funding from the Hauts-de-France region, several European funds from Feder (the European Regional Development Fund) and H2020 (e-balanceplus project), and funding from the Université Catholique de Lille foundation. It has received support and funding from the MEL (So Mel So Connected project funded by ADEME and Junia's Smart Buildings as Nodes of Smart Grids chair) and from the City of Lille. It has also received funding under the State-Region plan via the Sunrise project. Several companies are supporting the development of Live TREE.

Numerous actors are contributing to the development of these demonstration sites; the authors would like to thank them via this non-exhaustive list: Benoit Bourel, Yohann Rogez, Grégoire Destombes, Flovic Gosselin, Grégory Vangreveninge, Khaled Almaksour, Anthony Aouad, Arnaud Davigny, Nicolas Gouvy, Serkan Surkulu and Jad Nassar.

8.8. References

[ADE 12] ADEME, Le besoin d'eau chaude sanitaire en habitat individuel et collectif, Technical report, 2012.

[ADE 17] ADEME, L'Effacement de consommation électrique en France, Report, 2017.

[AOU 22] AOUAD A., ALMAKSOUR K., ABBES D., "Storage management optimization based on electrical consumption and production prediction in a photovoltaic system", *Conférence Electrimacs 2022*, Nancy, 2022.

[BEN 21] BENCHEKROUN A., Supervision énergétique des réseaux de distribution intégrant des véhicules électriques, des énergies renouvelables, du stockage et descharges passives pilotables, PhD thesis, University of Lille, 2021.

[BES 15] BESLAY C., GOURNET R., "Les professionnels du bâtiment face aux enjeux de la performance énergétique : nouveaux savoirs, nouveaux enjeux", *Revue Sociologies*, 2015.

[BRU 14] BRUCKNER S.S.T., FULTON L., HERTWICH E. et al., "Annex 3: Technology-specific cost and performance parameters", Climate Change 2014: Mitigation of Climate Change, Contribution of Working Group III to the Fifth Assessment Report of the Intergovernmental Panel on Climate Change, available at: https://www.ipcc.ch/site/assets/uploads/2018/02/ipcc_wg3_ar5_annex-iii.pdf, 2014.

[DOB 13] DOBROWSKI B., BLANC F. DIEMUNSCH V., "Le véhicule électrique : un vrai challenge induisant des réels changements d'habitudes", *REE*, no. 1, pp. 40–48, 2013.

[EAR 14] EARTO, The TRL scale as a research & innovation policy tool, Report, EARTO Recommendations, 2014.

[LAT 88] LATOUR B., AKRICH M., CALLON M., "A quoi tient le succès des innovations ? L'art de l'intéressement et le choix des porte-parole", *Gérer et Comprendre*, no. 11, 1988.

[ROB 15] ROBYNS B., FRANCOIS B., DELILLE G. et al., *Energy Storage in Electric Power Grids*, ISTE Ltd, London and John Wiley & Sons, New York, 2015.

[ROB 16] ROBYNS B., SAUDEMONT C., HISSEL D. et al., *Electrical Energy Storage in Transportation Systems*, ISTE Ltd, London and John Wiley & Sons, New York, 2017.

[ROB 23] ROBYNS B., LENGLET C., BARRY H. et al., *Smart Users for Energy and Societal Transition*, ISTE Ltd, London and John Wiley & Sons, New York, 2022.

[SAD 89] SADIN S.R.F., POVINELLI P., ROSEN R., "The NASA technology push towards future space mission systems", *Acta Astronautica*, vol. 20, pp. 73–77, doi: 10.1016/0094-5765(89)90054-4, 1989.

[SHA 12] SHAAD M., MOMENI A., DIDUCH C.P. et al., "Parameter identification of thermal models for domestic electric water heaters in a direct load control program", *25th IEEE Canadian Conference on Electrical and Computer Engineering: Vision for a Greener Future*, CCECE, pp. 1–5, 2012.

[SUB 15] SUBREMON H., AUBRY-BRECHAIRE M., BOUSQUET P. et al., "Coriolis à l'usage : un bâtiment performant sous tension", *Métropolitiques*, available at: http://www.metropolitiques.eu/Coriolis-a-l-usage-un-batiment.html, 2015.

[ZEL 15] ZELEM M.-C., BESLAY C. (eds), *Sociologie de l'énergie, gouvernance et pratiques sociales*, CNRS, Paris, 2015.

Postface

Acclaim for the *Prosp'Active* Citizen

This book is a gem in terms of its approach to the modern challenges posed by global warming and the changes in climate. Not only does it provide a precise, well-documented technical vision of new forms of energy production and energy cooperation, the existing opportunities and their limitations, but also it demonstrates that the major challenge is the initiation of a positive, creative socio-perceptive dynamic to transform the human *habitus*.

At the heart of today's foremost challenges is the distancing of people's truth from experts' truth. The analytical truth of the experts, supported by a segmented vision, is coming up against a more holistic "truth perceived by people", in the face of the widespread collapse of traditional hierarchies, including that of knowledge. The efficiency envisaged by forward-looking roadmaps clashes (analytically constructed) with real-world perceptions, and the attempt to integrate all factors related to the fulfillment of individual and collective desires.

The origins of this phenomenon are manifold. They are to be read as part of a complex system of interrelated determinants. Each interferes with the others, without any of them being able to assume the role of primary cause. We can list them in no particular order of importance:

– the collapse of traditional hierarchies, particularly that of knowledge and expertise;

– global disintermediation as a result of individualization, particularly on the Internet;

– the creation of ephemeral communities for sharing and reinforcing opinions of varying degrees of validity;

– the emergence of new epistemic processes, i.e. the creation of knowledge. This tends to become knowledge of links between knowledges of different epistemological status (founded knowledge, rumors, gurus, pseudoscientists, experts, etc.), rather than knowledge of place based on in-depth disciplinary study, subject to controversy and critical experience;

– the transition from an exo-distributive epistemic manufacturing process, in which the subject's intelligence is external to the object system, to an endo-contributive process, in which subjective intelligence is embedded in each component of the system, which have become subject-objects, hybrid network actors;

– the collapse of belief in science as a powerful lever for enduring progress;

– the maintenance and even growth of technological culture;

– social imaginaries that elaborate narratives whose main object of desire is no longer the discovery of truth, especially scientific truth, but the shaping of the world in accordance with the fulfillment of individual desires.

Many see this metamorphosis as a risk, stimulating concern about our humanization process. They see it as a major cultural and intellectual regression from the Enlightenment and emancipation through reason. But can we not also meet this upheaval with a positive spirit? Consider it as a still tentative, paradoxical adaptation to a profound process of reaction in the face of social, economic and ecological vulnerability that is concretely exploding in our individual and collective consciousness. This would be what Bernard Stiegler called an anti-entropy phenomenon, aimed at rebalancing the living in the face of entropy forces, that is, alteration and destruction. *The challenge is to get "real people" to embrace a desire to transform traditional conceptions of happiness and progress.*

However, this is no easy task, all the more so given that in the West there is a kind of contempt, or at best distrust, of what is considered to be based on incompetence, or even ignorance. Ever since the Greek philosophers, and more recently the Enlightenment and Descartes in particular, distrust of appearances has been at the heart of a hierarchy of knowledge that has become a social hierarchy for those who put it into practice. As a result, what people believe to be true has always been suspect in the eyes of the so-called experts. What people take to be true has almost always been considered by the experts (philosophers, scientists, politicians, etc.) at best as the initial moment in the discovery of truth, a partial element to be put in its rightful place, that is, to be overcome by integrating it. At worst, as something to be ignored, or even denigrated.

So, we think that to solve the problem, we need more reasoning, more rational arguments, more education of the people, as well as better argued, better evaluated, better planned roadmaps. We need to explain more and more. But is this the right way? The more rationality is demonstrated, the more holistic resistance finds a way to challenge it. This has been proven in the case of *fake news*, for example. For some years now, we have witnessed a rejection of truths by experts, in the name of real experiences and gut feelings. The result is a growing divide between truth and verity, which gives rise to mistrust on the one hand, condescension on the other, and is the cause behind the general lack of understanding.

The way out of this situation, which is damaging to both democracy and ecology, is to take people's truth more seriously, and to articulate it with the backing of experts. Truths, but not through the knowledge hierarchy. It would seem, then, that we should try to seize these phenomena in a positive way, to rebuild a transformed link between expertise, the rationalized management of change and concrete action, by the people.

This means setting the truth in motion, by going back and forth to the truth. Clearly, in the face of increasingly complex systems, political decision-making relies more and more on experts of all kinds as a source of authority. The media are constantly validating their relevance, to help fill the 24-hour news channels. Experts, and sometimes "supposed experts", cost less than field reports by investigative journalists. *This amounts to an expert assessment of the democratic act, and even of the functioning of democracy itself. It is thus more a question of democratizing expertise in collective action and projects, on a scale (geographical or thematic) accessible to the citizen, in an enlightened co-elaboration of citizen action. I call this* prosp'active.

Therefore, it is in the *agora*, in public discussion, that choices must be made. Again and again, over and over, because it is all about being attentive to the signs of a constantly changing timeline. Hence, with these lessons in mind, I am advocating a circular *prosp'active* approach, as opposed to the classic linear perspective. For the concept of *prosp'active* means acting with foresight along the way, constant reconstruction through action and experience in a circular movement between existential truth and verity. This path of self-reflexive action constantly challenges the permanence of truth. It would be a scenario in which time is reconciled through movement.

This would be a radical opening-up of possibilities, challenging the status quo intrinsic to which planning carries with it the curse of the arrow of time, when it is reduced to prediction. The end of linear prediction, which confines itself to observing a set of factors causing a probable future, would empower a *prosp'active* approach. In other words, "prediction in the making", opening the field to new narratives of the future, created as we walk. Disruptive, creative, artistic and experimental stories that would make sense to the greatest number of people in a viral, networked progression that would become exponential and out of control. It would explore fields of radical innovation. These narratives would give rise to trend-based trajectories, breaking with all continuities and linear extrapolations.

It would then be crucial to reintroduce scientific truth into the object of desire of social imaginary narratives, accepting its changed status and mourning its aspect of superiority.

Framework for a *prosp'active* politic

A famous quote in the military world, difficult to attribute to a specific author, is: "A battle plan that prepares for war often explodes at the first cannon shot". The Maginot Line is an obvious example of this. From a political point of view, the point here is not to delegitimize all predictions, especially those of climate catastrophe if nothing is done, but to instead put into perspective the effectiveness of top-down plans. Above all, it is about realizing that they implicitly do more of the same, that is, they reinforce the practices at the root of the problems we are trying to solve, without calling these practices into question.

Implicit in these predictions is a vision of the issues at stake and the solutions to be found, which limits the possibilities and reinforces the problem. The example of a categorical imperative such as the constant expansion in the size of megacities induces sustainable solutions that implicitly reinforce what then becomes an inevitability (the Réseau Express Régional (RER), for example, or centripetal networks). In this way, the sustainable treatment of mobility in megacities leads us to implement solutions that only reinforce the problem, because they are centrally conceived and prevent us from seeing the weak signals, that is, the attempts at disruptive solutions that challenge hyper-metropolization. This would be a concept based on horizontal, meshed networks for a different approach to the relationship between urban, suburban and rural areas.

These alternative models, more lateral and networked, more endo-contributive than exo-distributive, are unable to make inroads due to a lack of encouragement in terms of skills and funding. Furthermore, they are usually stifled as soon as they challenge the interests of various dominant actors (economic, political, state, corporatist, etc.). To implement them, we need to rebalance public policies and take them out of the realm of large-scale structural plans, which often do no more than reinforce the models at work in all decision-makers' representations of what needs to be done and how. This means admitting that we do not know much about what is going to happen socio-economically over the next 40 years, apart from the climate and ecological challenge. In fact, complexity is irreducible and hides, in weak and sometimes imperceptible signals, the dynamics of disruption or butterfly effects, of strange attractors that escape the futurists' roadmaps.

We are witnessing a radical change in the mindset of politicians. It is no longer a question of inventing solutions, having them voted for and financed in advance, and then reporting back. It is about creating the conditions for the adaptability and heightened awareness of people (experts, employees, managers, etc.) who, in their particular or diverse contexts (professional, personal, consumer), are sensitive to predictions from a global point of view, and experiment in community, both globally and locally, with different solutions and attitudes.

The idea, then, would be to build – or promote, where they already exist, alongside existing bodies for monitoring, coordinating and evaluating public action – a multitude of "initiative gardens" (third places, associations, organizations with a mission, etc.) in all sectors, where people, according to their different statuses, would come to bring, compare, analyze and connect their initiatives. A community of "initiative gardeners", made up of professionals, semi-professionals or volunteers, could be created in all organized communities (regions, CCIs[1], departments, towns, intermediary bodies, etc.). This community would be managed, trained and facilitated as a transversal network.

It is a question of considering people's growing awareness and need for initiative, which is often masked, bridled, by the requirement to fit into roadmaps. This highly biomimetic process involves thinking of the human community as a living organism that constantly reinvents itself in the face of all that tends to degrade or even destroy its environment.

It is therefore not a question of doing away with major structural plans, but rather of reinventing them, in part based on needs as witnessed by fieldwork, and seeking to disrupt rather than reproduce pre-existing schemes in people's social imaginations. The advantage of this model is its agility and, above all, the creation of a world that is highly connected to the needs and expectations of people in real-life situations, immersed in the new challenges they perceive. All this could create the conditions for active resilience in the face of issues perceived and experienced in an integrated way. A permanent active prosperity to which politics would be attentive at all levels of public good management. This would be done in three ways: precaution, adaptation and repair. This leads us to a political *prosp'activism* that would renew democracy and the exercise of citizenship.

1 CCI: French Chamber of Commerce and Industry.

Without wishing to cast aspersions on the authors of this book, my interpretation is that it invites us to take part in this radical transformation. In this respect, it is also a gem to be consumed without moderation.

Pierre GIORGINI
Essayist, associate researcher in technology ethics
Laboratoire ETHICS, Université Catholique de Lille

Index

M, N

P, R

S, T

U, V, W

Other titles from

iSTE

in

Energy

2024

BENALLOU Abdelhanine
Heat Exchangers: Design and Sizing Algorithms
(Energy Engineering Set – Volume 6)

MORY Mathieu
Mechanical Energy Conversion: Exercises for Scaling Renewable Energy Systems

2023

ALVAREZ-HERAULT Marie-Cécile, GOUIN Victor, CHARDIN-SEGUI Trinidad, MALOT Alain, COIGNARD Jonathan, RAISON Bertrand, COULET Jérôme
Distribution System Planning: Evolution of Methodologies and Digital Tools for Energy Transition

BEN KHADER BOUZID Monia, CHAMPENOIS Gérard
Electrical and Mechanical Fault Diagnosis in Wind Energy Conversion Systems

JANNOT Yves, MOYNE Christian, DEGIOVANNI Alain
Heat Transfer 1: Conduction and Convection
Heat Transfer 2: Radiation and Exchangers

ROBYNS Benoît, LENGLET Claude, BARRY Hervé, BOZZO-REY Malik
Smart Users for Energy and Societal Transition

2022

ABOUELATTA Mohamed, SHAKER Ahmed, GONTRAND Christian
Smart Power Integration

BOUTAUD Benoit
Energy Autonomy: From the Notion to the Concepts

GONTRAND Christian
Electromagnetism: Links to Special Relativity

2021

ROBYNS Benoît, DAVIGNY Arnaud, FRANÇOIS Bruno, HENNETON Antoine, SPROOTEN Jonathan
Electricity Production from Renewable Energies (2nd edition)

2020

BOISGIBAULT Louis, AL KABBANI Fahad
Energy Transition in Metropolises, Rural Areas and Deserts

SOUALHI Abdenour, RAZIK Hubert
Electrical Systems 1: From Diagnosis to Prognosis
Electrical Systems 2: From Diagnosis to Prognosis

2019

BENALLOU Abdelhanine
Energy Transfers by Convection
(Energy Engineering Set – Volume 3)
Energy Transfers by Radiation
(Energy Engineering Set – Volume 4)
Mass Transfers and Physical Data Estimation
(Energy Engineering Set – Volume 5)

LACHAL Bernard
Energy Transition

2018

BENALLOU Abdelhanine
Energy and Mass Transfers: Balance Sheet Approach and Basic Concepts
(Energy Engineering Set – Volume 1)
Energy Transfers by Conduction
(Energy Engineering Set – Volume 2)

JEMEÏ Samir
Hybridization, Diagnostic and Prognostic of Proton Exchange Membrane Fuel Cells: Durability and Reliability

RUFFINE Livio, BROSETA Daniel, DESMEDT Arnaud
Gas Hydrates 2: Geoscience Issues and Potential Industrial Applications

VIALLET Virginie, FLEUTOT Benoit
Inorganic Massive Batteries
(Energy Storage – Batteries and Supercapacitors Set – Volume 4)

2017

BROSETA Daniel, RUFFINE Livio, DESMEDT Arnaud
Gas Hydrates 1: Fundamentals, Characterization and Modeling

LA SCALA Massimo
From Smart Grids to Smart Cities: New Challenges in Optimizing Energy Grids
(Advanced Smart Grids Set – Volume 2)

MOLINA Géraldine, MUSY Marjorie, LEFRANC Margot
Building Professionals Facing the Energy Efficiency Challenge

SIMON Patrice, BROUSSE Thierry, FAVIER Frédéric
Supercapacitors Based on Carbon or Pseudocapacitive Materials
(Energy Storage – Batteries and Supercapacitors Set – Volume 3)

2016

ALLARD Bruno
Power Systems-on-Chip: Practical Aspects of Design

ANDRÉ Michel, SAMARAS Zissis
Energy and Environment
(Research for Innovative Transports Set – Volume 1)

DUFOUR Anthony
Thermochemical Conversion of Biomass for the Production of Energy and Chemicals

SOUSTELLE Michel
Phase Transformations
(Chemical Thermodynamics Set – Volume 5)
Thermodynamics of Surfaces and Capillary Systems
(Chemical Thermodynamics Set – Volume 7)

2015

MONCONDUIT Laure, CROGUENNEC Laurence, DEDRYVÈRE Rémi
Electrodes for Li-ion Batteries: Materials, Mechanisms and Performance
(Energy Storage – Batteries and Supercapacitors Set – Volume 2)

ROBYNS Benoît, FRANÇOIS Bruno, DELILLE Gauthier, SAUDEMONT Christophe
Energy Storage in Electric Power Grids

ROSSI Carole
Al-based Energetic Nanomaterials: Design, Manufacturing, Properties and Applications
(Nanotechnologies for Energy Recovery Set – Volume 2)

TARASCON Jean-Marie, SIMON Patrice
Electrochemical Energy Storage
(Energy Storage – Batteries and Supercapacitors Set – Volume 1)

SOUSTELLE Michel
Phases Modeling Tools: Application to Gases
(Chemical Thermodynamics Set – Volume 1)
Modeling of Liquid Phases
(Chemical Thermodynamics Set – Volume 2)
Thermodynamic Modeling of Solid Phases
(Chemical Thermodynamics Set – Volume 3)
Chemical Equilibria
(Chemical Thermodynamics Set – Volume 4)
Ionic and Electrochemical Equilibria
(Chemical Thermodynamics Set – Volume 6)

2014

DE LARMINAT Philippe
Climate Change: Identification and Projections

2012

BECKERS Benoit
Solar Energy at Urban Scale

FOULADGER Javad
Electrothermics

2011

GAO Fei, BLUNIER Benjamin, MIRAOUI Abdellatif
Proton Exchange Membrane Fuel Cell Modeling

MONMASSON Eric
Power Electronic Converters: PWM Strategies and Current Control Techniques

MULTON Bernard
Marine Renewable Energy Handbook

RAZIK Hubert
Handbook of Asynchronous Machine with Variable Speed

2010

BRUNET Yves
Energy Storage

JUFER Marcel
Electric Drives: Design Methodology

2009

SABONNADIÈRE Jean-Claude
Low Emission Power Generation Technologies and Energy Management
Renewable Energy Technologies

2008

MEUNIER Gérard
The Finite Element Method for Electromagnetic Modeling

Printed and bound by CPI Group (UK) Ltd, Croydon, CR0 4YY
28/07/2026
14931340-0005